Lecture Notes on Data Engineering and Communications Technologies

Volume 194

Series Editor

Fatos Xhafa, Technical University of Catalonia, Barcelona, Spain

The aim of the book series is to present cutting edge engineering approaches to data technologies and communications. It will publish latest advances on the engineering task of building and deploying distributed, scalable and reliable data infrastructures and communication systems.

The series will have a prominent applied focus on data technologies and communications with aim to promote the bridging from fundamental research on data science and networking to data engineering and communications that lead to industry products, business knowledge and standardisation.

Indexed by SCOPUS, INSPEC, EI Compendex.

All books published in the series are submitted for consideration in Web of Science.

Andriy Semenov · Iryna Yepifanova ·
Jana Kajanová

Editors

Data-Centric Business and Applications

Modern Trends in Financial and Innovation
Data Processes 2023. Volume 2

Editors
Andriy Semenov 🆔
Faculty of Information Electronic Systems
Vinnytsia National Technical University
Vinnytsia, Ukraine

Iryna Yepifanova 🆔
Faculty of Management and Information
Security
Vinnytsia National Technical University
Vinnytsia, Ukraine

Jana Kajanová
Department of Economics and Finance
Faculty of Management
Comenius University Bratislava
Bratislava, Slovakia

ISSN 2367-4512   ISSN 2367-4520 (electronic)
Lecture Notes on Data Engineering and Communications Technologies
ISBN 978-3-031-53983-1   ISBN 978-3-031-53984-8 (eBook)
https://doi.org/10.1007/978-3-031-53984-8

Preface

Information technologies are widely used in the decision-making process in business. Databases of important socio-economic characteristics exist mainly in developed countries. In developing countries, the relevant databases needed to optimize economic, financial, production, and innovation activities are often incomplete. This volume includes several chapters in which some statistical data characterizing the economy of Ukraine are given. Also, some chapters are devoted to the description of analysis methods used by people who make decisions at different levels of the hierarchy in business, finance, and innovation management in Ukraine. This allows international scientists to obtain information both about the state of Ukraine's economy, including the impact of the Russian attack, and about the direction of financing future economic aid. Materials of this volume are also important because they can serve as a powerful source of information to improve the effectiveness of communication between international financial donors and the Government of Ukraine when discussing the application of international aid by Ukraine.

Consequently, in the preface, we provide short sketches of every study included in the volume. Starting with the first chapter titled "Formation of Strategies for the Development of Startup Ecosystems as a Prerequisite for Sustainable Entrepreneurship", this study examines startup ecosystems in the world countries based on clustering as a prerequisite for sustainable entrepreneurship. The analysis of startup ecosystems was based on a database presented in the Global Startup Ecosystem Index 2022 from Startup Blink, and 100 countries were included. Two criteria were chosen for cluster formation: Total Score, which demonstrates an overall position in the ranking, and Rank Change (from 2021), which demonstrates the direction of movement of the ecosystem (growth or decline). The authors find four homogeneous clusters of startup ecosystems in considered countries. The first cluster includes only the United States (Total Score 195). The second cluster includes 14 countries (from the United Kingdom with 52 to Ireland with 15), the third cluster includes 46 countries, and the fourth cluster includes 35 countries. The paper proposes 4 types of strategies: absolute leader, leadership retention, considering mistakes, and gradual growth for the obtained classes, respectively. The obtained results allow the government to use the Global Startup Ecosystem Index and Startup Blink databases to form profitable strategies for the country's development. It will be especially useful for developing countries.

The next chapter presents a study on "The Strategy of Sustainable Development of Digital Business in the Conditions of the Variability of the Business Environment and European Integration". Based on a forecast of the sustainable development index for the EU countries until 2030, the authors identified the conditions that the economy of Ukraine, as a future member of the EU, should satisfy. Digital technologies are proposed as a powerful driver for the development of Ukraine's economy. Such technologies should be used in all spheres of the economy. The authors note that since the majority of Ukrainian enterprises are significantly behind the EU countries or the world, the effect of growth will be large at the beginning. The section presents the results of a survey of 50 experts from all sectors of the economy of Ukraine aged 30–55. According to their estimates: the growth of the economy of Ukraine will be at least 10–15% per year, the high-tech industry sector will grow by 15–20% per year, and there will be growth and attraction of development investments in the country both in production and in R&D centers, incubators, and technological companies. The results obtained by the authors will be of interest to a wide range of specialists in computer modeling of sustainable development in the EU countries. Also, these results will be important for modeling the economic state of the EU taking into account the acceptance of new members.

The next chapter titled "Business Development towards the Application of Innovative Customer Relationship Management (CRM) Technologies in the Context of Global Transformational Changes" ascertains that the world market of CRM systems is developing very quickly. In contrast, the Ukrainian market of CRM systems has not yet gained mass development, which is caused by several factors: insufficient business budgets for implementation, lack of understanding among a significant number of business owners and top management regarding the availability and necessity of CRM implementation, lack of qualified personnel, etc. The IT methods are very needed in the Ukrainian CRM systems market, where the number of clients exceeds hundreds or even thousands. In particular, CRM stores the entire history of communication with customers; CRM improves customer satisfaction by reducing response time and improving the quality of customer support; CRM increases customer retention. The paper has found that for the successful implementation of CRM, a necessary condition is the definition and specification of the company's goals in the short- and long-term perspectives and the definition and development of a strategy that determines relations with customers.

The subsequent chapter "Methodological Principles of Smoothing the Effect of Seasonal Fluctuations on the Components of Labor Intensity in Construction" develops statistical methods for the analysis of the influence of natural factors (season fluctuations) on components of labor intensity in construction in the conditions of Ukraine, and the effectiveness of their application is investigated. It has been proven that in the conditions of Ukraine, climatic conditions and seasonality have a moderate influence on the progress of construction, which can be expressed in the form of seasonal fluctuations in indicators of deviations during the execution of construction works and such parameters as construction terms, labor intensity, and estimated cost during the year. The deviation of construction terms from a plan for the objects being built in the cities of Kyiv and Chernivtsi in the years 2012–2019 was investigated to identify the impact on the terms of operation, labor intensity, and cost of the seasonality factor and justify correction coefficients to the construction parameters depending on a season. It was found that the level of deviation was 8.3% for labor intensity, and 14.1–16.0% for terms and cost of work for Ukrainian climate conditions.

In the next study titled "Data-Driven Public Budgeting: Business Management Approach and Analytics Methods Algorithmization", the authors investigated the formation of a data-driven government and public budgeting system based on quality conditions of digitized data collected and the support of interaction of all budgeting participants, including citizens. Predictive models, business management approaches to public budgeting efficiency, and data mining techniques were included in this budgeting system. The authors draw attention to the fact that multi-criteria decision-making (MCDM) methods should be used at all stages, especially for the tasks of state and public budgeting. It is in this way that the interests of various active budgeting actors can be reconciled. Both favorable and unfavorable prerequisites for the use of information in the process of forming state and public budgets based on collected data are removed in the section. The authors also discuss specific features of managerial decision-making in the formation of the Ukrainian state and public budgets in conditions of war, high population migration, and relocation of enterprises and businesses.

In the next chapter on "Research of Information Platforms and Digital Transformation Algorithms for Post-war Recovery of Ukrainian Business", the authors propose to introduce a set of scenarios for designing a financial architecture and a large-scale information platform that will allow Ukrainian businesses not only to adapt to the digital dimension in the existing conditions of martial law but also to involve modern tools of digital integration (engineering, technology transfer, cloud technologies, etc.) for post-war recovery. The authors present theoretical developments in the field of substantiating the importance, necessity, and possibilities of implementing modern information systems, developing theoretical models, substantiating stages and procedures of business digitalization, and outlining models and mechanisms separately for the management and accounting systems of companies. Modeling recovery strategies for business with the help of an information software product will allow us to take into account the full toolkit of the latest market tools, which were previously not used to their full extent due to the usual complexity of combining theory and practice. The results described in the section represent Ukraine's current approaches to the development of directions for post-war reconstruction with the involvement of information technologies.

In the following study, entitled "Diagnostics as a Tool for Managing Behavior and Economic Activity of Retailers in the Conditions of Digital Business Transformation", a toolkit for diagnosing the digital behavior and economic activity of retailers in the context of digital business transformation was developed based on modern scientific investigations. The authors offer a set of quantitative parameters characterizing states and processes in the retail trade business. The section offers methods for assessing the level of the enterprise's mastery of digital technologies and the level of the enterprise's digital readiness for digital transformation. They claim that it is the diagnosis and monitoring of the proposed parameters that is the key factor for fueling the efficiency of retailers. The study provides an opportunity to assess the level of awareness of the scientific community of Ukraine regarding requirements for the activities of retailers in terms of modern business functioning. Also, retail management methods in the conditions of Ukraine may be important for the international scientific community.

The subsequent chapter "Methodical Tools for Identification and Quality Control of Design Products" develops the methods for increasing rationality and quality control of design products, the influence of a material consumption factor on a coefficient of rationality of design products, and other factors are considered. The author's purpose is the development of a methodological toolkit for checking the quality of project documentation to match the designer's remuneration with the results of his work. A possible variant of instrumentation, which is based on dimensionless relative values, is proposed in the paper. The initial value of this type of model is a weighted average linear combination of input variables or certain constants. The fuzzy inference algorithms are chosen for application. The general structure of a microcontroller using fuzzy logic is shown. The main results are equations, which showed that the influence of material consumption and other factors on the quality and price of design products is multidirectional. The membership function of the material consumption of project products is presented, too. The obtained results may be interesting for the optimization of the elaboration of product design.

The next chapter presents a study on "Methodical Approach to Assessment of Real Losses Due to Damage and Destruction of Warehouse Real Estate". Based on information about the estimated value of 192 objects of warehouse real estate in the period 01.21.22–02.04.22 in Ukraine, the chapter has made the systematization of the main price-forming characteristics and their categories and clustering of objects' prices. To determine the market value of a real estate at the time preceding a damage, the paper proposes to use a comparative approach and to substantiate the impact of the main price-forming characteristics on the value of real estate, it makes sense to use statistical methods, as well as technologies of intelligent data analysis. For classification of the price-forming characteristics on the value of real estate, the model in the form of a two-layer perceptron with three neurons in the hidden layer was used. The output of the neural network is the growth force of the area unit value for the evaluated object of warehouse real estate against the average market indicator. This model takes into account the dynamic nature of the warehouse real estate price and is characterized by a high level of approximation reliability. The quantitative results of neural network classification were discussed in the chapter. The first example of the result of estimating the real losses from damage and destruction of warehouse real estate as a result of the war in Ukraine is presented in the chapter.

In the next study on "Development of Information Processes as a Prerequisite for the Sustainable Development of Agricultural Enterprises", the authors propose to introduce the priority factors that determine the specificity of the digital transformation of agriculture and determine the directions of informatization of agricultural production. The authors established that the informatization of the agricultural sector is characterized by fragmentation. This leads to the differentiation of the level of use of information technologies by both individual agricultural enterprises and branches of agriculture. The authors developed information technologies that can be successfully applied in agriculture to ensure digitization. The authors propose to increase the level of economic development due to the synergistic effect of informatization of agricultural enterprises of the agrarian sector. This will save labor resources in the village, ensure overcoming the digital divide, and contribute to the socio-economic revival of the village. The approaches presented in the section describe a road map for the development of agriculture in Ukraine in the post-war period so its results can be used to forecast the effectiveness of the post-war development of Ukraine's economy.

In the chapter titled "Development and Increasing the Value Added Scenarios for the Woodworking Industry of Ukraine in the Context of the Circular Economy", the contribution of individual types of products to the formation of the gross value added of the industry in 35 countries was examined. The series of linear models of the dependence of a gross value added on the natural indicators of the output of products in 2000–2018 was built in the paper. Based on the analysis, the authors identified and analyzed three possible scenarios for the period up to 2030 for the development of the woodworking industry in Ukraine regarding import substitution, development for domestic consumption, and expansion of exports for various types of products. Based on the results of the analysis of these scenarios, the authors conclude that Ukraine needs to implement a combination of scenarios. The results of this paper can be used to form a database for optimizing international aid for the development of Ukraine after the end of the war.

In the next study, titled "Methodological and Technological Solutions to Improve the Security of Ukraine's Accounting System During the Hostilities", the authors carried out a comprehensive analysis of threats in the information space of accounting associated with the start of full-scale hostilities on the territory of Ukraine. The most important threats in Ukrainian accounting were identified both at the enterprise level and the country level. Blockchain technology was chosen as a perspective for protection. The authors propose the use of blockchains at both enterprise level and macro level. The scheme for the interaction of two-level blockchains was proposed, and the choice of several characteristics of these blockchains was developed. At the enterprise level, the hierarchical structure of internal blockchain users was elaborated. In contrast, at the macro level, the external blockchain was chosen. The description of the steps that Ukrainian scientists are proposing as necessary for the successful integration, including considerations for data migration, system integration, and development of blockchain-based solutions, will be useful for international scientists.

The next chapter is titled "Modeling of the Strategy of Light Industry Enterprise Behavior under Crisis Conditions of Martial Law". The formulation of business behavior and strategies, especially in the light industry sector, faces unique challenges during times of crisis, particularly in the context of martial law. This chapter explores the complex dynamics and strategic maneuvers of industrial entities encircled by crises. The authors reveal the complexities of corporate cooperation as a strategic approach in a crisis. They carefully develop models that not only identify the challenges faced by light industry enterprises but also provide a structured methodology for formulating optimal strategies for these entities, both individually and in a collaborative coalition. The chapter emphasizes the complex interaction between various factors influencing company performance and emphasizes the need for consistency in the evaluation of utility matrices and the importance of coherent expert grades in decision-making processes. It also examines the complexity of assessing the optimal strategies of coalitions, both broad and incomplete, and outlines the criteria for companies to join or deviate from coalitions based on projected benefits. Overall, this chapter serves as an insight guide for enterprise management and provides a systematic framework for determining behavior strategies in crisis situations. It provides a roadmap for decision-makers to assess the formation of coalitions, maximizing collective profitability while traversing the complex landscape of crisis-related industries.

The next chapter is titled "Innovative Technologies to Make Effective Business Decisions at Every Stage of a Mining Company's Development". In the current era of rapid technological progress and digitalization, industries around the world are witnessing fundamental changes in their operational landscape. The profound integration of digital technology and seamless automation of business processes are crucial factors for informed decision-making and the stability and growth of enterprises. In the midst of this wave of digital evolution, however, the metals and mining industry has not been at the forefront as a "digital leader". Comparative studies indicate that the mining sector lags behind, exhibiting digital maturity levels between 30 and 40% lower than similar industries such as pharmaceuticals, chemicals, and logistics. Based on real examples and using contemporary technology platforms such as Micromine, this study presents a detailed roadmap for the application of innovative technologies and describes their role in improving the efficiency, accuracy, and efficacy of mining operations. As the mining industry is on the verge of digital transformation, this chapter serves as a beacon for industry professionals, stakeholders, and decision-makers to take full advantage of the potential of digital technologies. The author's in-depth research lays the foundations for a paradigm shift, advocating the adoption of advanced technologies to strengthen the mining industry's journey toward informed decision-making and sustainable development.

In the forthcoming chapter "Innovative Method of Forecasting the Manifestation of Dangerous Properties of Coal Seams", the authors examine the multifaceted nature of factors that contribute to the emergence of hazardous properties within coal cables. Their analysis highlights the tripartite blocks that comprise natural conditions during geological processes, mining and geological conditions of coal deposits, and metamorphic transformations that are crucial to understanding the hazardous manifestations. By exploring these blocks, they emphasize the interaction between these blocks and highlight the complexity of the prediction and prevention of emergencies in mining operations. This chapter combines rigorous research, critical analysis, and pioneering technological advances to pave the way for a more comprehensive under-standing of forecasts and the prevention of hazardous phenomena in coal seams. The conclusions contained in this report not only promise to strengthen the safety protocols within the mining industry, but also to significantly mitigate the economic and human costs of accidents. This scientific contribution is a valuable asset for researchers, policy makers, and stakeholders in the industry who have invested in the safety and sustainability of coal mining around the world.

The next study is titled "Assessment of the Efficiency of Decentralization Trans-formations in the Rural Areas of Ukrainian Western Polissia: Current Trends and Challenges Under the Conditions of Martial Law". The dynamic landscape of rural development in Ukraine has been strongly influenced by decentralization efforts, particularly in the western region of Ukraine named Western Polissia. In the midst of these changes, it is essential to assess the effectiveness of these trans-formations in order to identify the current trends and tackle the challenges arising, especially in the complex field of martial law. The chapter emphasizes the urgent need to establish effective cooperation between the State and the local authorities and to ensure the principle of subsidiarity, even when there is a tendency to centralize under martial law. It supports coordinated efforts to address urgent needs while promoting sustainable development at the local level. In conclusion, the organiza-tional and economic mechanisms for sustainable development in rural areas depend on sound budgetary and fiscal policies. These policies are linked to the wider goals of social, ecological, and economic progress and form the foundations of effective governance and comprehensive community development. This chapter, therefore, offers a nuanced exploration of the multifaceted landscape of rural decentralization in the midst of the complex background of martial law. It serves as a testimony to the resilience of rural communities and the necessity of adaptation management strategies in times of turbulence.

The subsequent study, "The Technological and Environmental Effect on Marketing of Children's Food", explores the complex interaction between tech-nology, environmental impacts, and their effects on the marketing dynamics of children's food. The influence of these factors on the production and marketing of vegetable products for children is a central point of contemporary discourse. The essence of marketing in children's food has transcended traditional paradigms because of the convergence of technology, environmental concerns, and the decisive role they play in shaping consumers' preferences. Research in this chapter is not only an academic exercise, but an innovative effort to provide meaningful insights for stakeholders in the food industry, policymakers, and marketers. By carefully studying the complex chains of technological advances, environmental impacts, and consumer behavior, the authors seek to chart a way forward for a more informed, sustainable, and consumer-centered approach to the marketing of children's food. This chapter serves as a fundamental contribution to scientific discussions, offering a comprehensive understanding of the evolving dynamics in the area of children's food marketing. The insights obtained from this research promise to guide future strategies, foster innovation, and promote a holistic approach to ensuring healthier and more sustainable consumption patterns among the younger generation.

In the final chapter, titled "Model for Universal Classification of Social Agents' Activity/Behavior in Hierarchical Systems", the authors present the results of elaborating the developed apparatus that allows modeling the activity/behavior of social systems consisting of people and artificial agents. The activity/behavior of a person or artificial agents is described within the universal framework of those environmental changes that a person/object makes. It is shown that there is a finite number of classes of operators that can be used to describe human activity/behavior. The study argues that artificial systems of various origins, such as technical systems, robots, drones, expert and learning systems, and computer bots, can be described in the same terms as human activity/behavior. The obtained results can form a basis of methods for increasing the efficiency of the use of specially structured databases and developed information technologies to accompany and support communication in the process of international negotiations. These results can also form a basis for building new methods of restoring Ukraine's infrastructure in conditions of shortage of highly qualified workers when some people are replaced by robots or robotic complexes.

Vinnytsia, Ukraine   Andriy Semenov
                                                                        semenov.a.o@vntu.edu.ua

Vinnytsia, Ukraine   Iryna Yepifanova
                                                                          yepifanova@vntu.edu.ua

Bratislava, Slovakia   Jana Kajanová
                                                                      jana.kajanova@fm.uniba.sk

Contents

Formation of Strategies for the Development of Startup Ecosystems as a Prerequisite for Sustainable Entrepreneurship

Valentyna Smachylo⑩, Olena Dymchenko⑩, Olha Rudachenko⑩, Iryna Bozhydai⑩, and Yana Khailo

Abstract The goal of this paper is the formation of strategies for the development of startup ecosystems in the country based on clustering as a prerequisite for sustainable entrepreneurship. The analysis of startup ecosystems was based on the data presented in the Global Startup Ecosystem Index 2022 from StartupBlink. This allowed us to form a sample of 100 countries. Two criteria were chosen for cluster formation: Total Score, which demonstrates an overall position in the ranking, and Rank Change (from 2021), which demonstrates the direction of movement of the ecosystem (growth or decline). As a result of clustering, homogeneous groups of countries have been formed that have similar characteristics of startup ecosystems and, therefore, common approaches to further development, which allows the form of strategies for developing startup ecosystems of countries. Four homogeneous clusters of startup ecosystems of countries have been formed by the Startup Blink rating based on cluster analysis using the k-medium method, and their characteristics are given. The first cluster includes only the United States, with a maximum Total Score of 195.37. The second cluster includes 14 countries; with a Total Score of 52.555 (United Kingdom) to 15.914 (Ireland) and minor positive +3 (Singapore) and negative −4 (Switzerland) changes in rank, including without rank changes. The third cluster includes 46 countries; with a Total Score of 0.2600–14.4810 and mostly negative Rank Change changes (maximum −16 positions). The fourth cluster includes 35 countries characterized by: a Total Score level of 0.2750–14.193 and positive Rank Change changes from 1 to 18. Thus, the countries of the first cluster should apply the strategy of absolute leader, generating constant growth and rupture with other countries; the countries of the second cluster—a leadership retention strategy; the countries of the third cluster choose a strategy for correcting (taking into account)

V. Smachylo · O. Dymchenko (✉) · O. Rudachenko · Y. Khailo
O.M. Beketov National University of Urban Economy in Kharkiv, 17, Marshal Bazhanov Street, Kharkiv 61001, Ukraine
e-mail: polkin87@ukr.net

I. Bozhydai
State Biotechnological University, 44, Alchevskikh Street, Kharkiv 61002, Ukraine

errors in order to change the negative value of Rank Change to positive; It is advisable for the countries of the fourth cluster to choose a strategy for gradual growth and demonstrate annual positive growth. Four types of strategy (absolute leader, leadership retention, mistake accounting, gradual growth) have been identified, which are based on four defined clusters and are appropriate for cluster countries to use. The use of strategies will improve the country's startup ecosystem, which will contribute to the sustainability of entrepreneurship.

Keywords Sustainability · Innovation · Entrepreneurship · Startup ecosystem · Development strategy

1 Introduction

The modern world faces a significant number of challenges, which are due to both objective factors and factors that are caused by the activity of the person himself. The international community has recognized that humanity is following the path of self-liquidation through its own lack of awareness and lack of awareness of the impact on the environment, which is manifested in all spheres of life. Entrepreneurship as a separate specific sphere of human activity, which ensures economic growth and forms the GDP of countries, is no exception. Accordingly, international institutions pay a lot of attention to entrepreneurship in concepts and documents that are aimed at ensuring sustainable development as a reconciliation of the interests of society, business, and ecology. Accordingly, there is a need to transfer entrepreneurship to sustainable principles of activity, which allows us to talk about sustainable entrepreneurship, which combines economic growth with sustainable principles of management.

Interesting in this aspect is the theory of Ilian Mikhov, professor of economic and business transformation dean of one of the leading European business schools INSEAD, who developed a theory on the key factors that contribute to economic growth. It is called 4I: innovation, initial conditions, investment, and institutions [1]. He notes that the growth of these factors should occur taking into account each other's condition. Simply making investments and/or increasing them in an unfavorable environment will not improve the situation. Therefore, we should talk about an ecosystem approach to ensuring sustainable entrepreneurship, which will provide for taking into account the interests of all participants in the process and their interaction, the formation of a friendly environment for the entrepreneur, and encourage him to develop on the basis of innovation. A generally recognized driver of innovation development is startups, which are part of the entrepreneurial ecosystem, which are based on the introduction of innovations in all areas of management. And, accordingly, the startup ecosystem is a component of the entrepreneurial ecosystem. Therefore, there are questions that need to be addressed to ensure sustainable entrepreneurship in relation to the development of strategies for the development of startup ecosystems of countries as a prerequisite for economic and sustainable growth, which will ensure the impact on the achievement of sustainable development goals and the implementation of the concept of sustainable entrepreneurship in Ukraine.

Goals formation of strategies for the development of startup ecosystems of the country on the basis of clustering as a prerequisite for sustainable entrepreneurship.

2 Literature Survey

For the first time, the startup ecosystem became known by James Moore [2], who proved that the ecosystem includes the internal and external environment, and the author first introduced the concept of "entrepreneurial ecosystem". Thus, Humenna and Hanushchak-Yefimenko [3] see the ecosystem as a set of institutions that effectively interact in the economic system. Agnihotri [4]. He thoroughly described the modern startup ecosystem, where he substantiated that the creation of startups makes the "business" of a career that deserves attention outside the traditional trading community, a political path for creating startups and success in large-scale improvement. The study of factors that make it possible to evaluate the startup ecosystem was also engaged by Ramesh Menon [5].

The author has developed a comprehensive valuation model for startups that are still not generating revenue, including qualitative factors that help venture capital investors assign them value. The relationship between venture capital and the startup ecosystem is also represented by the authors Burstrom et al. [6], where they present that the ecosystem architecture in a segmentation matrix of investor types and roles, including Active Hubs and Complementors. The entrepreneurial startup ecosystem as a tool for measuring its elements and using it to further compile an index of the entrepreneurial ecosystem in the Netherlands was investigated by Stam and van de Ven [7].

In the entrepreneurial aspect, the startup ecosystem was also explored by Jacobides et al. [8], where the authors believe that ecosystems enable organizations to coordinate their multilateral dependence through a set of roles that are guided by similar rules so that it is possible to enter into individual contractual or non-contractual agreements with each of the partners. Van Vulpen et al. [9] investigate management activities through an ecosystem that creates incentives that take into account the following factors: mutual trust, knowledge absorption, lifecycle flow, openness to business discussion, orchestrator leadership, and external alignment. A more detailed description of the startup ecosystem is also presented by the authors in the work [10], where the startup ecosystem means an environment in which people and organizations (enterprises) are involved that create and support scientific developments and the possibility of their entry into the market. Also, in [11], the authors understand the startup ecosystem as "an interacting and interdependent set of institutions whose activities create an environment for the qualitative and quantitative growth of startups as subjects of innovative development of entrepreneurship."

This brings us to the point, which is that is advisable to consider the ecosystem as an interacting set of various agents of the national economy (commercial and non-commercial nature) between which there are relationships that allow the creation of a favorable environment. We will interpret a startup as an initial specific form of entrepreneurial activity, which is based on innovative approaches to solving problems and meeting needs forms economic and social value through a scalable and high-risk business model. Thus, the startup of ecosystems is determined by the interacting and interdependent set of institutions whose activities contribute to the qualitative and quantitative growth of startups as subjects of innovative development of entrepreneurship [11]. As a result, the basis of the startup ecosystem is the environment, favorable development of startups, as well as their qualitative and quantitative characteristics. This vision corresponds to the components by which the Startup Blink (2022) rating is built [12] and is determined by the Global Startup Ecosystem Index [13]. Therefore, it is advisable to put such structuring (quantitative component, qualitative component, business environment) into the basis of the study.

Thus, the modern startup ecosystem is an important element in ensuring sustainable entrepreneurship. Thus, authors Watson et al. [14] believe that sustainable entrepreneurship can contribute to the sustainable development of the country by finding synergies between social, environmental, and economic results. The conceptual model of entrepreneurship and the promotion of its economic growth, taking into account the preservation of ecology and the development of society, was investigated by the author—Agrawal [15], which invites existing entrepreneurs, potential entrepreneurs, politicians, academics and researchers to promote the cause of sustainable development in all aspects of business and research. However, it is worth noting that the basis for ensuring sustainable entrepreneurship is the right management approach, ensuring the achievement of strategic and operational goals at different levels of management.

Scientific and theoretical studies of the basic principles for determining the essence of strategic management were devoted to the fundamental works of both domestic and foreign scientists. It is believed that, nevertheless, the founders of strategic management are Ansoff [16], Cassels [17], Pierce and Robinson [18] and others. The concept of "strategy" is quite broad and can refer to many systems and phenomena. Thus, the analysis of most scientific works on the study of the essence of the concept of "strategic management" showed that there is no single approach to its definition. The process of strategic management itself is a method by which a strategy is developed and implemented that can lead to a sustainable competitive advantage for an enterprise, community, region, or country. However, in general, under the strategy, we propose to consider a plan, or rather a document, to achieve the goal. The development of a long-term strategic document is necessary to identify current trends and patterns of local, regional, and state development, to form on this basis scenarios for promising socio-economic and environmental development, and to determine the stages and terms of achieving priorities. To ensure the implementation of certain priority areas of the strategy, targeted programs for the development of individual areas of activity are being developed and implemented, which should become an effective tool for implementing long-term policy. Strategy formulation is an analytical process of choosing the most suitable course of action to achieve goals.

It is worth noting that the formulation of the strategy should cover three levels:

- strategic plan (why?);
- tactical plan (what?);
- operational plan (how?).

Thus, in Ukraine, the main document ensuring national interests in the sustainable development of the economy, civil society and the state to achieve an increase in the level and quality of life of the population, observance of constitutional rights and freedoms of man and citizen is the Decree of the President of Ukraine of September 30, 2019 No. 722/2019 "On the Sustainable Development Goals of Ukraine for the period up to 2030" [19], which provides for ensuring and complying with the 17 Sustainable Development Goals (SDG) of Ukraine for the period up to 2030. Also currently being approved by the Cabinet of Ministers of Ukraine (CMU) is the "Draft Strategy for Sustainable Development of Ukraine until 2030" [20], which defines the strategic directions of Ukraine's long-term development. The principal aspect of its development is to take into account the SDGs adapted for Ukraine until 2030 and take into account the main provisions of the Updated EU Sustainable Development Strategy. The developed Strategy for Sustainable Development of Ukraine until 2030 and the National Action Plan, subject to their approval at the state level, can later become an effective tool for implementing the principles of balanced development in the regions and settlements of Ukraine by taking into account the main provisions in their development strategies. This document will be especially important for writing strategies for the development of territorial communities. As well, both at the regional and state levels, it is necessary to adhere to the strategic and operational goals of the transition to integrated economic, social and environmental development, taking into account all the envisaged goals approved by the CMU Resolution.

It should be noted that one of the main tasks of the Government of Ukraine to ensure the Sustainable Development Goals in the announced documents of the Cabinet of Ministers of Ukraine is to create a favorable environment for doing business, developing small and medium-sized businesses, attracting investment and improving the efficiency of the labor market. Accordingly, the development of strategies for the startup ecosystem of the country has a direct impact on the achievement of sustainable development goals and the implementation of the concept of sustainable entrepreneurship in Ukraine.

3 Methods

The clustering method allows to representation of a significant sample of indicators in the form of generalized homogeneous groups constructed in accordance with and within a set of criteria or features [21, 22].

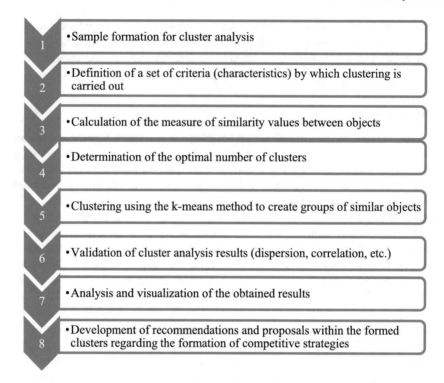

Fig. 1 Algorithm for cluster analysis. *Source* [23–25]

In accordance with Fig. 1, we have the following sequence of clustering [23–25]: the formation of a sample of objects; definition of criteria; conducting a hierarchical cluster procedure in order to form a hypothesis regarding natural clustering; hypothesis testing by the k-mean method; verification of the reliability of the results.

Data analysis is carried out using the SPSS Static software product.

The analysis of the characteristics of startup ecosystems of countries was based on the data presented in the Global Startup Ecosystem Index 2022 from StartupBlink. This allowed us to form a sample of 100 countries.

The choice of this particular rating is due to the presence of a structured startup ecosystem, which is described by components containing relevant indicators and descriptions.

The methodology for calculating the startup ecosystem index Startup Blink is based on 3 components (Fig. 2).

Two criteria have been chosen for cluster formation: Total Score, which demonstrates an overall position in the ranking, and Rank Change (from 2021), which demonstrates the direction of movement of the ecosystem (growth or decline) compared to the previous year (2021).

To solve the problem of cluster analysis, it is necessary to define the concepts of similarity and heterogeneity.

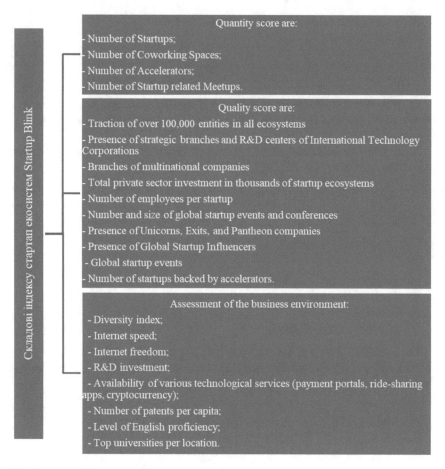

Fig. 2 Startup blink ecosystem index components. *Source* [12]

Objects *i*th and *j*th would fall into the same cluster when the distance between points X_i and H_j is quite small and would fall into different clusters when this distance is large enough. Thus, the involvement in one or different clusters of objects is determined by the concept of the distance between X_i and H_j with Er, where Er is the r-dimensional Euclidean space.

Clusters and their optimal number are determined using different methods and the most used integral measures—the square of the Euclidean distance and the Euclid distance.

In order to establish the significance of the selected factors and the interdependence between the defined sampling criteria, a variance analysis was carried out, which allows, using Fisher statistics (F-relation), to determine the proportion of the distribution between group and intragroup variability and the corresponding statistical significance of this difference [23–25].

We determine the statistics of Fisher by the formula.

$$F = \frac{S_1^2}{S_2^2},\tag{1}$$

S_1^2, S_2^2—values of estimates of greater and smaller variances, respectively. The numbers of degrees of freedom to search for a critical value are chosen equal $n_1 - 1$ and $n_2 - 1$.

To exclude the interdependence between given indicators, we use the Pearson correlation coefficient [23–25], which is determined by the formula:

$$r = \frac{\sum_{i=1}^{n}(x_i - \overline{X})(y_i - \overline{Y})}{(n-1)s_x s_y},\tag{2}$$

where

x_i, y_i the value of two variables,
$\overline{X}, \overline{Y}$ Their average values,
s_x, s_y Standard deviations,
n number of pairs of values.

The Euclidean metric is the most popular. The Euclidean distance between two points x and y is the shortest distance between them. In the case of k variables, the Euclidean distance is calculated by the formula [23–25]:

$$d_2(X_i, X_j) = \left[\sum_{k=1}^{p}(x_{ki} - x_{kj})^2\right]^{\frac{1}{2}}\tag{3}$$

In the case of cluster analysis of objects, most often the measure of divergence is the square of the Euclidean distance

$$d_{ij}^2 = \sum_{h-1}^{m}(x_{ih} - x_{jh})^2\tag{4}$$

where x_{ih}, x_{jh}—the value of the hth attribute for the ith and jth objects, and m—is the number of characteristics.

Compared to Euclidean distance, this measure attaches more serious importance to large distances.

Clustering using the k-mean method distributes the input set of vectors by k clusters $S_i (i = 1, 2, ..., k)$, each of which is associated with a centroid c_i. Denote the set of input vectors $S = \{x\}$, $|S| = n$. Нехай $D(x, c)$ the distance between the vector x and the centroid c:

$$D^2(x, c) = \sum_{i=1}^{d} (x_i - c_i)^2 \tag{5}$$

Denote the set of centroids obtained on iteration t, $SC_t = \{c_i\}$. The algorithm for clustering k-averages in its usual version is described as follows:

1. Set $t = 0$ and set the initial location of the centroids SC_0.
2. For a given set of centroids, SC_t perfom actions, specified in paragraphs 2.1 and 2.2, and we obtain an improved set of centroids SC_{t+1}:

   2.1 We fund such a partition of S, which distributes S by k clusters S_i (i=1, 2, ..., k) and satisfies the condition.

   $$S_i = \{x | D(x, c_i) \leq D(x, c_j) \forall_j \neq i\} \tag{6}$$

   2.2 Calculate the centroid c_i for each cluster S_i (i=1, 2, ..., k), to get a new set of centroids SC_{t+1}:

   $$c_{ij} = \frac{1}{m_i} \left(\sum_{l=1}^{m_i} x_{lj} \right), j = 1, 2, ..., d, \tag{7}$$

   where m_i—number of vectors belonging to the cluster S_i.

3. Calculate the total distortion $E^2 = \sum_{x \in S} D^2(x, c)$ для SC_{t+1}. If it differs from that obtained on the previous iteration by a sufficiently small value, we stop the process. Otherwise, we assign $t \leftarrow t + 1$ and return to step 2.

The algorithm is guaranteed to coincide with a finite number of iterations. The clustering error and the number of iterations depend on the initial choice of centroids, so it is common practice to run the k-average several times with different initial candidates for centroids [20, 23, 24].

It is advisable to check the obtained results of clustering for significance using the same indicators of variance, correlation, two-stage cluster analysis, etc.

On the basis of clustering, it is possible to form homogeneous groups of countries that have similar characteristics of startup ecosystems, and, therefore, common approaches to their further development, which will allow to development of appropriate development strategies.

Using the matrix approach allows you to develop 4 types of strategies that correspond to the 4 quadrants of the matrix and are defined within the Total Score and Rank Change axes (from 2021).

4 Discussion

Hierarchical cluster analysis allows you to justify the optimal number of clusters. For this purpose, values were standardized in the range from 0 to 1 of the Rank Change, since it includes both negative and positive numeric values. The most commonly used interval measures in this case are the Euclidean distance or the square of the Euclidean distance, which we use to compare the results obtained. The number of clusters obtained as a result of hierarchical clustering by different methods and measures of distance is presented in Table 1.

Based on the data in the table, it is advisable to form 4 corresponding clusters. We check the obtained results by the method of two-stage cluster analysis (Fig. 3), which confirms a good silhouette measure of connectedness and distribution of clusters.

After confirming the hypothesis of the feasibility of forming 4 clusters, we will carry out clustering using the K-mean method.

Using the SPSS Statistic software product, we define the initial centers of the clusters (Table 2), and conduct the required number of iterations to achieve minimization of the distance of each sample element to the center of the cluster.

Similarity is achieved by the criterion of small value. The maximum absolute coordinate change of any cluster is 0.000. 10 iterations were made. The minimum distance between the initial centers is 34,248. In accordance with the research methodology and the obtained data, we form in tabular form the final centroids of the defined clusters (Table 3).

Table 1 The number of clusters depending on the clustering method

The Hierarchical clustering method	The options	
	Euclidean distance square	Euclidean distance
1. The Intergroup relationships	4	4
2. The Intragroup relationships	5/4	4
3. The closest neighbor	3	3/5
4. The distant neighbor	4	4
5. The centroid clustering	4	4
6. The median clustering	4	4
7. The ward method	4	4

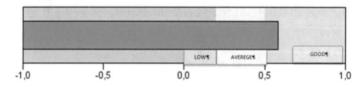

Fig. 3 Cluster quality based on the results of two-stage cluster analysis

Table 2 Initial cluster centers

	Cluster			
	1	2	3	4
Rank change (from 2021)	0	0	−16	18
Total score	195.3700	52.5550	4.3890	0.2750

Table 3 Cluster end centers

	Cluster			
	1	2	3	4
Rank change (from 2021)	0	0	−4	5
Total score	195.3700	26.2145	4.7666	3.5426

The procedure allowed to minimize discrepancies in the indicators of certain groups of clusters and to form the most homogeneous data within their limits. To test the hypothesis of the relationship of two variables after the previous ranking, we use the Spearman's rank correlation coefficient and Kendall rank correlation coefficient (Table 4).

Table 4 Spearman's rank correlation coefficient and Kendall rank correlation coefficient

The title of the correlation	Indicator	Meaning	Rank change (from 2021)	Total score
Kendall rank correlation coefficient	Rank change (from 2021)	Correlation coefficient	1.000	0.071
		Meaning (2-sided)		0.338
		N	96	96
	Total score	Correlation coefficient	0.071	1.000
		Meaning (2-sided)	0.338	
		N	96	100
Spearman's rank correlation coefficient	Rank change (from 2021)	Correlation coefficient	1.000	0.090
		Meaning (2-sided)		0.383
		N	96	96
	Total score	Correlation coefficient	0.090	1.000
		Meaning (2-sided)	0.383	
		N	96	96

For variables belonging to the ordinal scale, the Spearman's rank correlation coefficient is calculated; when calculating it, the indicators of the compared variables (arithmetic mean and variance) that are not related to the distribution are taken into account but ranked.

The Kendall rank correlation coefficient (concordation coefficient) is an independent original method based on calculating the ratio of pairs of values of two samples that have the same or different trends. Applying the Kendall rank correlation coefficient is preferable if the source data has emissions.

Correlation analysis of data allows us to conclude that the relationship between the signs is insignificant.

Generalized information on the number of countries of the maximum and minimum values of indicators in each cluster is given in Table 5.

Thus, as a result of cluster analysis, 4 homogeneous clusters were formed.

Only the United States belongs to cluster 1, with a maximum Total Score of 195.37. The analysis of this cluster shows that for the period from 2021 to 2022, United States still dominates the ranking of global startup ecosystems. In 2022, the United States maintained a significant gap between itself and the rest of the world.

At the same time, the gap in the overall score has slightly decreased between the United States and lower-ranking countries.

The second cluster includes 14 countries representing the leaders of the startup movement, which are in the top 15. Their rates range: Total Score from 52.555 (United Kingdom) to 15.914 (Irland) and are characterized by minor positive +3 (Singapore), +1 (Sweden, Australia), and negative −4 (Switzerland), −1 (Germany) changes in rank, including without changes in rank (United Kingdom, Israel, Canada, The Netherlands, Estonia, Finland). The lack of rank change indicates a great effort by countries that are aimed at developing all components of the country's startup ecosystem, but this mainly concerns the quantitative and qualitative components. After all, according to [11], the level of the business environment in the countries of the 2nd cluster is of high importance, sometimes even higher than that of the leader of the USA rating.

The third cluster includes 46 countries; with Total Score from 0.2600 to 14.481 and, mostly, negative changes in Rank Change (maximum −16 positions—Ukraine; −15—Rwanda, Jamaica; −12—Russia; −11—North Macedonia, Moldova; −10—Bosnia and Herzegovina). Without changes in the ranking of Bangladesh (93rd position), Czechia (32nd position). Also in the cluster, there is a startup ecosystem of countries that have minimal positive change (+1)—Belgium, Taiwan, India, Japan.

The fourth cluster includes 35 countries characterized by: Total Score level from 0.2750 to 14.193 and positive Rank Change changes from 1 (Serbia, Malta) to 18 (Angola), 16 (Morocco), 14 (Iceland), 13 (Senegal). Verification of the reliability of the results obtained is carried out on the basis of the indicators of dispersion analysis (**ANOVA**), which are given in Table 6.

Table 6 shows the values of Fisher's statistics indicators (F-criterion), which in this case is used only as an indicator since clusters were chosen in such a way as to maximize the discrepancy between the indicators of different clusters. At the same time, the trend of significance indicators at and $\rightarrow 0$ is positive.

Table 5 Generalized information on clustering startup ecosystems of countries

Cluster	Rank	Country	Rank change (from 2021)		Total score		Number of countries
			Min.	Max.	Min.	Max.	
1	1	United States	0	0	195.37	195.37	1
2	2–15	United Kingdom, Israel, Canada, Sweden, Germany, Singapore, Australia, France, China, The Netherlands, Switzerland, Estonia, Finland, Ireland	−4	3	15.9140	52.5550	14
3	16, 17, 19, 20, 21, 22, 25, 26, 27, 28, 29, 31, 32, 33, 36, 42, 43, 45, 46, 47, 49, 50, 51, 53, 56, 57, 58, 62, 63, 66, 69, 70, 72, 76, 77, 82, 83, 84, 86, 87, 88, 93, 94, 95, 98, 99	Spain, Lithuania, India, Japan, South Korea, Belgium, Taiwan, Brazil, United Arab Emirates, Portugal, Russia, Italy, Czechia, Poland, Bulgaria, Malaysia, Latvia, Croatia, Turkey, Slovenia, South Africa, Ukraine, Hungary, Thailand, Uruguay, Philippines, Slovakia, Kenya, Peru, Jordan, North Macedonia, Belarus, Saudi Arabia, Pakistan, Lebanon, Ghana, Tunisia, Rwanda, Qatar, Ecuador, Moldova, Bangladesh, Jamaica Bosnia and Herzegovina, Somalia, Kuwait	−6	11	0.2600	14.481	46

(continued)

Table 5 (continued)

Cluster	Rank	Country	Rank change (from 2021)		Total score		Number of countries
			Min.	Max.	Min.	Max.	
4	18, 23, 24, 30, 34, 35, 37, 38, 39, 40, 41, 44, 48, 52, 54, 55, 59, 60, 61, 64, 65, 67, 71, 73, 74, 75, 78, 79, 80, 81, 85, 90, 91, 92, 97	Denmark, Austria, Norway, New Zealand, Chile, Mexico, Argentina, Indonesia, Romania, Luxembourg, Iceland, Colombia, Greece, Serbia, Vietnam, Cyprus, Malta, Armenia, Nigeria, Bahrain, Egypt, Costa Rica, Mauritius, Georgia, Kazakhstan, Albania, Panama, Morocco, Cape Verde, Mongolia, Azerbaijan, Sri Lanka, Namibia, Senegal, Angola	1	18	0.2750	14.193	35

Table 6 Results of dispersion analysis (**ANOVA**)

	Cluster		Accuracy		F	Meaning
	Mean square	Degrees of freedom	Medium square	Degrees of freedom		
Rank change (from 2021)	509.849	3	17.086	92	29.840	,000
Total score	13,577.090	3	29.989	92	452.733	,000

Thus, we will formulate descriptive characteristics of startup ecosystems of countries depending on clusters and present them in Table 7.

Graphically, the dispersion of startup ecosystems of the countries of the world is presented in Fig. 4.

In accordance with Fig. 4 we will present the placement of clusters in the coordinate system, where we will put the Total Score along the abscissa axis and Rank Change—the ordinate axis (Fig. 5).

Table 7 Generalized characteristics of startup ecosystems of countries depending on the cluster

Cluster	Characteristics of the startup ecosystem of the cluster countries
1	High level of development startup ecosystem of the country, a leader among 100 countries of the world. The gap with the next startup ecosystem is slowly shrinking from 4.3 in 2021 to 3.7 in 2022. The advantage is formed by the Quality Score (164.15) and, partly, the Quantity Score (27.56), while the Business Score of 3.66 is even lower than that of lower-ranked countries
2	High level of development of the startup ecosystem of the country; Top 15 countries in the world. There are no significant changes in the rating, only within the cluster both upwards (+3 positions) and downwards (−4 positions). Total Score ranges from 52.5550 (2nd position in the overall ranking) to 15.9140 (15th position in the overall ranking). Cluster ecosystems have a developed Business Score (2.24–3.80), which corresponds to the level of the leader—cluster 1. More weight in the overall result is occupied by the Quality Score, which corresponds to the trends of the cluster of the leading country
3	The startup ecosystem of the countries of this cluster is characterized by a drop in the rating during the year (maximum −16 positions); a decrease in the value of all components of the index compared to the values in previous clusters (up to 0.07); changes in the weight of the components of the startup ecosystem index—the Quantity Score begins to prevail
4	The startup ecosystem of the countries of this cluster is characterized by growth in the ranking during the year (maximum 18 positions); in the specific gravity of the formation of Total Score, the role of the components varies

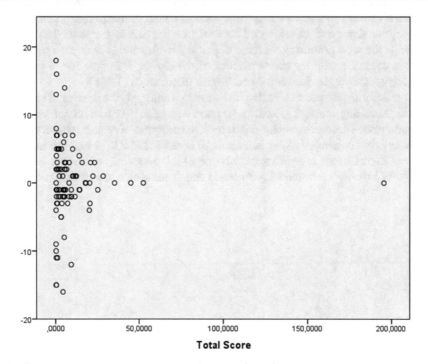

Fig. 4 Scatter diagram startup ecosystems of the countries of the world

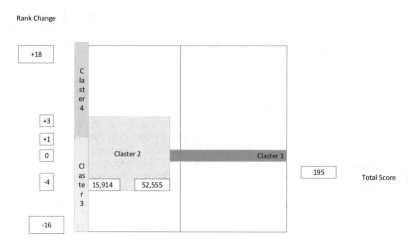

Fig. 5 Clusters of startup ecosystems in countries

As can be seen from Fig. 5, cluster 1 is in the middle of all quadrants and heads towards the maximum value. Cluster 2 covers two quadrants with both positive and negative changes in Rank Change. Cluster 3 is located in the lower corner of the matrix, which is characterized by both low Total Score values and negative growth rates. This cluster contains a startup ecosystem of countries that have negative trends and require significant changes and decisions on the part of the state. The fourth cluster is the most promising, as it has high growth rates and growth potential.

In accordance with the characteristics of the cluster, it is possible to propose development strategies for startup ecosystems of countries (Table 8).

Consequently, the countries of the first cluster should apply the strategy of absolute leader, generating constant growth and rupture with other countries; countries of the second cluster—leadership retention strategy; the countries of the third cluster choose a strategy for correcting (taking into account) errors in order to change the negative value of Rank Change to positive; it is advisable for the countries of the fourth cluster to choose a strategy for gradual growth and demonstrate annual positive growth.

Table 8 Characteristics of development strategies for startup ecosystems of cluster countries

Cluster	The strategy	Characteristics of the strategy
1	The Absolute leader Be faster!	Holding and increasing the gap; enhanced development of Business Score and Quantity Score; further development of Quality Score
2	Retaining leadership or building up "Catch up and get ahead Follower strategy"	Maintaining/strengthening the rate of narrowing the gap with the country leader. Enhanced development of Quantity Score, Quality Score; retention and Business Score build-up
3	Error correction "Find and fix the fault"	Analysis and identification of the causes of the fall; development of measures to correct errors; build-up at a higher rate and retention of the Quantity Score, Quality Score, and Business Score
4	Gradual growth "Do what you do"	Retaining/increasing growth rates; increasing at a higher rate and retaining Quantity Score, Quality Score, and Business Score

5 Conclusion

1. A detailed description of the ecosystem, entrepreneurship and strategy is given, unresolved problems regarding the development of startup ecosystems of countries are highlighted, which enabled the authors to form their own vision of scientific issues with ensuring sustainable development of entrepreneurship. The author's team determined that it is advisable to consider the ecosystem as an interacting set of various agents of the national economy (commercial and non-commercial nature) between which there are relationships that allow the creation a favorable environment; startup as an initial specific form of entrepreneurial activity, which is based on innovative approaches to problem-solving and meeting needs, forms economic and social value through a scalable and high-risk business model. Accordingly, the startup ecosystem is defined as an interacting and interdependent set of institutions whose activities contribute to the qualitative and quantitative growth of startups as subjects of innovative development of entrepreneurship.

2. 4 homogeneous clusters of startup ecosystems of countries have been formed in accordance with the Startup Blink rating based on cluster analysis using the k-medium method, and their characteristics are given. The first cluster includes only the United States, with a maximum Total Score of 195.37. The second cluster includes 14 countries; with a Total Score of 52.555 (United Kingdom) to 15.914 (Irland) and minor positive +3 (Singapore) and negative −4 (Switzerland) changes in rank, including without rank changes. The third cluster includes 46 countries; with a Total Score of 0.2600 to 14.4810 and mostly negative Rank

Change changes (maximum −16 positions). The fourth cluster includes 35 countries characterized by: a Total Score level of 0.2750 to 14.193 and positive Rank Change changes from 1 to 18.

3. There are 4 types of strategy (absolute leader, leadership retention, taking into account mistakes, gradual growth), which are based on 4 defined clusters and are appropriate for use by cluster countries. The countries of the first cluster should apply the strategy of absolute leader, generating constant growth and rupture with other countries; the countries of the second cluster—a leadership retention strategy; the countries of the third cluster choose a strategy for correcting (taking into account) errors in order to change the negative value of Rank Change to positive; It is advisable for the countries of the fourth cluster to choose a strategy for gradual growth and demonstrate annual positive growth. The use of strategies will improve the country's startup ecosystem, which will contribute to the sustainability of entrepreneurship.

References

1. To restore post-war Ukraine, it is necessary to create a comfortable environment for business and specialists. (2023) https://mim.kyiv.ua/news/article/dlya-vdnovlennya-povonno-ukrani-neobhdno-stvoryuvati-komfortne-dlya-bznesu-fahvcv-seredovische-dekan-insead-lan-mhov [in Ukrainian]
2. Moore, J.F.: Predators and prey: a new ecology of competition. Harv. Bus. Rev. **71**(3), 75–86 (1993)
3. Humenna, O.V., Hanushchak-Yefimenko, L.M.: Formuvannia spozhyvchoi tsinnosti znan v innovatsiinii ekosystemi (2016). https://er.knutd.edu.ua/bitstream/123456789/2684/1/201 61021_103.pdf. Accessed 20 Jan 2022
4. Agnihotri, D.: Startup ecosystem. Manag. Forum 1–13 (2018)
5. Menon, R.: Estudio Empírico de los Factores Relevantes en la Metodología de Valuación de Startups. Valuación de Startups InnOvaciOnes de NegOciOs **19**(37), 56–88 (2022). https://doi.org/10.29105/revin19.37-392
6. Burstrom, T., Lahti, T., Parida, V., Wartiovaara, M., Wincent, J.: A definition, review, and extension of global ecosystems theory: trends, architecture and orchestration of global VCs and mechanisms behind unicorns. J. Bus. Res. **157**, 113605 (2023). https://doi.org/10.1016/j.jbusres.2022.113605
7. Stam, E., Van de Ven, A.: Entrepreneurial ecosystem elements. Small Bus. Econ. **56**(2), 809–832 (2021). https://doi.org/10.1007/s11187-019-00270-6
8. Jacobides, M.G., Cennamo, C., Gawer, A.: Towards a theory of ecosystems. Strateg. Manag. J.. Manag. J. **39**(8), 2255–2276 (2018). https://doi.org/10.1002/smj.2904
9. Van Vulpen, P., Jansen, S., Brinkkemper, S.: The orchestrator's partner management framework for software ecosystems. Sci. Comput. Program. **213**(1–31), Article 102722 (2022). https://doi.org/10.1016/j.scico.2021.102722
10. Kyzym, M., Dymchenko, O., Smachylo, V., Rudachenko, O., Dril, N.: Cluster analysis usage as prerequisite for implementing strategies of countries startup ecosystems development. In: Arsenyeva, O., Romanova, T., Sukhonos, M., Tsegelnyk, Y. (eds.) Smart Technologies in Urban Engineering. STUE 2022. Lecture Notes in Networks and Systems, vol 536, pp. 290–301. Springer, Cham (2023). https://doi.org/10.1007/978-3-031-20141-7_27

11. Dymchenko, O., Smachylo, V., Rudachenko, O., Palant, O., Ye, K.: Modeling the Influence of startup ecosystem components: entrepreneurial aspect. Sci. Horiz. **25**(11), 131–140 (2022). https://doi.org/10.48077/scihor.25(11).2022.131-140
12. Startup Blink (2022). https://lp.startupblink.com/report/. Accessed 20 Jan 2022
13. Global Startup Ecosystem Index (2022). https://www.startupblink.com/startupecosystemre port. Accessed 20 Jan 2022
14. Watson, R., Nielsen, K., Wilson, H., Macdonald, E., Mera, C., Reisch, L.: Policy for sustainable entrepreneurship: A crowdsourced framework. J. Clean. Prod. **383**, 135234 (2023)
15. Agrawal, R.: Conceptual framework for sustainable entrepreneur-ship. J. Appl. Manag. Jidnyasa **14**(1), 61–74 (2022)
16. Ansoff, H.I.: Implanting Strategic Management, p. 510. Prentice/Hall International, Englewood Cliffs, N.J. (1984)
17. Cassels, E.: Book 1. Introduction. Strategy, p. 432. The Open University, Walton Hall, Milton Keynes (2000)
18. Pearce II, J.A., Robinson, R.B. Jr.: Strategic Management, 2nd edn. Homewood, III: Richard D. lrwin (1985)
19. Decree of the President of Ukraine.: On the Sustainable Development Goals of Ukraine until 2030 (2019). https://zakon.rada.gov.ua/laws/show/722/2019#Text. Accessed 20 Jan 2022 [in Ukrainian]
20. Law of Ukraine.: On the Strategy for Sustainable Development of Ukraine until 2030 (2018). https://ips.ligazakon.net/document/JH6YF00A?an=332. Accessed 20 Jan 2022 [in Ukrainian]
21. Kryvinska, N., Auer, L., Strauss, C.: Managing an increased service heterogeneity in a converged enterprise infrastructure with SOA. IJWGS **4**, 440 (2008). https://doi.org/10.1504/IJWGS.2008.022546
22. Kryvinska, N., Auer, L., Strauss, C.: The place and value of SOA in building 2.0-generation enterprise unified versus ubiquitous communication and collaboration platform. In: 2009 Third International Conference on Mobile Ubiquitous Computing, Systems, Services and Technologies (2009). https://doi.org/10.1109/UBICOMM.2009.52
23. Bogiday, I.: Clusterization of agro-industrial enterprises of Ukraine as the basis of effective strategic management. Agric. Resour. Econ. **5**(2), 86–98 (2019). https://doi.org/10.22004/ag.econ.290315
24. Sotska, Y.I.: Methodological basis of cluster analysis of the competitiveness of Ukrainian banks. Financ. Credit Act. Probl. Theory Pract. **2**(19), 177–185 (2015). https://doi.org/10.18371/fca ptp.v2i19.57261 [in Ukrainian]
25. Scitovski, R., Sabo, K., Martínez-Álvarez, F., Ungar, S.: Cluster Analysis and Applications, p. 271. Springer Cham (2021). https://doi.org/10.1007/978-3-030-74552-3. https://doi.org/10.1007/978-3-030-74552-3#bibliographic-information

The Strategy of Sustainable Development of Digital Business in the Conditions of the Variability of the Business Environment and European Integration

Oksana Polinkevych⦿, Olena Kuzmak⦿, and Oleh Kuzmak⦿

Abstract In the conditions of the development of digital technologies, the business environment faces new challenges to ensure sustainable development. Accordingly, processes arise that require the development of management mechanisms. One such tool for building a digital business towards the goals of sustainable development and European integration is the construction of a strategy for the sustainable development of digital business in the conditions of the variability of the business environment. Variability is one of the characteristics of the digital economy. The purpose of the study is to develop a strategy for the sustainable development of digital business by the declared indicators of the development of countries, territories, and communities. The research aims to identify changes in the business environment and assess trends in sustainable development and digital business. On this basis, the sustainable development index for individual EU countries and Ukraine has been determined. This index for Ukraine proved the need to develop a strategy for the sustainable development of digital business as a tool for achieving the goals of sustainable development. The dependence is deduced that digital business is developing rapidly and, by 2030, will prevail in all economies of the countries. Accordingly, the business environment in Ukraine should adapt such that it will ensure competitiveness not only in domestic but also in international markets. The strategy of sustainable development of digital business in the conditions of a changing environment should include objects, subjects, tools, and methods of achieving the planned goals. It will become decisive. In the post-war transformation of the economy of Ukraine and the development of the state based on the principles of social partnership. The main strategic directions of development are the development of business based on sustainable development through the activation of entrepreneurial activity, the use of digital tools in the activities of

O. Polinkevych (✉) · O. Kuzmak · O. Kuzmak
Lutsk National Technical University, Lvivska Street, 75, Lutsk 43018, Ukraine
e-mail: o.polinkevych@lutsk-ntu.com.ua

O. Kuzmak
e-mail: o.kuzmak@lutsk-ntu.com.ua

O. Kuzmak
e-mail: oleh_kuzmak@lutsk-ntu.com.ua

21

business structures through the formation of a digital business model, the formation of an eco-friendly space for interaction and the improvement of the management system, and the activation of development processes with using lean production. The consequences of the implementation of the strategy of sustainable development of circular business in Ukraine and its absence from the economic development of the country are determined.

Keywords Sustainable development · Digital business · Changing environment · European integration · Digitalization · The index of sustainable development · Industry 4.0

1 Assessment of the Business Environment of Digital Business in Conditions of Variability

1.1 Trends of Changes in the Business Environment of Digital Business in Conditions of Variability

The world economy develops cyclically. A rapid economic rise causes a deep decline, which is expressed in a crisis. In 2008–2009, the global financial crisis took place, which negatively affected the competitiveness of domestic product manufacturers. In 2014, an anti-terrorist operation began in eastern Ukraine. In 2019, businesses faced the challenges of the COVID-19 pandemic. The year 2022 was marked by the fact that war began in Ukraine due to armed aggression on the part of Russia. Due to these factors, a recession occurs in business systems, which involves a decrease in business activity, a decrease in consumer spending, bankruptcy of enterprises, and an increase in the unemployment rate. Economists consider it part of a development cycle, where there is growth, reaching a peak, falling, the lowest level, and the next level of growth. Since 1948, 11 recessions have been recorded in the USA [1]. Therefore, the business system needs changes and quick adaptation to new business conditions.

Thus, these processes in the economy affect business systems, and it is quite difficult to predict the vector of development of both a business entity and a region or a state in general [2, 3].

The business environment of the twenty-first century can be characterized as changing, uncertain, complex, and ambiguous [4]. Variability is the nature and dynamics of change, as well as the nature and speed of changing forces and catalysts. Uncertainty is a lack of predictability, the prospect of surprise, and a sense of awareness and understanding of problems and events [5]. Complexity describes the multitude of factors that define a problem surrounding a business organization [6]. Ambiguity is associated with an uncertain reality, the possibility of misunderstanding, and determining cause-and-effect relationships. Joyner V. defined four management styles [2]:

(1) 1st generation management—management through actions: performing tasks independently. Society has the necessary knowledge, skills, abilities, and technologies. All this contributes to the satisfaction of consumer requirements;

(2) 2nd generation management—management through leadership: employees can expand their capabilities if superior managers tell them how to do it;

(3) 3rd generation management—management by results: the manager rewards or punishes employees based on how well they perform tasks;

(4) 4th generation management—a systemic approach: in contrast to the first three generations of management, which characterizes a traditional organization (i.e. an organization closed to stakeholders) and is characterized by bureaucracy and the old role of the manager, which creates barriers between him and his people, this is a holistic an approach that allows you to manage the business as a system (that is, as a whole) by doing it open to stakeholders (involving stakeholders) but only customer-oriented (quality-oriented) [7, 8].

Fourth-generation management is the first management style for enterprises that have moved from a traditional organization to a modern one. In the 1990s, Joyner set out to pioneer quality-oriented business generation through a customer-centric systems approach. Fourth-generation management is a significant step forward through the interplay of effective management and increased productivity. The key control elements of the 4th generation are [2]:

1. Quality is the understanding that the client determines quality. It is necessary to satisfy the client's requests, to exclude things that cause irritation, egoism, anger, and aggression from the process of interaction, and to offer a product or service in an interesting form. All this is done to draw his attention to those characteristics that are important to him and solve his needs. Not only employees of one department but also all employees of the enterprise should work on this.

2. Scientific approach—learn to manage the business as a system, develop process thinking (systems or systems thinking in a business environment), make data-driven decisions, and understand variation.

3. All are one team—faith in people, mutual respect, trust, and mutual support in the team. Teamwork should be aimed at benefiting all stakeholders, such as customers, employees, shareholders, suppliers, and the communities in which we live. The main goal of working in a team should be aimed at constantly increasing the level of customer satisfaction in various areas: economic, social, psychological, and legal.

The business environment can be defined as a set of factors within which a business functions to make a profit or achieve a sustainable socio-psychological effect. From the point of view of a specific enterprise, the business environment is a set of uncontrollable (external environment) and controllable (internal environment) factors that change approaches to management and development of business structures [9, p. 194]. Accordingly, for an enterprise, the external environment is a set of system-forming factors [10, 11]: business communication factors (unifying parameters of the environment), resources (auxiliary facilities), and business processes (ordered actions), and subjects (legal entities and individuals). The internal environment consists of such factors as enterprise owners (businessmen, entrepreneurs, shareholders, investors, founders), business communication factors (intra-corporate formalized and informal structure-forming norms, rules, values), resources (auxiliaries controlled by owners), business processes (ordered actions within the enterprise that are under the control of the owners) [12, 13]. The simplified structure of the business environment in terms of variability and European integration is shown in Fig. 1.

The main tools for assessing the business environment in terms of variability and European integration are the most well-known and widespread: SWOT analysis, and PEST (STEP) analysis. SNW-analysis, GRID-matrix, EFAS-analysis, ETOM-analysis, and others [8]. Accordingly, in the conditions of the digital economy and sustainable development, entrepreneurship should be considered as a system capable of giving the maximum effect for all interested parties: the state, territorial communities, businesses, employees, managers, and shareholders, ensuring the safety of the environment, etc. [14, 16].

Most experts agree that the global security environment is currently characterized by a high degree of uncertainty and unpredictability [16]. The US National Intelligence Council's report Global Trends 2040: A More Contested World emphasized

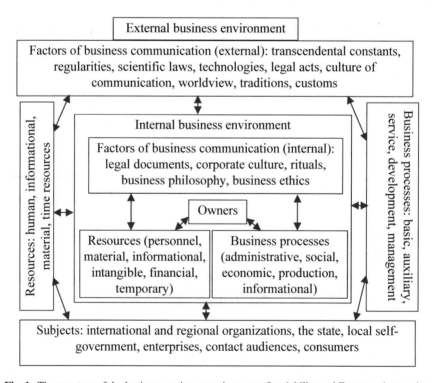

Fig. 1 The structure of the business environment in terms of variability and European integration

that the COVID-19 pandemic has reminded the world of its fragility and demonstrated the high interdependence of various risks. According to experts' conclusions, in the coming years and decades, the world may experience more intense challenges of various origins with cascading effects, which will be a test of stability and adaptability for societies, states, and the international system in general [17]. The Russian Federation's hybrid aggression against Ukraine highlighted the European security crisis [18]. This confirms the conclusions about the existence of a conflict of interests in the international arena and the strengthening of rivalry between states. Ukraine is fully aware of the changes taking place in the world today.

1.2 Changes in Business Under the Influence of the Digital Economy

In the future, there will be changes in the work process under the influence of digital technologies [19, 20]. Types of work organization in the digital economy will be determined by two parameters: technological capabilities (from low to high) and the level of democratization (stability and duration of labor relations). This was the conclusion reached by John Boudreau [3]. The digital economy contains a core that is the primary sector (software, IT consulting, information services, telecommunications), the secondary sector (digital services and digital platforms), and the tertiary sector (e-business, commerce, industry 4.0, smart agriculture, algorithmic economy). Li et al. [4] explore the digital economy that promotes the development of intra- and inter-regional industrial sectors using the example of China. The digital economy generates a positive impact and influences the development of industry and technological innovation. The digital economy and technological innovation have a threshold effect. Zhou et al. [5] define the digital economy at the county level. It affects the sustainable development of enterprises. The impact of the digital economy on sustainable development should be multi-criteria. The digital economy is being affected by the COVID-19 pandemic. It changes the trajectory of sustainable development of enterprises.

Gregory and Holzmann [6] researched sustainable business models and entrepreneurship, showing that digital technologies allow the creation of new configurations of sustainable business model components: mixed value proposition, integrative value creation, and multidimensional value coverage. They noted that digital technologies can significantly contribute to achieving the goals of sustainable development. However, how this potential can be realized in practice is still largely unknown, especially for entrepreneurs seeking to create socio-environmental value through financially viable business models. Hajiheidari et al. [7] proposed an innovative business model (BMI) to achieve sustainability and adapt to changes in the digital economy, following the principles of dynamic capabilities. It consists of four levels, including approach, aspect, dimension, and component. Based on the quantitative results, the 16 parameters were divided into four main levels: "sustainable comput-ing", "sustainable execution", "sustainable engagement", and "sustainable results". Considering sustainability and digital transformation as the main drivers of change for modern business, a new structure in the digital economy is proposed. However, the question of forming a business model of a digital economy of sustainable devel-opment in terms of its instrumental content has remained practically unexplored, which does not allow for to achievement of positive results in the transformation of the economy of Ukraine and restrains the transition of domestic enterprises to innovative development.

In the economy, processes are changing under the influence of digital technologies, and flexibility to changing environmental conditions is increasing. All this affects the labor market and employment, management, and coordination of labor resources. For example, in the last decade, Western countries have faced a shortage of certain STEM (science, technology, engineering, and mathematics) occupations. Recently, online marketplaces and crowdsourcing [10] for business services have proliferated. These marketplaces can act as intermediaries between employers and freelancers and have enabled on-demand contract hiring between project initiators (employers) and freelancers (employees) on a global scale [11]. They create new dynamics among STEM and intellectual workers, and firms develop new strategies for hiring and managing intellectual capital [12].

Digitization trends are actively being introduced to the national economy of Ukraine. However, the pace of digitization is somewhat behind the advanced coun-tries of the world. Unfortunately, in 2022, these processes in several s, sectors of the national economy began to slow down due to colossal losses and losses that domestic businesses will suffer from a full-scale war in the country [14].

In Ukraine, there is a steady trend towards an increase in the number of enterprises that have access to the Internet: from 43,303 enterprises in 2018 to 44,508 in 2021. These are mainly enterprises of the processing industry (11,323 in 2021), wholesale trade, repair of motor vehicles and motorcycles (10,630), wholesale trade (7112), and construction (5141). The percentage change in the number of enterprises with access to the Internet shows a decrease from 88% in 2018 to 86.6% of the total number of enterprises in 2021. 28% of the total number of employed workers in enterprises have access to the Internet. The digital economy is characterized by the activity of enterprises in the online environment. One of the forms of such activity is the presence, support, and regular updating of the company's website. On average, slightly more than 35% of enterprises in Ukraine had a website in 2021. We can note that with an increase in the number of employees, enterprises are more inclined to digitize their activities and expand access to the Internet, which is explained by the possibility of simplifying communication, speeding up the movement of information flows, and reducing the time it takes to deliver orders and orders from management to final executors. However, the larger the number of employees at the enterprise, the more difficult it is to provide them with access to the Internet (in 2021: only 22.6% of employees of large enterprises; 32.6% of employees of medium-sized enterprises; 39.2% of employees of small enterprises) [13]. From Fig. 1 we can see the obvious leadership of large enterprises with more than 250 employees. in digitalization of activities. This is explained by the presence in the structure of such enterprises of defined marketing departments, which professionally deal with the positioning of the company, its brand, and products on the Internet and above all on the official website. For financial (personnel) reasons, small businesses are usually unable to fully provide technical, commercial, information, and marketing aspects of permanent website support. In the future, an increasing number of enterprises, regardless of their size and types of economic activity, will have websites with wide functionality—this is an indispensable condition for the development of the digital economy.

Digitization should be perceived as a tool that helps to stimulate and develop the informational openness of society, which is one of the main factors in increasing labor productivity, economic growth, and competitiveness of national enterprises, creating jobs, and overcoming poverty and social inequality. In the technological aspect, the digital economy is defined by four trends: mobile technologies, business analytics, cloud computing, and social media; globally—social networks such as Facebook, YouTube, Twitter, LinkedIn, Instagram, etc. In recent years, such Internet-dependent markets as tourism, games and e-sports, media, and banking services have also received active development.

The gradual digitization of the mass segment is changing the nature of online consumption. Digitization is aimed at the development of tools and mechanisms for the operation of virtual digital coworking centers, cross-platforms with the digital industry, digital hubs-studios, hubs-associations, and hackathons in Ukraine using "cloud" technologies and software-defined architecture (Fig. 2).

According to McKinsey, in China, up to a 22% increase in GDP by 2025 can occur due to digital technologies, and in the USA—up to 10% [15]. In addition, research by the Center for Economic Strategy states that the benefit of the digital sector of China's economy will increase to 10% of GDP by 2025 from 7.8% of GDP in 2020, and by 2035, it will become the world leader in the digital economy [16].

1.3 The Role of Sustainable Development in the Formation of Digital Business in Conditions of Variability

Modern development can be defined as critical, which is caused by the global inconsistency between its economic, ecological, and social components. The world community drew attention to the critical ecological state of the planet at the end of the 60s of the twentieth century. Before that, sustainable development was associated only with economic growth. The new paradigm of social development, called sustainable development, was developed based on the analysis of the causes of the catastrophic degradation of the natural environment on the scale of the biosphere and the search for ways to overcome threats to the environment and human health [19, p.15].

The concept of sustainable development is based on five main principles [20]:

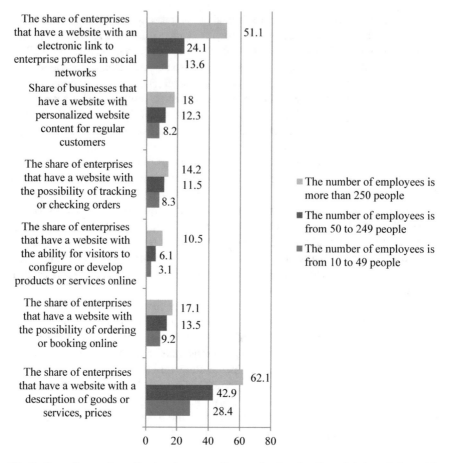

Fig. 2 Share the number of enterprises according to the functionality of their website, with distribution by the number of employed workers in Ukraine, 2021

1. Humanity can indeed give the development a sustainable and long-term character so that it meets the needs of people living now, while not losing the opportunity for future generations to meet their needs.
2. Limitations that exist in the field of exploitation of natural resources are relative. They are connected with the modern level of technology and social organization, as well as with the ability of the biosphere to self-renew.
3. It is necessary to satisfy the elementary needs of all people and allow everyone to realize their hopes for a better life. Without this, sustainable and long-term development is simply impossible. One of the main causes of environmental and other disasters is poverty, which has become a common phenomenon in the world.

4. It is necessary to improve the living conditions of those who use excessive means (monetary and material) and ecological opportunities for the planet, in particular, the use of energy.
5. Volumes and rates of population growth must be coordinated with the production potential of the Earth's changing global ecosystem.

The application of the stated principles involves the interrelationship of three components of development: economic, social, and environmental. However, in the conditions of the development of the digital economy, it is worth adding a digital one to these components [21, 22].

The economic component consists of the optimal use of limited resources and the application of nature-, energy- and material-saving technologies to create a flow of aggregate income that would ensure at least the preservation (not reduction) of the aggregate capital (physical, natural or human) with the use of which this aggregate income is created [23, 24].

The social component is focused on human development, maintaining the stability of social and cultural systems, reducing the number of conflicts in society, and forming a philosophy of communication and culture. A person should not become an object, but a subject of development [25].

The ecological component of sustainable development must ensure the integrity of biological and physical natural systems and their viability, on which the global stability of the entire biosphere depends. Of particular importance is the ability of such systems to self-renew and adapt to various changes, instead of remaining in a certain static state or degradation and loss of biological diversity [26].

The digital component determines the rapid development of digital technologies and the formation of a safe information space of interaction, where the acts of buying and selling goods and services and exchanging ideas, take place. This world is the opposite of reality, where e-business takes place through communication and self-motivation.

The Institute of Applied System Analysis of NTUU "KPI" proposed to assess the level of sustainable development using the appropriate index of sustainable development, which is calculated as the sum of indices for three dimensions: economic (index of competitiveness and index of economic freedom), ecological (ESI index) and social (index of quality and life safety, human development index and society index) with the appropriate weighting coefficients (ISR = 0.43 × IED + 0.37 × ISD + 0.33 × IED) [27]. Each of the indices is calculated using indices and indicators known in international practice. We propose to introduce the I-DESI index, which will show the index of the digital economy and will be separated from the social index. The calculation formula will take shape:

$$ISR = 0.43 \times (IFI + GSI) + 0.25 \times (HDI + SPI)$$
$$+ 0.12 \times (I - DESI) + 0.33 \times (ESI) \qquad (1)$$

The methodology for calculating indices defines the period of indices 2022, 2021, and 2020, which are placed in the relevant reports [21–25]. In the absence of data

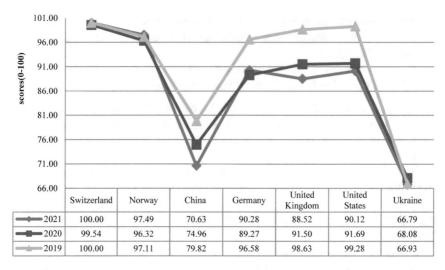

	Switzerland	Norway	China	Germany	United Kingdom	United States	Ukraine
2021	100.00	97.49	70.63	90.28	88.52	90.12	66.79
2020	99.54	96.32	74.96	89.27	91.50	91.69	68.08
2019	100.00	97.11	79.82	96.58	98.63	99.28	66.93

Fig. 3 Index of sustainable development of Ukraine and certain countries of the world in 2020–2022 [21–25]

for the relevant period for the country, data for the previous period were taken. This situation is typical for the index ESI. In Fig. 3 presents the index of sustainable development for 2022–2020 in the countries of Switzerland, Norway, China, Germany, the USA, Great Britain, and Ukraine. The best index is in Switzerland and Norway. Ukraine has the lowest index of sustainable development among the compared countries. In most of the studied countries, there is an increase in the index.

Thus, in the USA it increased in 2022 against 2020 to 99 points, and in Great Britain to 98 points. However, in Norway, it decreased, albeit slightly.

Accordingly, Ukraine needs to develop a strategy for the sustainable development of digital business in conditions of variability.

2 The Strategy of Sustainable Development of Digital Business in Conditions of Variability

2.1 Elements of the Strategy of Sustainable Development in the Formation of Digital Business in Conditions of Variability

The digital transformation of business involves the restructuring of the management system of a business organization and several management decisions—the choice of a method of digital transformation. Boston Consulting Group researchers believe that firms use one of two paths [26].

The first is the path of gradual transformation, which is considered ineffective because changes in digital technologies are happening too quickly.

The second is leapfrogging and riskier, during which firms build up their internal technological capabilities as quickly as possible.

A distinction is made between an online strategy and a firm's digital strategy. An online strategy is a strategy for using a company's digital (online) assets to maximize its business results. The task of the consultant is a diagnosis of business problems, for the solution of which the company's online assets can be used; development of ways to use online assets to meet needs, and achieve company and stakeholder goals; determination of priority online initiatives of the firm. The main thing is to determine the client's ability to implement digital strategies with the identification of sources of threats and opportunities for business, and already on this basis—to justify the allocation of resources and investments in traditional and digital strategies. Digital strategy is the strategy of transforming a firm into a digital one, in which digital communication ensures the interaction of a business organization with its customers, individualized and personalized product/service offers, decision-making based on data, under the influence of changes in the external and internal environment of the firm.

Digital strategies also include strategies for digital operations, the use of cloud technologies, and application management. A digital strategy builds business relationships through digital networks supported by enterprise-class technology platforms used by the business organization to support critical business functions and services. The digital firm emphasizes digital support of business processes and services using modern technologies and information systems. Thanks to this, digital firms have the opportunity to decentralize operations, increase market readiness and responsiveness, and improve interaction with customers. The goal of a digital firm is cost savings, competitive advantage, business continuity, and efficiency. Managers' attention should be drawn to the digital firm's use of technological platforms, including CRM—customer relationship management; SCM—supply chain management; ERP—company resource planning; KMS—knowledge management system; ECM—company content management, etc. The purpose of these technology platforms is to enable digital integration and exchange of information within the organization with employees and beyond with customers, suppliers, and other business partners.

Therefore, the core of a digital firm and the most effective tool for its management are information technologies. All areas of business are changing under the influence of digital strategizing:

First, marketing strategies, and interactions with consumers (clients). It is primarily about digital marketing, which uses digital channels to attract and retain customers. Digital marketing includes (a) Internet marketing (SEO promotion of the site, context, webinars, etc., i.e. all channels that are available to the user only on the Internet), (b) promotion of goods/services/activities on any digital media, digital methods, and digital channels, using ICT outside the network (mobile devices, local networks, digital television, interactive screens, POS terminals). That is, digital marketing makes it possible to reach both online and offline consumers, to have clear and detailed data about goods and buyers, which are recorded by analytical systems; apply a flexible approach to the target audience.

Secondly, competitive strategies. In the digital economy, the "pioneers" and their "closest followers" who can provide additional value to the consumer will win. "Pioneers" are pioneers and open the market for consumers based on innovation.

Thirdly, personnel strategies. The main focus is on the smart working model, which uses new technologies to improve productivity and employee satisfaction from work, including remote, and digital workplaces. The latter are virtual, mobile, flexible, and do not require the constant presence of employees. Companies have the opportunity to use limited resources more effectively, people—to work remotely (at home or in other convenient places), using e-mail, instant messaging, and tools for conducting virtual meetings. This saves time and improves their quality of life.

Fourth, innovative strategies. Digital technologies can become the basis of a new type of competitive advantage, for example, a new quality of customer service due to access to global digital platforms, gaining access to new customer databases and new technologies. In addition, companies try to use all the advantages of innovative business conduct: not only to offer new products and use new technologies for their products but also to change the forms and methods of conducting business.

McKinsey specialists compiled a list of five problems that company managers should be aware of [26]. Including:

1. Insufficient understanding of the essence of the digital economy.
2. Lack of a clear understanding of the digitalization process, without which the manager cannot form a strategy that would combine digitalization and his business.
3. Lack of attention to digital ecosystems that are replacing the concept of industry. At the same time, platforms that allow digital players to cross the boundaries of one sphere destroy the traditional model make it possible to attract an almost infinite number of customers, and use artificial intelligence and other tools to provide exceptional service. Business models that previously seemed unimaginable are becoming a reality. For example, grocery stores in the US now have to take into account Amazon's actions, not just their neighborhood stores. In China, Tencent and Alibaba are platform companies that combine traditional and digital businesses (and their suppliers) in insurance, healthcare, real estate, and others. Facebook has become the biggest media player without creating content. Uber and Airbnb sell global mobility and accommodation without owning cars or hotels. A McKinsey study shows that digital ecosystems could reach $60 trillion in revenue by 2025 (more than 30% of the global corporate sector's revenue) [28].
4. Over-focus on the consumer prevents us from seeing the growing importance of digital products in B2B markets.
5. Ignoring the duality of digitization. For most companies, it is very difficult, and sometimes impossible, to move away from the existing business. They need to digitize the current business and implement innovative models.

According to MIT experts, "a digital platform is a high-tech business model that creates value by facilitating exchanges between two or more interdependent groups of participants" [29]. The platform allows you to receive income from communication relationships; protect ownership of information, provide communication opportunities, and make them a factor of production. The following types of platforms are distinguished:

A transactional is a technology, product, or service that acts as a channel (or intermediary) to facilitate exchanges or transactions between different users, buyers, or suppliers.

Innovation is a technology, product, or service that serves as a foundation on which other firms (loosely organized into an innovation ecosystem) develop complementary technologies, products, or services.

Integrated is a technology, product, or service that is both a transaction platform and an innovation platform. This category includes companies such as Apple, which has both platforms, an App Store, and a large three-way ecosystem that supports the creation of content (information content) on the platform.

Investment—consists of companies that have developed a portfolio strategy and act as a holding company, an active platform investor or combine both functions.

The basis of the development of a strategy for the sustainable development of digital business in conditions of variability should be digital competencies, which are presented in Figs. 4 and 5.

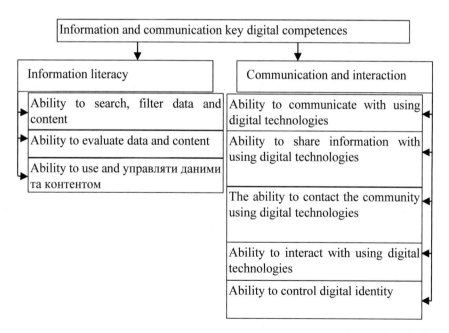

Fig. 4 Information and communication are key digital competencies in the strategy of sustainable growth of digital business

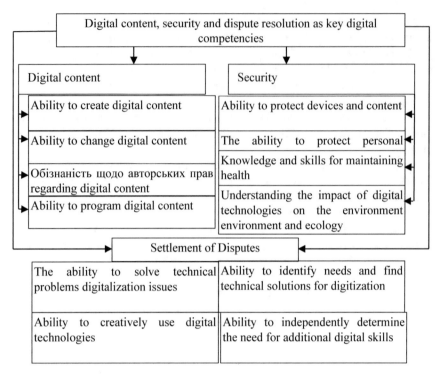

Fig. 5 Digital content, security, and dispute resolution as key digital competencies in the strategy of sustainable digital business growth

All of them should form the core of a strategy for the sustainable development of digital business in a changing environment.

2.2 Stages of Developing a Strategy for the Sustainable Development of Digital Business in Conditions of Variability

There are 3 stages of developing a strategy for the sustainable development of digital business in conditions of variability:

1st stage—analysis. Research and strategic analysis, digital business profile, social biological research.

2nd stage—is planning. Forecasts and scenario modeling. Defining the mission and vision of the process, SWOT analysis, strategic choice, action plan (goals and tasks), projects, and measures to achieve goals.

3rd stage—implementation. Public discussion, adoption of decisions by relevant bodies, development of target programs, management structure, monitoring and evaluation, revision, and adjustment.

We will develop a strategy for the sustainable development of digital business in Ukraine in 3 stages.

Digital business is developing rapidly all over the world [30, 31]. Digitization is one of the main factors in the growth of the world economy in the next 5–10 years. In addition to the direct productivity gains that companies get from digital technologies, there is a chain of indirect benefits of digitalization, such as saving time, creating new demand for new goods and services, and new quality and value [32]. Digitization will be the main tool for achieving Ukraine's strategic goal of increasing GDP by 8 times, up to $1 trillion. in 2030, and ensuring the well-being, comfort, and quality of life of Ukrainians at a level higher than the average in Europe. The share of the digital economy in the GDP of the world's largest countries in 2030 will reach 50–60%. In Ukraine, this indicator, according to our estimates, can be even higher—65% of GDP (under the implementation of a forced scenario of the development of the digital economy in Ukraine).

A digital business profile includes:

1. Digital technologies: Internet of things, robotics and cyber systems, artificial intelligence, big data, paperless technologies, additive technologies (3D printing), cloud and fog computing, unmanned and mobile technologies, biometrics, quantum technologies, identification technologies, and blockchain.
2. Consumers of digital technologies: the state, business, and citizens.
3. Areas of digital business: all types of economic activity, not only in the information and telecommunications sector but also in basic industries, agriculture, and construction.
4. Sectors of the economy: public and private; real, non-productive, and financial; mining, processing, and service sector.
5. Digital trends: data that become the main source of competitiveness; development of the Internet of things sphere; digital transformations of both individual businesses and entire sectors; sharing economy; virtualization of physical infrastructure IT systems; artificial intelligence; digital platforms.
6. Problems of digital business development in Ukraine:

   6.1 Institutional:
       6.1.1 Low inclusion of state institutions regarding the implementation of the Concept of Development of the Digital Economy and Society (Digital Agenda of Ukraine).
       6.1.2 Inconsistency of specialized legislation with global challenges and opportunities (progressive drafted bills have not yet become laws).
       6.1.3 Inconsistency of national, regional, and sectoral strategies and development programs with digital capabilities.
   6.2 Infrastructural:
       6.2.1 Low level of coverage of the territory of the country with digital infrastructures (for example, the EU goal is to cover 100% of the

territory with broadband Internet access by 2020, in Ukraine, this indicator is about 60%).

  6.2.2 Lack of separate digital infrastructures (for example, the infrastructure of the Internet of Things, electronic identification and trust, etc.).

  6.2.3 Unequal access of citizens to digital technologies and new opportunities (digital divides).

 6.3 Ecosystem:

  6.3.1 Weak state policy regarding incentives and incentives for the development of the innovative economy.

  6.3.2 Immature investment capital market.

  6.3.3 The outdated education system, teaching methods, lack of focus on STEM education, soft skills, and entrepreneurial skills, imperfect models of technology transfer, and consolidation of knowledge and skills.

  6.3.4 Shortage of highly qualified personnel for the full development of the digital economy and digitalization in general.

 6.4 In the field of electronic government and governance ("the state in a smartphone"):

  6.4.1 Low level of automation and digitalization of public services due to weak motivation of government institutions (there is no full understanding of the potential benefits of total digitalization).

There are two scenarios for the development of digital business in Ukraine, depending on the assessment of criticality and the need for rapid and deep changes in the traditional economic structure: inertial (evolutionary) and targeted (forced) (Table 1).

The mission of the digital business is to rethink approaches to doing business based on lean production with a combination of digital technologies and processes, as well as managing them through a smartphone.

Table 1 Comparative characteristics of digital business development scenarios in Ukraine

Indication	Inertness	Targeted
Basic elements in business	Traditional production	Digital platforms
Core	Production technologies, Automated systems	Information technology
The roles in the scenario	Manager and subordinate, Initiator and executor	Leader, Experimenter, regulator and defender, Popularizer
Methods and tools in business	A system, E-accounting, Cloud technologies, Advertising, Bank ID, Mobile ID	CRM, SCM, ERP, KMS, ECM

The vision of a digital business is to achieve a clean environment and ensure the sustainable development of territorial communities, and the creation of "smart systems" in all spheres of activity.

In SWOT analysis, it is recommended to use the methodology of Kucharczyk and Kardas [33]. The SWOT analysis matrix is shown in Fig. 6.

The strategic choice should be based on the strategic directions of digital business development. The main areas of development are:

1. Business development based on sustainable development.
2. The use of digital tools in the activities of business structures.
3. Formation of an environmentally safe interaction space.
4. Improvement of the management system and activation of development processes.

In Table 2, we present the main measures, thanks to which the strategic goals of the sustainable development of digital business will be achieved in the conditions of a changing environment.

Strong points	• Weak sides
• development of new information technologies; • improving communication with clients, partners, and employees; • new opportunities for customer service; • increasing the level of competitiveness in the business market by optimizing the organization's processes	• institutional; • infrastructural; • ecosystem; • in the field of electronic government and governance ("the state in a smartphone")
Opportunities	Threats
• expansion of payment methods; • use of electronic money; • online lending; • smart logistics and related services; • improvement of citizens' digital skills; • development of cross-border electronic trade	• the escalation of the conflict in the East and the end of the war in Ukraine; • political instability in Ukraine; • deepening of the economic crisis in EU countries; • deterioration of the demographic situation; • inflation growth, instability of national banks, high credit rates; • deterioration of the standard of living of citizens, decrease in the purchasing power of the population; • high level of cybercrime

Fig. 6 Matrix of SWOT analysis of sustainable development of digital business in conditions of variability

Table 2 Strategic goals and measures for sustainable development of digital business in Ukraine

Strategic direction	Strategic goal	Actions
Business development on this basis enable development	Activation of business activities	Promotion of entrepreneurial initiatives Approval of development programs Development and implementation of innovations Building a business based on the principles of social responsibility
The use of digital tools in the activities of business structures	Formation of a digital business model	Introduction of Internet marketing tools Development and distribution of its products on digital platforms Expansion of communication channels with stakeholders
Formation of an eco-safe interaction space	Ecological space of interaction between stakeholders	Introduction of waste-free production Compliance with the principles of environmental sustainability in production Landscaping of territories Development of the concept "Healthy nutrition and adherence to the principles of a healthy lifestyle" Introduction of environmentally friendly transport
Improvement of the management system and activation of development processes	Production management according to the principle of lean production	Production and supply of goods according to specific terms Reduction of labor costs Shortening the terms of development of new products Reduction of product creation time Reduction of production and warehouse areas Guarantee of product supply to the customer Maximum quality at minimum cost

To the strategic goals and measures of sustainable development of digital business in Ukraine, it is worth developing an action plan for their achievement (Table 3).

The proposed list of actions should be considered and discussed by all interested parties, suggestions should be made to it. This is possible by observing the principles of transparency and publicity. They can be implemented through the coverage of

Table 3 Action plan for achieving strategic goals and measures for sustainable development of digital business in Ukraine

Strategic goal	Actions	Source	Expected results
Activation of business activities	Popularization of business initiatives Approval of development programs Development and implementation of innovations Building a business based on the principles of social responsibility	Funds of local authorities, own funds of business structures, grant funds	Business development using digital technologies, increasing the level of competitiveness, innovative development
Formation of a digital business model	Introduction of Internet marketing tools Development and distribution of its products on digital platforms Expansion of communication channels with stakeholders	Grant funds, funds of business structures	Implementation of digital tools in the activities of business structures, the attraction of new customers from the Internet space, creation of prerequisites for international business
Ecological space of interaction between stakeholders	Introduction of waste-free production Compliance with the principles of environmental sustainability in production Landscaping of territories Development of the concept "Healthy nutrition and adherence to the principles of a healthy lifestyle" Introduction of environmentally friendly transport	Funds of interested parties, funds of business structures, funds of local authorities	Reduction of emissions into the atmosphere, ggreen in theg of territories, reduction of the greenhouse effect, reduction of waste

(continued)

Table 3 (continued)

Strategic goal	Actions	Source	Expected results
Production management according to the principle of lean production	Production and supply of goods according to specific terms Reduction of labor costs Shortening the terms of development of new products Reduction of product creation time Reduction of production and warehouse areas Guarantee of product supply to the customer Maximum quality at minimum cost	Funds of business structures, grant funds	Increasing competitiveness, reducing the terms of order fulfillment, improving the quality of products (services)

information about the sustainable development of digital business on their Internet pages, and social networks, through advertising and promotional videos.

In the last stage, strategic directions and goals should be coordinated with the strategy of territorial communities. Develop clear indicators of interaction between them and control over the process of their achievement. The main indicator of achieving the goals of sustainable development of digital business is the volume of sales, which should increase dynamically at constant costs [34, 35]. This can be achieved with the use of digital tools for attracting customers to the business [34, 35], constantly searching for them, and turning potential customers into regular ones by drawing attention to new products, social projects, and company tours.

Monitoring and control according to the action plan for the sustainable development of digital business should be carried out every decade, monthly, quarterly, and annually. Moreover, it should be conducted not only by the governing bodies of business structures but also by local self-government bodies to develop and adjust the development strategy for the relevant period. Such monitoring should take place annually.

Strategies for the sustainable development of digital business should become the basis of an innovative management mechanism for the reconstruction of Ukraine's economy in the post-war period.

The basis of the development and implementation of the strategy of sustainable development of digital business is software and information support. A strategy is a complex process that contains an analysis of components and has an analytical structure. A successful strategy should include building blocks such as resources, business processes, and growth potential.

The role of a digital business sustainability strategy is related to strategies such as:

- product strategy—technical and technological characteristics of digital business on the market;
- competitive strategy—competitive advantages that are formed by digital business in conditions of sustainable development and European integration;
- resource strategy—personnel, information, and resource potential of digital business;
- knowledge management strategy—aimed at introducing the idea of learning into the production process (learning-by-doing), introducing dual education, and industrial training.

In general, we can talk about the combination of all strategies into one, which will form a strategy for the sustainable development of digital business in a changing environment and global integration. The strategy for sustainable development of digital business is presented in Fig. 7.

The strategy of sustainable development of digital business provides for the creation of a favorable business environment that would ensure the observance and spread of the principles of sustainable development and post-war transformation of

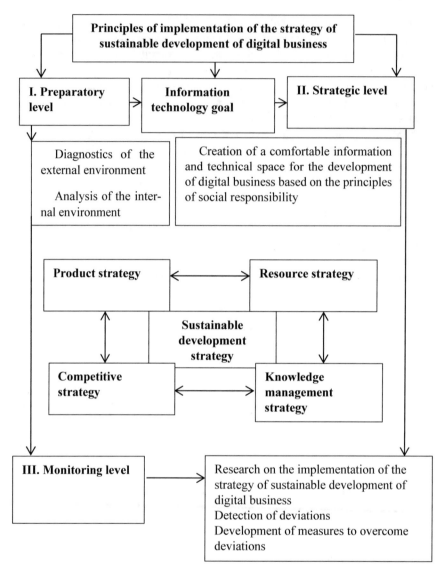

Fig. 7 Strategies for sustainable development of digital business in changing conditions

Ukraine's economy. Such an environment should be based on the principles of openness, integrity, environmental friendliness, social responsibility, and balance. Implementation of the strategy of sustainable development of digital business involves three levels—preparatory, strategic, and monitoring. At the preparatory level, the factors of the internal and external environment that shape the development of the circular business are determined. The strategic level involves the consistent implementation of product, competitive, resource strategies, and knowledge management strategies for the formation of an effective sustainable development strategy. The main goal of the strategy is to create comfortable information and technical space for the development of digital businesses based on social responsibility. At the monitoring level, control of indicators (productivity, innovativeness, quality and time of existence of the enterprise on the market, cost price), diagnosis, and development of measures to overcome negative phenomena are ensured.

3 Conclusion

The strategy of sustainable development of digital business in conditions of variability and European integration is based on improvement indicators (productivity, quality and time of existence of the enterprise on the market, innovations, customer experience, and cost); the main emphasis should be on values (optimization of processes, asset management, productive service, digital design, remote monitoring, Supply Chain Optimization, energy efficiency. This can be achieved with established cooperation between businesses the orit the customers based on a formed culture of interaction and business philosophy according to the principles of social responsibility.

According to these developed measures, digital business in Ukraine will achieve the following indicators, which are defined and characterize Industry 4.0 [36]:

- increase in production capacity—up to 60%;
- increase in the number of orders fulfilled on time—up to 95%;
- reduction of stocks—up to 20%;
- an increase in the overall efficiency of the installed equipment—up to 15%;
- reduction of equipment downtime—up to 22%;
- savings on purchase costs—up to 30%.

The growth of industrial production will be ensured by at least 7–10% per year. In addition, the majority of Ukrainian enterprises are significantly behind the EU countries or the world, which means that the initial effect of growth will be much greater.

According to the expert assessment conducted among 50 respondents from all sectors of the economy aged 30–55, the Ukrainian economy will receive:

1. The growth of the industrial sector is not less than 10–15% per year.
2. Preservation and anticipatory growth of high-tech industrial segments up to 15–20% per year. Significant growth in exports of these segments.
3. Additional growth and attraction of development investments to the country both in production and in R&D centers, incubators, and technology companies.

If the proposed strategy for the sustainable development of digital business for Ukraine is not implemented, then by 2030, high-tech segments on which competitiveness depends will disappear from the market, primarily mechanical engineering, electrical machines and equipment, instrument engineering, biopharmaceuticals, and energy [37]. The number of scientific and educational institutions will also decline, which will lead to a sharp reduction in the educational, engineering, and scientific potential of the country [34]. High import dependence will increase not only in construction but also in engineering. There will be a fixation of the economic structure on natural resources and raw materials.

References

1. Kovalenko, Yu.O.: Approaches to assessing the sustainable development of regions based on socio-ecological and economic indicators. DonDUU Manag **3**(88), 45–55 (2020). https://doi.org/10.35340/2308-104X.2020.88-3-04
2. Joiner, B.L.: Fourthh Generation Management: The New Business Consciousness. McGraw Hill, New York (1994)
3. Boudreau, J.: Work in the future will fall into these 4 categories. Harvard Bus. Rev. **3** (2016). https://hbr.org/2016/03/work-in-the-future-will-fall-into-these-4-categories
4. Liu, L., Ding, T., Wang, H.: Digitl economy, technological innovation and green high-quality development of industry: a study case of China. Sustainability **14**, 11078 (2022). https://doi.org/10.3390/su141711078
5. Zhou, Z., Liu, W., Cheng, P., Li, Z.: The impactt of the digital economy on enterprise sustainable development and its spatial-temporal evolution: an empirical analysis based on urban panel data in China. Sustainability **14**, 11948 (2022). https://doi.org/10.3390/su141911948
6. Gregori, P., Holzmann, P.: Digitl sustainable entrepreneurship: A business model perspective on embedding digital technologies for social and environmental value creation. J. Clean. Prod. **272**, 122817 (2020). https://doi.org/10.1016/j.jclepro.2020.122817
7. Hajiheydari, N., Shouraki, M.K., Vares, H., Mohammadian, A.: Digital sustainable business model innovation: applying dynamic capabilities approach (DSBMI-DC) Foresight, Vol. ahead-of-print No. ahead-of-print (2022). https://doi.org/10.1108/FS-02-2022-0012
8. Shtal, T., Buriak, M., Ukubassova, G., Amirbekuly, Y., Toiboldinova, Z., Tlegen, T.: Methos of analysis of the external environment of business activities. Espacios **39**(12), 22 (2018)
9. Naboka, Yu.V.: Business environment: characteristics, structure, development, diagnosis. Econ. Space **138**, 192–200 (2018). https://doi.org/10.30838/P.ES.2224.231018.192.257
10. Maskell, P.: Accessing remote knowledge–the role of trade fatheir, pipelines, crowdsourcing, and listening posts. J. Econ. Geogr. **14**(5), 883–902 (2014)
11. Massini, S., Caspin-Wagner, K., Chilimoniuk-Przezdziecka, E.: Global sourcing and the unbundling of innovation: challenges and opportunities for emerging countries. In: Lewin, A.Y., Kenney, M., Murmann, J.P. (eds.) Building China's Innovation Capacity: Overcoming the Middle-Income Trap. Cambridge University Press (2016)
12. Research Policy. Special Issue: Innovation and Skills in the Digital Economy (2018). https://www.journals.elsevier.com/research-policy/special-issues-in-progress/-special-issue-innovation-and-skills-in-the-digital-economy
13. The use of information and communication technologies at enterprises: the use of the Internet, cloud computing services, robotics. http://ukrstat.gov.ua
14. Kotelevets, D.: Development trends of the digital economy in Ukraine. Probl. Modern Trans. Ser. Econ. Manag. **5** (2022). https://doi.org/10.54929/2786-5738-2022-5-03-01
15. Matviienko-Biliaieva, G.: Development trends of the digital economy as an economic system. Економічний простір **181**, 115–119 (2022). https://doi.org/10.32782/2224-6282/181-20
16. Grygorenko, Y.: Twilight of the dragon: problems in China's economy threaten the whole world GMK Center, 14 січня (2022). https://gmk.center/ua/posts/sutinki-drakona-problemi-v-ekonomici-kitaju-zagrozhujut-usomu-svitovi/

17. US National Intelligence Council. Global Trends 2040: A More Contested World (2021). https://www.dni.gov/files/ODNI/documents/assessments/GlobalTrends_2040.pdf
18. Reznikova, O.O.: National stability in the conditions of a changing security environment: monograph. Kyiv: NISD (2022)
19. Kharazishvili, Y.M.: Systemic security of sustainable development: assessment tools, reserves, and strategic implementation scenarios: monograph/NAS of Ukraine. Inst. Indus. Econ. Kyiv (2019)
20. Daly, H.E.: The perils of free trade. Sci. Am. **269**(5); Springer Nat. **5**(11), 24–29. (1993)
21. The 2021/2022 Human Development Report. https://hdr.undp.org/system/files/documents/glo bal-report-document/hdr2021-22pdf_1.pdf
22. Social Progress Index. https://www.socialprogress.org/
23. Index of economic freedom. https://www.heritage.org/index/ranking
24. Environmental Sustainability Index. https://sedac.ciesin.columbia.edu/data/collection/esi
25. Digital Economy and Society Index (DESI). http://digital-strategy.ec.europa.eu/en/library/dig ital-economy-and-society-index-desi
26. Digital transformation of business: changing strategies and development models (2020). https://ndipzir.org.ua/wp-content/uploads/2020/02/Strizhkova19Mono/Strizhkova19 Mono%20(4).pdf
27. Zgurovskyi, M.: Ukraine in global dimensions of sustainable development. Dzerkalo tyzhnya newspaper (2006). https://kpi.ua/620-7
28. Why do digital strategies fail? http://open.kmbs.ua/digital-strategies-fail/
29. Evans, P.C., Gawer, A.: The rise of the platform enterprise. Glob. Surv. **9** (2016)
30. Polinkevych, O., Glonti, V., Baranova, V., Levchenko, V., Yermoshenko, A.: Change of business models of Ukrainian insurance companies in the conditions of COVID-19. Insur. Markets Co. **12**(1), 83–98 (2021). https://doi.org/10.21511/ins.12(1).2021.08
31. Polinkevych, O., Kamiński, R.: Corporate image in behavioral marketing of business entities. Innov. Mark.. Mark. **14**(1), 33–40 (2018). https://doi.org/10.21511/im.14(1).2018.04
32. Polinkevych, O.M., Leshchuk V.P.: Structuring of factors influencing the external environment on innovative business processes of enterprises according to the degree of risk. Actual Probl. Econ. **1**, 210–213 (2014). http://nbuv.gov.ua/UJRN/ape_2014_1_26
33. Kucharczyk, A., Kardas, E.: Ocena potencjału wybranego przedsięwzięcia za pomocą analizy SWOT/TOWS. Archiwum wiedzy inżynierskiej **3**(1), 3–7 (2018)
34. Rauer, J.N., Kroiss, M., Kryvinska, N., Engelhardt-Nowitzki, C., Aburaia, M.: Cross-university virtual teamwork as a means of internationalization at home. Int. J. Manag. Educ. **19**, 100512 (2021). https://doi.org/10.1016/j.ijme.2021.100512
35. Kryvinska, N., Kaczor, S., Strauss, C., Greguš, M.: Servitization—its raise through information and communication technologies. In: Lecture notes in business information processing, pp. 72–81 (2014). https://doi.org/10.1007/978-3-319-04810-9_6
36. Kuzmak, O., Kuzmak, O., Pohrishchuk, B.: Sustainable development: Trends and realities of Ukraine. In: E3S Web of Conferences 255, 2021. International Conference on Sustainable, Circular Management and Environmental Engineering (ISCMEE 2021), p. 01035 (2021). https://doi.org/10.1051/e3sconf/202125501035
37. Hrozniy, I., Kuzmak, O., Kuzmak, O., Rusinova, O.: Modeling of diversification of foreign economic interactions. Probl. Perspect. Manag. **1**, 155–165 (2018). https://doi.org/10.21511/ ppm.16(1).2018.15

Business Development towards the Application of Innovative Customer Relationship Management (CRM) Technologies in the Context of Global Transformational Changes

Oleh Kuzmak⬤, Olena Kuzmak⬤, and Serhii Voitovych⬤

Abstract The article summarizes issues related to the key tasks of modern business owners, in particular, ensuring a competitive positioning on the market and sustainable development of enterprises in the conditions of the development of digital transformation of economic relations. The main goal of the work is to research and determine the points of contact between customer loyalty and satisfaction and business efficiency, to assess the state of application of CRM systems and their impact on business indicators of activity. The actuality of solving the established problem consists in the fact that the effective use of CRM systems should become one of the tools for ensuring the preconditions of competitive advantage of the business entity, strengthening the image by improving the management of relationships with existing and potential customers, attracting and retaining consumers, developing dialogue and gaining loyalty. This publication highlights the aspects of transformational changes, globalization of economic relations, and their impact on the formation of modern business relations with clients. An attempt was made to give a systematic presentation of the main advantages of using information technologies as a tool for the effective management of relationships with clients. It has been found that for the successful implementation of CRM, a necessary condition is the definition and specification of the company's goals in the short- and long-term perspective and the definition and development of a strategy that determines relations with customers. The study concludes that in modern business conditions, it is necessary for companies to use CRM systems, and managing the successful implementation of CRM

O. Kuzmak (✉) · O. Kuzmak · S. Voitovych
Faculty of Business and Law, Lutsk National Technical University, Lvivska Street, 75, Lutsk 43018, Ukraine
e-mail: oleh_kuzmak@lutsk-ntu.com.ua

O. Kuzmak
e-mail: o.kuzmak@lutsk-ntu.com.ua

S. Voitovych
e-mail: gnidawa@ukr.net

© The Author(s), under exclusive license to Springer Nature Switzerland AG 2024 47
A. Semenov et al. (eds.), *Data-Centric Business and Applications*, Lecture Notes on Data Engineering and Communications Technologies 194,
https://doi.org/10.1007/978-3-031-53984-8_3

requires a complex integrated, and balanced approach to technologies, processes, and personnel.

Keywords Business development · Globalization · Competition · Customer relations · CRM system · Customer loyalty · Communication · Customer relationship management

1 Introduction

Ensuring competitive positions on the market and sustainable business development in conditions of uncertainty in the development of economic relations caused by the pandemic, as well as the frequent irrational behavior of consumers, is a key task of company owners. In recent years, the sphere and environment of modern business functioning have been extremely changeable under the influence of global transformational changes and the dynamic development of markets. The changes taking place have an extremely huge impact on the structure of the environment and relationships between companies, and in particular with customers.

It is recognized that the world economy has become more integrated due to the process of globalization [1]. Redding defines globalization interprets the integration of the market of goods and services, the integration of the capital market, cultural exchange, migration policy, and various combinations of these elements [2]. Globalization due to the increased liberalization of trade and the development of modern technologies has led to a sharp increase in the trade and economic turnover of goods and services, as well as an increase in the exchange of financial resources. In modern conditions, mass production and mass marketing technologies have significantly changed the competitive environment, by increasing product availability for consumers, the origin of goods has become secondary, and geographical distance is no longer an obstacle for many services.

Globalization changes the rules of competition and the ability of businesses to compete with each other. Success depends not so much on the competitive capabilities of an individual firm, but on the success of its interaction with clients. However, it is worth noting that the process of commerce, which allowed the seller and the client to interact personally, communicating with each other "face-to-face", so to speak, has also fundamentally changed and acquired the character of a global business market. It can be said that customers, to some extent, have lost their uniqueness, as they have become a "serial number", and it has become difficult for sellers to carry out the accounting and assessment of the individual needs of their customers, as the market has become overflowing with goods and services.

In addition, research in the field of behavioral economics for many years indicates that the preferences of modern consumers and their decision-making abilities are far from being as stable and rational as producers would like. Due to various circumstances, such as time constraints, frugality, exalted state, instability of the external environment, and pandemic, people often behave irrationally. "Our cognitive responses take into account intuition, emotions, colors, norms, availability, and a huge number of other prejudices that make it almost impossible to instantly make a 100% rational purchase decision" [3].

However, the fact that people can behave irrationally does not mean that their behavior is accidental. Irrationality is quite predictable, so studying how and why, and most importantly, considering this consumer behavior and its impact on it can be extremely useful for business owners in the process of forming a customer relationship strategy.

Even though today there are significant developments in the field of marketing research aimed at effective interaction with customers, however, there are still many problems in the practice of using these results in the work of modern business. In particular, the saying of marketing classic F. Kotler "marketing is now in a miserable state, not the theory, but the practice of marketing…" [4] is still relevant today. We can agree with the author's opinion that the "deadly sins of marketing" are the reasons why marketing does not work, namely: insufficient attention of companies to market concentration, orientation on clients, and understanding of their target consumers (as a rule, marketing strategy is based on taking into account rational behavior of customers, and attention to their irrationality is lost at the same time); incorrect construction of the customer relationship management system; the inability of companies to find new opportunities; the imperfect process of marketing planning and implementation of the policy regarding products and services; insufficient development of brand building and communication skills; insufficient organization for effective and skillful marketing.

In the modern market, has aroused the situation when traditional, so-called "marketing formulas of success" are increasingly ceasing to be the key that opens the door to the company's success with the help of higher quality; more perfect service; lower prices; higher market share; continuous improvement of products; product innovation [5], instead, how and in what way the company's relationship with consumers is built, namely whether the customer relationship management system is properly built, is becoming more and more important and weighty.

To correct the situation, F. Kotler suggests replacing traditional approaches with elements of new marketing approaches: focusing on values that exist throughout the lifetime of consumers; orientation to the satisfaction of several groups of shareholders; managing so that all employees are involved in the marketing activities of the enterprise; to form a brand through all the company's activities; focus on "serving" consumers; promise less, give more; make the value chain the unit of analysis [6]. To the list, you can add: focusing on the feelings, emotions, and culture of consumers, while skillfully using this information and effectively interacting with customers.

As noted by M. Biedenbach and H. Marell, the use of marketing communications directly affects the formation of the brand, recognition, image, and, therefore, the value of the brand [7]. The modern consumer receives huge amounts of various information about companies and their services, however, often buyers do not pay attention to the sources from which they get it. As a result, information received from a large number of sources (television, operational information systems, etc.) merges into a single entity in the minds of consumers and creates a general impression of the enterprise. If the information coming from different sources is contradictory, it provokes distrust of the company and its services. Quite often, there is a situation where companies are not able to coordinate the work of all their communication channels. As a result, the consumer cannot make sense of the message mix. Advertisements say one thing, the price level indicates something else, something different is written on the label, sales agents tell something of their own, and the Web node of the company, the enterprise, seems to be completely unrelated to anything. Therefore, the presence of good effective channels and methods of communication adds value to the company's product and also strengthens its competitive position, as customers are confident in their purchase and are more likely to stop their option on the chosen brand.

The importance and necessity of forming an effective customer relationship management system are confirmed by global trends in business development. In particular, studies indicate that interaction with a significant number of existing customers (up to 50%) of many companies does not bring enough profit due to ineffective interaction with them. Customer dissatisfaction is the reason for frequent changes in suppliers. For example, a large part of companies on the Fortune 500 list lose up to 50% of their customers every 5 years. The decrease in customers has a strong impact on the company's image since customers who are dissatisfied with the quality of service disseminate information about their negative experience significantly more widely than those who are satisfied with a positive interaction [8].

Today, many companies are trying to restore relations with new and existing customers, considering the stability and loyalty of the customer base as key factors in their viability. Because more consumers shop online than ever before, customer service is now a key differentiator for companies, is the main question for customers, and is an independent source of income. Today's consumers expect service to be fast, easy, and effective, and they're willing to look elsewhere if they don't get it.

The situation that has developed on the internal market and orientations in the direction of European integration requires a modern business to find new effective tools for further development and increasing business competitiveness, one of which is the use of technologies aimed at ensuring communication with market segments, attracting and retaining existing loyal customers, focusing on factors of variability and uncertainty of the market environment. Effective management of relationships with clients requires an individual approach to each client and an analysis of relationships with them to identify the most promising ones. For this, it is necessary to collect and process large amounts of information from the history of relations with each client.

Accordingly, an effective customer relationship management policy, which would take into account as much as possible various factors influencing consumer behavior, as well as the specifics of integration processes, should be one of the tools for ensuring the prerequisites for obtaining a competitive market advantage and sustainable business development both at the national and world levels.

The purpose of the work is to research and determine the relationship between customer loyalty and satisfaction and business efficiency, to assess the state of application of CRM systems and their impact on business performance indicators.

2   Research Methods

In the process of writing, the following methods of scientific research were used: analysis and synthesis (to study and research the features and state of the CRM market, the formation of factors and conditions for the use of CRM systems, the formation of customer loyalty, calculation-constructive when developing proposals for their improvement), dialectical and abstract- logical (for theoretical generalization and formulation of conclusions).

The authors analyzed the scientific publications of the scientometric database for the period 2000–2022 in terms of their reflection on aspects of transformational changes, globalization of economic relations, and their impact on the formation of modern business relations with clients. Also, the authors analyzed publications regarding the reflection of the essence, conceptual foundations, and features of the implementation and use of CRM systems in the activities of business structures. This allowed us to determine the factors and effectiveness of using CRM systems and their advantages for the formation of long-term loyal relationships with customers. For deepened analysis, the authors used case studies. The work uses methods of generalization and comparison to determine the level and effectiveness of CRM systems application in the activities of Ukrainian businesses and other countries of the world.

3   Business, Customers, CRM: Points of Contact

Considering business development in conditions of uncertainty and variability of the market environment, globalization of economic relations, and structural changes caused by global quarantine restrictions, there is an opinion that attracting new customers is the key to success. Attracting new customers is great, but it is not always as convenient and profitable as it is usually thought. This is because the process of identification, qualification, and training before turning them into paying customers requires significant resources for the company. Therefore, it is not necessary to make extraordinary efforts to attract new customers but to invest more in keeping the most valuable existing customers.

Research points to quite interesting facts about business effectiveness due to the maintenance of existing loyal and regular customers [8]:

- Costs for attracting a new client are, on average, 5–25 times higher than for maintaining an existing one;
- Relationships with the majority of clients begin to bring a steady profit only a year after starting work with them. Therefore, if a new client works with the firm for less than a year, the costs of attracting him are not paid off and the firm suffers losses;
- Concluding an agreement with an existing client is significantly easier and 5–10 times cheaper than with a new buyer;

- An increase in the number of regular customers by 5% increases sales volumes by more than 25%, and profits by 50–100%;
- About 80% of the company's income is provided by 20% of its clients;
- Customers who are dissatisfied with the interaction with the company reproduce a negative opinion about it significantly more widely than satisfied—positive ones.

According to research by Zendesk, an American company that develops software as a service related to customer support, sales, and other communications with customers online, more than 60% of business respondents indicate that they have higher standards of customer service today after the crisis of 2020, 73% of business leaders point to a direct link between customer service and business performance. Additionally, research shows that 61% of customers would switch to a competitor after just one bad experience, a 22% increase from 2020 [9]. The importance of customer satisfaction cannot be denied because happy customers are like free advertising for a company [10].

The results shown in Table 1 confirm the importance of effective communication and indicate that 90% of customers increase their loyalty to a particular business when close and personalized communication is provided.

As we can see, in today's reality, consumers value customer experience more and more. And what is customer experience (CX)? It is the total experience of customers before and after the sale, that is, from the moment when a potential customer learns about the company's product until all his interactions with the company as a customer [11]. In other words, if customers like the company, they will use its services for a long time and cooperate, and most importantly recommend it to others. But, company management should understand that the definition of CX is that the interaction between the company and the client throughout the term of their business relationship

Table 1 Quality service can drive sales (based on [9])

Quantity customers (%)	Customer opinion	Increase/decrease loyalty
93	Will spend more on companies that offer a preferred customer contact option (such as chat)	
92	Will spend more on companies that guarantee they won't have to repeat information	
90	Will spend more with companies that personalize the customer service they receive	
89	Will spend more on companies that allow them to find answers online without contacting anyone	

requires a creative approach, awareness, and the desire to open, develop, and create advertising campaigns, establish sales channels, communications, and service. At the same time, the difficulty lies in the fact that the customer experience depends on many starting points (points of interaction). That is, it is familiarization with web content, a request for a necessary product or service, a free trial subscription, submitting an application for a service, and finally, re-applying to the company's services. All this should lead the company to constant interaction, monitoring of the client's actions, and, to some extent, the formation of information related to the individual needs of clients. This will allow you to direct customers to purchase the product. In today's environment, effective CX management can reduce customer churn, increase revenue, and encourage repeat business with existing customers.

According to Esteban Kolski, 72% of customers will share a positive experience with 6 or more people. On the other hand, if a customer is unhappy, 13 of them will share their experience with 15 or even more. The problem is that in most cases, customers do not say they are not satisfied, they simply leave, and only 1 in 26 dissatisfied customers can express their dissatisfaction [12]. It is important to understand that customer service experience and customer experience are completely different things.

Figure 1 shows where they intersect and their differences. Assuming that a customer happens to come to the company in person or makes an order for the first time over the phone, of course, an experienced manager will use the opportunity to make an impression and provide excellent customer service.

The next time the same customer comes or calls, another, perhaps less experienced manager, does not find a common language with him, makes a bad impression, or does not understand the customer; in this case, the concept of customer experience

Fig. 1 Commonalities and distinctive features of customer service and customer experience (based on [13])

immediately comes into play: the situation clearly shows that customer service is only one aspect of the overall customer experience.

That is, in the new realities of the business environment, customer service becomes a key factor. And whether a company exceeds or falls short of consumer expectations is often directly related to business success. For businesses that succeed in attracting and retaining consumers, there are huge opportunities for growth. Not only will they claim the growing number of first-served customers, but they will also have real opportunities to expand, retain, and deepen their customer base.

Today, among marketing specialists, there is an opinion that if a brand is not on social networks, it does not exist as a company. The same applies to business management: if operations and finances are not optimized, the company has no chance to effectively develop and compete, since it can be said with confidence that competitors are engaged in the optimization of business processes.

If we consider the effective management system of a modern enterprise, the following scheme can be drawn: the business owner trusts the top management, which is prepared and oriented to the management, and expects results from department heads who control their subordinates: marketers, managers, SMM. At the same time, none of them can know everything about everyone, including clients. Mistakes occur often, and there is no panacea for them. However, if one mistake can make the consumer smile, then another can destroy the relationship with him forever.

Today, some businesses compete effectively and win the battle for the consumer by implementing relationship-marketing principles through strategic and technological applications for customer relationship management (CRM). The CRM system supports the strategy according to which the customer is at the center of everything the company does. This customer-centric strategy must be based on clear goals and an understanding of the experience (Fig. 2). It is an integrated approach to relationship management focused on customer retention and relationship development.

Fig. 2 The CRM system is customer-oriented [14]

How CRM marketing technology provides the ability to collect, analyze and store contact information about existing and potential customers at all stages of the relationship (attraction, retention, loyalty), customer history and the status of their orders, identify sales opportunities, fix service problems and manage marketing processes and make information about every customer interaction available to anyone in the enterprise who might need it.

From the point of view of information technologies, the CRM system is a set of applications connected by a single business logic and integrated into the company's corporate information environment based on a single database. The software used at the same time allows you to automate business processes in the implementation of marketing campaigns, sales, and services.

CRM programs help organizations measure customer loyalty and profitability by metrics such as repeat purchases, dollars spent, and longevity. CRM applications help answer questions such as "What products or services are important to our customers? How should we communicate with our customers? What are my client's favorite colors or sizes?". In particular, customers benefit from the belief that they are saving time and money, as well as receiving better information and special treatment [15]. In addition, regardless of the channel or method of communication with the company, be it the Internet, call centers, sales representatives, or resellers, customers receive the same consistent and efficient service [16].

In research by N. C. Krämer and S. Winter notes that the business performance of companies that intensively use Internet networks for sales, as a rule, strongly depends on customer satisfaction. The ability to improve customer satisfaction through CRM may be the single most important marketing practice that allows companies to improve satisfaction with their performance. This pattern is also supported by studies that found that the benefits of information technology (IT) capabilities for company performance tended to be more pronounced in companies with high IT intensity than in companies with low IT intensity [17]. As the cliche goes, tech is the way of the future. While marketing revenue only grew by 9.5%, digital commerce revenue expanded by 17.1% [18]. At the same time, global IT spending is expected to total $4.5 trillion in 2022. That is a 5.5% rise from 2021.

So, the main feature of the CRM system is that with the help of applications connected by a single business logic and integrated into the corporate information grid, the business can collect, systematize, and manage information about its customers. With its help, you can monitor and control all stages of the pipeline movement of each of the leads, as well as personalize communication with them and improve the quality of service. And all this in one interface. The numbers speak for themselves; accordingly, below are the statistics on CRM implementation and application effectiveness, which we have collected by conducting various studies:

- CRM can increase the conversion rate by 300% [19].
- CRM increases customer retention and satisfaction by 47% [19].
- Return on investment in a CRM system, if properly planned, can exceed 245% [20].

- The average return on investment in a customer relationship management system is $30.48 for every dollar spent [21].
- Using CRM can increase business profitability by up to 29% [22].
- Companies using CRM forecast sales 42% more effectively [22].
- Sales teams that implemented CRM increased productivity by 34% [22].
- CRM programs can increase the income of each member of the sales team by up to 41% [23].
- 50% of entrepreneurs believe that the CRM system helped to increase their productivity [11].
- 65% of business owners say that CRM has allowed them to increase their bottom-line sales [11].
- 40% of entrepreneurs confirmed that CRM helped to reduce personnel costs [11].
- 74% of business owners say CRM has improved customer relationships [13].
- 74% of CRM users claim that the system has made it easier for them to access customer data [24].
- 65% of sales representatives work more productively and achieve sales targets due to the use of mobile CRM. Only 22% of sales representatives who do not use mobile CRM do the same [25].
- Mobile CRM users saw an 87% increase in sales, a 74% increase in customer satisfaction, and a 73% increase in overall business efficiency [11].

4 Integration of CRM into Business Processes

Considering the functional advantages and performance indicators of the application, the CRM system looks like an advantage for any business. But before implementing automated customer relationship management and choosing a specific CRM system, you need to understand whether your business needs it. It often happens that someone tells the business owner about the existence of such systems, or software vendors try to impose their product.

If your business is not interested in increasing the number of customers at a certain stage of development, if the loyalty of regular customers is based on long-term contracts, and all contracts with new customers are based on personal meetings, even the best CRM system will not give anything.

On the other hand, if incoming calls or inquiries (leads) from new customers are important in the business, if the business makes efforts to obtain and retain new customers, if the business has a long sales cycle, if it is no longer possible to remember all customers, and potential customers communicate with several people in the company, this is a clear sign that it is time to integrate business processes with CRM. For example, an online store, a wholesale company, or a beauty salon will not be able to work effectively without careful attention to leads (incoming inquiries and calls). Some of the biggest companies that use CRM are Coca-Cola, Tesco, Apple, McDonald's, KFC, and Lufthansa. Currently, for most companies that work with a large number of customers, CRM is a prerequisite for a successful business. After all, in each of these types of business, all orders must be fulfilled on time, buyers of goods and services remain satisfied, and customer loyalty increases.

But even if the business aims to automate business processes and use software products to manage relationships with customers, it is necessary to understand that, first of all, you need to build logic and sequence of actions. In particular, the first step should be setting strategic goals that determine the content of the action program and determining what benefits the company plans to receive from the implementation of CRM. It is necessary to set goals from the short-term and long-term perspectives of the company.

The second step should be to specify the goals that the company plans to achieve. The process of implementation and further use of CRM systems and their effectiveness is difficult to control if the defined quality parameters are not clearly defined in advance. Accordingly, it is worth focusing attention on the development of a mechanism for measuring and controlling results. At each stage, it is necessary to set certain target values, for example: obtaining information about 75% of visitors to the company's Website at the first stage; converting 50% of their number into buyers and obtaining the necessary information from them; obtaining information about the purchasing preferences of all customers to increase the number of repeat purchases.

The next step should be the development of a strategy that will clearly define the relationship with customers. Before starting changes in the structure, business processes, culture, and technology, the company must clearly define what results it expects as a result of the formation of loyal relations with customers. Such a strategy, for example, can be a competitive advantage due to targeting the largest customers or due to an increase in sales volumes through the most profitable sales channels. On the other hand, the company may consider the most appropriate choice of strategy for attracting new customers.

Even if the company has set strategic goals, defined and specified its goals, and developed a relationship strategy, it is also necessary to take into account the current trends in consumer behavior. In contrast to those times when communication with the client took place "face-to-face", or by telephone or fax, today, modern platforms of interaction with clients can support several channels: Internet, voice communication, digital communication, and e-mail (Fig. 3).

Companies are forced to interact with their customers through multiple channels, website forms, live chat, social media, and more. Taking into account modern communication capabilities, a multi-channel customer experience is formed for customers, which consists of separate points of interaction through various channels, which are smoothly connected, allowing them to continue from the place where they stopped on one channel, and continue work on to another. This is where the complexity comes in because while customers may be positive and accept different levels of service from different channels, they also expect consistent communication with them. Accordingly, the company's personnel must possess the skills of supporting multiple channels in one interaction.

Even with the fact that today many channels of interaction have been developed, still, communication-using telephones remain at a significant level, and therefore the level of customer experience due to this channel is of great importance. The client's

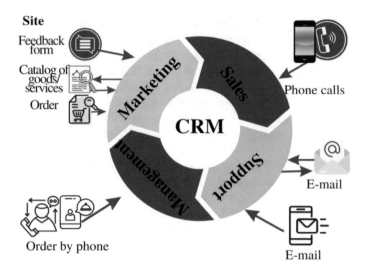

Fig. 3 CRM system and flow of information about sales, marketing, and service

first call to the company is often called the "bottleneck" of the sales funnel because many people do not make it through it. The reason for this can be many factors: long wait for the operator, inconvenient voice menu, very frequent voice waiting "your call is very important to us". All this can lead to the loss of a client before establishing cooperation with him. The integration of CRM and IP telephony will be able to help solve this issue and not lose (lose less) customers at the start. Of course, the integration won't add more operators or even change the voice menu. But it can affect such a problem as the human factor. If the company does not have the integration of telephony and the CRM system, all customer contacts are entered manually. As a result, a certain percentage of contacts are not entered into the customer base at all, and some are filled in with errors, without specifying important information. Studies show that even if the company monitors the process of forming and processing the customer base, still up to 20% of customer contacts may be lost or recorded with errors during manual processing.

In addition, today, it is undeniable that Internet users around the world are moving from desktop computers to mobile devices. People spend an average of 3 h 37 min on the mobile internet, and the average time people spend on social media is 2 h 27 min; search engines or web portals, chat and messaging platforms, and social networks are the most used websites and apps, people use social media to fill their free time, read news and stay in touch with friends and family, people on social media follow the accounts of acquaintances, friends, and relatives, famous people, actors, 72.0% of users Internet worldwide 16–64-year-olds use social media to find brands, 67.1% of the world's population are mobile phone users, with 5.29 billion, 4.55 billion (57.6%) active social media users [26]. Mobile business app users experienced an 87% increase in sales, 74% in customer satisfaction, and 73% in business process efficiency [27].

The fact that today customers have unlimited opportunities to communicate with companies through digital channels, accordingly, the customer experience takes on the features of digitalization. Companies that make an effort to optimize these interactions will have a significant advantage over those that do not.

Data highlighted in the results of a Salesforce study indicate that as recently as 2016, 58% of consumers agreed that technology has fundamentally changed their expectations of how companies should interact with them. In addition, the modern generation of consumers is becoming more and more dependent on mobile gadgets. Millennials are almost three times more likely than baby boomers to agree that they "manage their lives from their mobile devices" [28]. And the generation born after 2005 (Generation Z) generally came into the world with gadgets in their hands. While other generations are discussing the "technologies of the future", for this generation, the "future" has already arrived. They do not read printed newspapers, do not imagine life without the Internet, and were born with pages in social networks.

Compared to those doing business in person, digital customers are much more likely to expect instant or near-instant gratification. This means that any factors that can slow down the digital shopping process can end up turning potential customers away.

For example, about 5% of organizations had users leave their websites after only a one-second delay, and the longer the delay, the higher the percentage of customers who start looking elsewhere [28]. For example, 57% would not recommend a company with a poorly designed mobile website. In addition, if a website is not optimized for mobile devices, 50% of customers will stop visiting it, even if they like the company and its services [12].

Taking into account the latest trends in the development of multi-channel communication, mobile communications, and mobile and social networks, to successfully communicate with customers, it is worth targeting your business model for the application of CRM systems. In particular, one of the directions of development of CRM systems to strengthen the personal approach to customers is their integration with social networks, that is, through SCRM. Social CRM is a tool that promotes better, more effective interaction with the customer and uses the collective mind of a wider customer space with a predictable improvement in the contact between the company and its potential and real customers.

5 Modern Trends in the CRM Market and Application of CRM Systems in Business

In 2021, a large share, nearly 59.4% of the global CRM application market was held by the top 10 CRM software vendors. In 2020, Salesforce led the way with a market share of 31.3%, driven by a 12.6% jump in CRM revenue. Adobe ranked second, followed by Oracle, SAP, and Microsoft [29]. That is, the global market of CRM systems shows rapid growth every year. This is because customers constantly demand a better customer experience (CX) year after year, and companies constantly need to improve operational efficiency, which can be achieved through business process automation (BPA).

In a global CX study conducted by one of the major players in the CRM market, Oracle Corporation, 74% of senior executives believe that customer experience influences the customer's desire to be a loyal advocate [30]. BPA is expected to double over the next four years, growing from $9 billion in 2020 to over $19 billion in 2026. Companies see the real value of BPA, as they can increase their revenues by 39%, and get 60% more profitability. In addition, it is an effective way to increase productivity and enjoy work. And companies that ignore business process automation lose 20–30% of revenue every year due to manual, inefficient processes. Despite the positives, 42% (almost half) of all businesses still do not have an indicative automation strategy [31]. That is, if a company wants its customers to remain loyal and its staff to increase productivity and enjoy their work, it must invest in customer experience and business process automation [32, 33].

Research and global practice of CRM application indicate interesting facts, in particular [34]:

- 82% of organizations use customer relationship management systems for sales reporting and process automation.
- The importance of interacting with a salesperson whom they consider a trusted advisor at the time of purchase is a top priority for 79% of business buyers.
- It is predicted that 91% of data in CRM systems will be incomplete, outdated, or duplicated every year.
- In 74% of companies aimed at increasing their competitive status, converting potential customers into existing ones is their main priority.
- The highest CRM market shares are concentrated in Salesforce, Microsoft, and Oracle.
- 74% of respondents claim that CRM solutions give them better access to customer data, providing more personalized service.
- More than 91% of companies with 10 or more employees use CRM to manage customer interactions, 82% of them use it only as a sales tool, while just over 50% use it for calendar management and email marketing.
- During the first 5 years after creation, as a rule, 65% of enterprises consider it necessary and implement CRM technology (Fig. 4).

The evidence suggests that businesses that implement CRM will benefit from increased customer loyalty and long-term profitability. In addition, a fully functional CRM system can be considered a competitive advantage, which applies to both global and small and medium-sized enterprises.

World studies (Fig. 5) show that in 2021, the global customer relationship management market volume was estimated at 57.83 billion dollars. Interestingly, the impact of the pandemic in the same year significantly accelerated the demand for CRM. Digitization and digitization of society have prompted companies to look for new ways of communicating with customers. The CRM market is projected to grow from

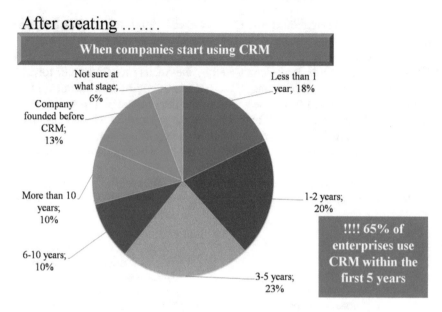

Fig. 4 Implementation of CRM after the start of the company's activities (based on [34])

$63.91 billion in 2022 to $145.79 billion in 2029, a CAGR of 12.5% over the next 7 years [34].

Scientific research by Nucleus Research showed (Fig. 6) that, on average, for every dollar spent on CRM technology in 2014, an average of $8.71 was returned [35]. Although this number has not been updated since then, Dynamic Consultants

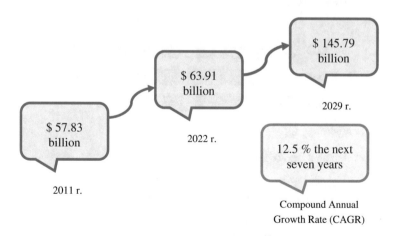

Fig. 5 Global customer relationship management market size

commented on the evolution of the CRM market in 2021 and estimated that this number increased to an average of $30.48 for every dollar spent in 2021 [21].

According to the results of the survey of the CRM market in Ukraine for 2018–2020 (Fig. 7), which was conducted by the Society for Consumer Research (Gesellschaft für Konsumforschung (GFK)) of 1,030 companies in the field of large, small and medium-sized businesses, 68% of the interviewed entrepreneurs do not know and have never heard of a CRM system. At the same time, 86% said that they do not plan to spend resources on implementation shortly. More than 90% of surveyed companies do not use specialized CRM systems for working with customers [36, 37].

Ukrainian businesses' distribution and use of CRM systems are related to the development trends of the information technology market itself, which is currently

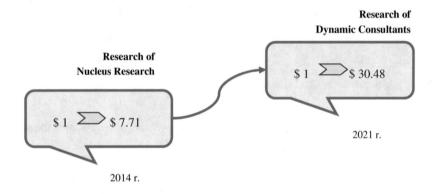

Fig. 6 Cost recovery for CRM

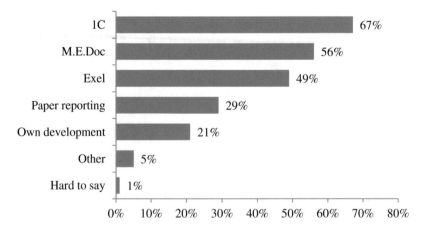

Fig. 7 Tools used by Ukrainian businesses for reporting when working with clients

several years behind the Western one. And while the IT industry is actively developing in the West, in particular, the CRM industry, the Ukrainian market is just beginning to "ripen up" and offer these systems. To collect reports on work with clients 49% of companies use Excel, and 29% of companies record key points on paper. 1C and MEDoc services are used by 67 and 56% of companies. At the same time, only 8.4% of surveyed companies use CRM or plan to do so [36, 37].

Of course, each business at its discretion chooses the features and mechanism of communication with customers, and even if the company keeps a history of calls and contacts on paper or in Excel, it can be considered a CRM system if the developed accounting and control scheme works and allows you to control all options for interacting with customers. Today, such methods of collecting information are a thing of the past, especially in industries where companies work with a very large number of customers, and a large amount of information, and without effective automation, it is difficult to imagine the operation of any business.

Respondents identified Bitrix24, Terrasoft, RetailCRM, AmoCRM, OneBox, Megaplan, Zoho, and Salesforce as the main CRM systems in the Ukrainian market (Fig. 8). At the same time, up to 50% of these companies used Russian products, such as Bitrix24, amoCRM or RetailCRM [38].

In Kyiv, the percentage of companies that use CRM systems in their work is slightly higher—12%. Most often, CRM systems are used by companies operating in the field of IT services and social services (53%), trade (23%), repair (6%), and industry (14%) (Fig. 9). At the same time, 47% of companies working with CRM note that the implementation of the system significantly increased their work efficiency, and another 45% noted that the efficiency increased, but slightly [37].

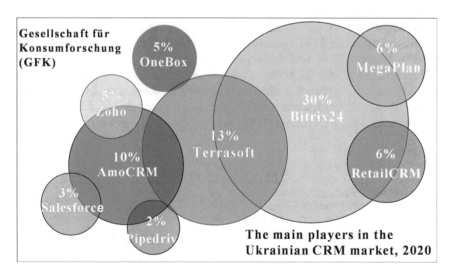

Fig. 8 Software products of SRM, which Ukrainian business mainly uses in its activities (based on [34])

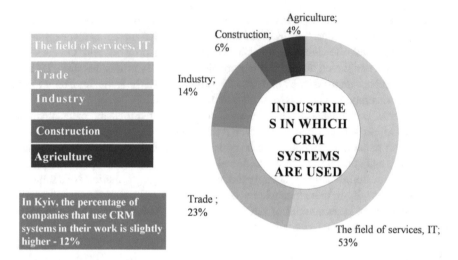

Fig. 9 Industries in which SRM systems are used to automate business operations

6 CRM and National Security of the Ukrainian Economy

As noted earlier, based on research conducted by the Society for Consumer Research (Gesellschaft für Konsumforschung (GFK)), in 2020 more than 50% of companies used Russian CRM products.

There is also another interesting piece of information, using the BuiltWith tool and open data, you can calculate how many Ukrainian business sites use Russian software products. Using the capabilities of BuiltWith, you can determine that the list of Russian services used by Ukrainian online stores includes: AmoCRM, RetailCRM, Bitrix24, Jivochat, Tilda, and Insales (Table 2) [39].

This is the only public data that can be found. At the same time, companies can use some closed programs it is difficult to detect from the outside. In addition, it is not rare to find a situation where a rather interesting situation develops that is difficult to understand. That is Ukrainian stores operating on Russian platforms, at the same

Table 2 Russian software products used by online stores on the Ukrainian market, as of 12 September 2022 (based on [39])

The name of the Russian service	Number of sites, BuiltWith
Tilda	13,453
Insales	199
Bitrix	3379
RenailCRM	87
AmoCRM	406
JivoChat	6344

time intensively using patriotic symbols. In this case, the question arises—is this a sincere position or marketing?

Also, after studying the tariffs on Russian products, you can calculate how much Ukrainian e-commerce pays to companies from the aggressor country. These calculations are approximate because it is impossible to make an exact calculation for many reasons. In addition, some companies offer custom solutions, the cost of which is calculated according to business requests. Also, there are services with a free tariff, but with reduced capabilities. So, as shown in Table 3, approximately 10,425 Ukrainian companies use Russian services and spend $122472, and this is UAH 4,479,380 per month at today's rate of the National Bank of Ukraine [39].

Today's trends in the development of the CRM market in Ukraine indicate that after recovering from the shock of the war, many Ukrainian companies decided to change the CRM system to a domestic one.

In particular, those businesses that were able to maintain their operations and had the budget to change CRM have already started, and some have even completed this process. On the other hand, businesses that have suffered more from the war and currently do not have additional budgets for data migration are forced to postpone this issue until their financial situation improves.

In addition to financial problems, many users still for various reasons have not been able to completely replace Russian CRM systems with an alternative Ukrainian or foreign product because the subscription to the Russian product is paid in advance in advance and has not yet expired; there is no possibility to switch to an alternative product because a large amount of data is lost; there are no specialists who could carry out the transition and migration of data; a large number of users are simply used to it and believe that they can continue to use it, despite a full-scale invasion.

Also interesting is the fact that Russian services transfer their domains, hiding their origin. For example, Bitrix24 moved the domain to the EU zone, AmoCRM is formally registered in San Francisco but belongs to 1C Company, which is a Russian business (Fig. 10).

Evaluating the situation in the CRM market and the behavior of Ukrainian businesses, it is possible to predict that in the future there will be 4 groups of companies:

Table 3 How much do Ukrainian companies spend on Russian servers (based on [39])

The name of the Russian service	Tariff	The approximate number of consumers	Amount of expenses, $
Tilda	8.20 $/month	6700	54940
Insales	27.35 $/month	100	2735
RenailCRM	25.27 $/month	45	1137
AmoCRM	16.38 $/month	200	3276
JivoChat	18.87 $/month	3200	60,384
	Total	10,245 companies	$122,472
	Total/year		$1,469,664

Fig. 10 How Russian servers disguise themselves as western ones [40]

Companies that will use Ukrainian CRM software; Companies that will use European and American CRM software; Companies that will use Russian CRM software no matter what; Companies that will operate without using CRM software (Fig. 11).

Ukrainian businesses should finally understand that using a Russian product is unpatriotic and dangerous for the following reasons. First, various bonuses in the form of special free tariffs will eventually expire, and the side effects will remain.

Second, it does not matter whether the servers are located in Russia or another country, the personal data of the company and its customers can be used by Russian companies. This is especially dangerous for online stores that receive information about their names, addresses, phone numbers, and emails from thousands of

Fig. 11 Forecasts of the use of CRM systems in Ukraine

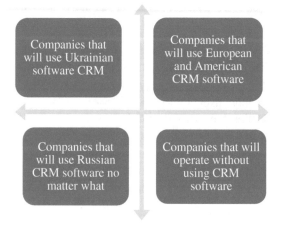

customers. The collected information can fall into the hands of those Russians who will use it for their own selfish, negative purposes.

Third, even if the money for using the product does not enter the Russian economy in the form of taxes, due to the intricacies of the legal process, it goes to the salaries of employees living in Russia. Opening an office in another country does not solve this problem, because companies rarely move the entire state and the bulk of people remain in Russia.

Fourth, in addition to issues of security and ethics, Russian services are not technically suitable for Ukrainians. At the beginning of the full-scale invasion, many users of Russian platforms did not have access to their admin accounts, and pages did not open. Ukrainian entrepreneurs should take a clear position regarding cooperation with the Russians and refuse any of their services.

7 Conclusion

So, summarizing the above, we can conclude that the world market of CRM systems is developing very quickly. CRM systems are often updated, and it happens that in the past the system did not solve all the issues that arise before it, but now it is necessary to pay attention to it, because its functionality has become much better. It is difficult to say that one CRM is the best and will suit all companies regardless of the industry. On the contrary, as many businesses exist, as many unique solutions can be chosen and built with different CRM systems for your business process and closing of your own needs.

Unfortunately, the Ukrainian market of CRM systems in Ukraine has not yet gained mass development, which is caused by several factors: insufficient business budgets for implementation; lack of understanding among a significant number of business owners and top management regarding the availability and necessity of CRM implementation; lack of qualified personnel, etc. At the same time, Ukrainian businesses need to understand that regardless of whether the company is well organized or just starting to scale and grow, the implementation of CRM systems and a successful customer service team can help attract new customers, increase retention and increase sales among the existing customer base. Without a doubt, relationship marketing is one of the most effective and profitable marketing strategies today. By winning the trust of customers, their loyalty increases, as a result, regular customers are more likely to become "rand advocates". However, it is difficult to build quality relationships when the number of clients exceeds hundreds or even thousands.

Today, customers almost no longer need to be told what they need or want. They already understand and know what they want, because they received this information through the Internet, social networks, forums, etc. Today, customers usually want to be treated as important persons, to feel that their requests are important to the company. Customers who feel valued are happy, and happy customers mean good feedback and repeat sales.

So, CRM can improve and ensure effective customer relations. In particular, CRM can be used to automate processes to congratulate customers on important anniversaries or holidays; CRM stores the entire history of communication with customers; CRM improves customer satisfaction by reducing response time and improving the quality of customer support; CRM increases customer retention. Automated workflows never leave the customer unattended, assisting at the best time.

References

1. Neuland, E., Hough, J.: Globalization of the world economy: The need for global strategies and mindsets for South African management. In: Paper Delivered at the EBM Conference Port Elizabeth (1999)
2. Redding, S.: Dynamic comparative advantage and the welfare effects of trade. Oxf. Econ. Pap.. Econ. Pap. **51**(1), 15–39 (1999)
3. Fastenau.: The Irrational Consumer: What it Means for Behavioral Economics (2019). Accessed from https://blog.crobox.com/article/the-rational-consumer-debunked. Accessed 25 Jan 2023
4. Kotler, P.: Ten Deadly Marketing Sins: Signs and Solutions, 1st edn. Wiley, New York (2004)
5. Joseph, B., Jimmie, B.: The Guru Guide to Marketing. A Concise Guide to the Best Ideas from Today's. Wiley, Hoboken, New Jersey (2004)
6. Kotler, P., Kartajaya, H., Setiawan, I.: Marketing 3.0: from products to customers to the human spirit. Management for professional. In: Kompella, K. (ed.) Marketing Wisdom, pp. 139–156. Springer, Berlin (2019). https://doi.org/10.1007/978-981-10-7724-1_10
7. Reid, M., Johnson, T., Ratcliffe, M., Skrip, K., Wilson, J.: Integrated marketing communications in the Australian and New Zealand wine industry. Int. J. Advert. **20**(2), 239–262 (2001). https://doi.org/10.1080/02650487.2001.11104889
8. Chornukha, O.: Implementation of the CRM System: The Role of CRM Technologies in Increasing Business Efficiency (2021). Accessed from https://tqm.com.ua/ua/likbez/crm-systemy/rol-vprovadzhennia-crm. Accessed 25 Jan 2023
9. Zendesk.: CX Trends 2022: Improve Your Bottom Line by Putting Customers at the Top (2022). Accessed from https://cx-trends-report-2022.zendesk.com/opportunity. Accessed 28 Jan 2023
10. Kotler, P., Keller, K.: Marketing Management, 14th edn. Pearson Education (2012)
11. Gilbert, N.: 75 Basic CRM Software Statistics: 2023 Data Analysis and Market Share (2023). Accessed from https://financesonline.com/crm-software-statistics/. Accessed 20 Jan 2023
12. Stattin, N.: 32 Customer Experience statistics You Need to Know For 2023 (2022). Accessed from https://www.superoffice.com/blog/customer-experience-statistics/. Accessed 10 Jan 2023
13. Robinson, C.: How to Improve Customer Experience Using a Helpdesk Software (2022). Accessed from https://financesonline.com/how-to-improve-customer-experience-using-a-helpdesk-software/. Accessed 10 Jan 2023
14. Lund, J.: Way Customer Relationship Management (CRM) is Key to Unlocking Business Potential (2022). Accessed from https://www.superoffice.com/blog/what-is-crm/. Accessed 10 Jan 2023
15. Kassanoff, B.: Build loyalty into your e-business. In: Proceedings of DCI Customer Relationship Management Conference. Boston, MA (2000)
16. Creighton, S.: Partnering for success to the e-business world. In: Proceedings of DCI Customer Relationship Management Conference. Boston, MA (2000)
17. Krämer, N., Winter, S.: Impression management 2.0. J. Media Psychol. Theor. Methods Appl. **20**(3), 106–116. https://doi.org/10.1027/1864-1105.20.3.106
18. Poulter, J., Dharmasthira, Y., Gupta, N., Amarendra, B.C.: Market Share: Customer Experience and Relationship Management, Worldwide, 2020 (2021). Accessed from https://www.gartner.com/en/documents/4001096. Accessed 05 Jan 2023

19. Spajic, D.: 31 Sizzling CRM Statistics to Help Your Business Soar (2022). Accessed from https://www.smallbizgenius.net/by-the-numbers/crm-statistics/. Accessed 07 Jan 2023
20. IBM Institute Business for Value.: 5Trends for 2023: Embracing chaos, taking charge (2023). Accessed from https://www.ibm.com/downloads/cas/JLKJK1ZP. Accessed 15 Jan 2023
21. Dynamic Consultants Group.: CRM–Return on Investment (2021). Accessed from https://dynamicconsultantsgroup.com/resources/crm/roi-of-crm-2021/. Accessed 25 Jan 2023
22. Salesforce.: Deliver success now with Salesforce Customer 360 (2023). Accessed from https://www.salesforce.com/
23. Traskvia.: Connect and Track Everything That Matters. https://trackvia.com/. Accessed 28 Jan 2023
24. Taylor, M.: 18 CRM Statistics You Need to Know ford 2023 (and beyond) (2022). Accessed from https://www.superoffice.com/blog/crm-software-statistics/. Accessed 28 Dec 2022
25. MacDonald, S.: Best CRM Software: How To Find The Right CRM For You (2021). Accessed from https://www.superoffice.com/blog/best-crm-software/. Accessed 20 Dec 2022
26. Kemp, S.: Digital 2021 October Global Statshot Report—DataReportal—Global Digital Insights (2021). Accessed from https://datareportal.com/reports/digital-2021-october-global-statshot. Accessed 14 Jan 2023
27. Lund, J.: CRM Mobile App: The Rise of CRM [Infographic] (2020). Accessed from https://www.superoffice.com/blog/infographic-the-rise-of-the-mobile-crm-app/. Accessed 11 Jan 2023
28. State of the Connected Customer.: (2016). Accessed from https://www.salesforce.com/content/dam/web/en_us/www/images/form/pdf/socc-2016.pdf. Accessed 18 Jan 2023
29. Apps Run The World.: Top CRM Software Vendors (2021). Accessed from https://www.appsruntheworld.com/cloud-top-500/CRM/. Accessed 25 Jan 2023
30. MacDonald, S.: How a customer experience strategy helps scale revenue growth (and achieve profitability) (2022). Accessed from https://www.superoffice.com/blog/customer-experience-strategy/. Accessed 12 Jan 2023
31. Super-Ofise.: 9 Real-World Examples of CRM Automation (For Sales, Marketing and Customer Serwise Teams) (2023). Accessed from https://www.superoffice.com/blog/crm-automation/. Accessed 16 Jan 2023
32. Ahmed, W., Rasool, A., Javed, A.R., Kumar, N., Gadekallu, T.R., Jalil, Z., Kryvinska, N.: Security in next generation mobile payment systems: a comprehensive survey. IEEE Access **9**, 115932–115950 (2021). https://doi.org/10.1109/ACCESS.2021.3105450
33. Kryvinska, N., Kaczor, S., Strauss, C., Greguš, M.: Servitization—its raise through information and communication technologies. In: Lecture Notes in Business Information Processing, pp. 72–81 (2014). https://doi.org/10.1007/978-3-319-04810-9_6
34. Carter, R.: The Ultimate List of CRM Statistics for 2022 (2023). Accessed https://findstack.com/resources/crm-statistics/. Accessed 17 Jan 2023
35. Nucleus Research.: CRM pays bask $8.71 for every dollar spent (2014). Accessed from. https://nucleusresearch.com/wp-content/uploads/2018/05/o128-CRM-pays-back-8.71-for-every-dollar-spent.pdf. Accessed 25 Jan 2023
36. Auspex.: Results of CRM market research in Ukraine (2018). Accessed from https://auspex.ua/articles/biznes-sovety/rezultaty-doslidzhennya-rynku-crm-v-ukrayini/. Accessed 20 Dec 2022
37. Business.: Only 8.4% of surveyed companies use specialized CRM systems (2020). Accessed from. https://biz.nv.ua/ukr/markets/90-ukrajinskih-kompaniy-ne-vikoristovuyut-crm-novini-ukrajini-50102087.html. Accessed 17 Jan 2023
38. Finance.: CRM systems and business: what to expect from the Ukrainian market (2022). Accessed from https://finance.ua/ua/saving/crm-systemy-ta-biznes. Accessed 16 Jan 2023
39. How Ukrainian business cooperates with the Russians despite the war—Horoshop study. Accessed from https://uaspectr.com/2022/10/04/biznes-spivpratsyuye-z-rosiyanamy-nezvazhayuchy-na-vijnu/. Accessed 20 Jan 2023
40. amoCRM—Crunchbase Company Profile and Funding. Accessed from https://www.crunchbase.com/organization/amocrm. Accessed 17 Jan 2023

Methodological Principles of Smoothing the Effect of Seasonal Fluctuations on the Components of Labor Intensity in Construction

Yevheniia Novak⬤, Viktoriya Tytok⬤, Oleksandr Kazmin⬤, Denis Dubinin⬤, and Olena Emelianova⬤

Abstract Tools and methods of mathematical statistics are used to determine and measure seasonal fluctuations in construction, especially such an indicator as labor intensity. Seasonality is a factor affecting failures and deviations of construction processes from planned indicators. Climatic conditions have a moderate influence on construction terms, which can be expressed in the form of seasonal fluctuations in deviations in the course of the construction process or the supply of resources from the plan. Determination of corrective components for smoothing the level of seasonality plays an important role in planning the organization of construction production. The research established seasonality indices for 2012–2019, which require further analysis for practical optimization at any construction enterprise. The factors that directly affect the size of the fluctuations have been determined, and the main stages of forecasting and methodological principles of adjusting labor intensity at the construction enterprise under the influence of seasonal fluctuations have been determined based on the decomposition of numerical series with the formation of a system of economic-mathematical models of labor intensity at the construction enterprise. Cyclical fluctuations in labor intensity in the construction industry force construction enterprises to use modern scientific methods of seasonality analysis.

Y. Novak
Yuriy Fedkovych Chernivtsi National University, 2 Kotsiubynsky Str., Chernivtsi 58012, Ukraine

V. Tytok (✉) · O. Kazmin · D. Dubinin · O. Emelianova
Kyiv National University of Construction and Architecture, 31 Povitroflotskyi Ave., Kyiv 03037, Ukraine
e-mail: tytok.vv@knuba.edu.ua

O. Kazmin
e-mail: kazmin_oh-2022@knuba.edu.ua

D. Dubinin
e-mail: dubinin.dv@knuba.edu.ua

O. Emelianova
e-mail: Iemelianova.om@knuba.edu.ua

Keywords Smoothing · Seasonal fluctuations · Intensity · Seasonality index ·
Trend · Dynamic series

1   Introduction

High instability, uncertainty of changes in the economic situation in Ukraine requires
the use of various methods and tools for forecasting strategic trends in the development of the construction industry of Ukraine. There is a constant need to take into
account not only already formed trends, but also the analysis of corrective factors
that have an external and internal influence, including force majeure circumstances
and reflect the real situation for the optimization and organization of construction
production at a construction enterprise [1–4]. The application of adaptive models of
economic smoothing [5] best shows the evolution of characteristics and their ability to
represent dynamic values, taking into account seasonal fluctuations that are inherent
in the construction industry.

The question of estimating seasonal fluctuations in itself is not new and is often
studied in various economic spheres [6, 7] and proves the need for correction coefficients. Similar scientific works related to the forecasting of sales volumes of building
materials based on expert assessments [8].

2   Literature Review and Methodology

Construction is one of the sectors of the national economy, which has clearly
expressed seasonal trends, which must be taken into account in the strategic forecasting of the economic activity of a construction enterprise [9–11]. The works
[12, 13] established a regularity regarding the presence of accidents at work in the
construction industry, namely, seasonality and cyclicality during the year. To date, a
number of methods and models have been developed, which should ensure the minimization of refusal and malfunctions of the construction process, namely models
and methods of organization, management, and economic evaluation of technological processes of construction production based on the fractal characteristics of a
series [14, 15], taking into account factors seasonality [16], mathematical modeling
of construction work parameters by means of fuzzy sets [17, 18].

Using the algorithm for studying the effect of seasonal fluctuations on the cost of
rolled metal [11, 19], we will apply it to the study of the most important and significant
indicator in the optimization and organization of a construction enterprise—labor
intensity and its components.

The application of correlation-regression analysis to determine the mathematical
model, form, strength and tightness of the correlation between the cost and labor
intensity of construction works and obtained the regression equation with the determination of the influence of each of the studied factors on the resulting indicators.

The obtained regularities can be used to calculate the projected costs of resources for repair work to restore the properties of thermal insulation and equipping facades [20].

Special attention is paid to modern processes of digital transformation in construction, which have a huge impact on the organization and optimization of construction processes in the construction enterprise [14, 15, 21].

The creation of information models of building objects (BIM-models) makes it possible to obtain comprehensive information at all stages of the life cycle of a construction object: from design and construction, maintenance, repair, and finally to its liquidation [22–24]. The use of BIM technology will improve quality, make work on the project transparent, the project itself more manageable, which will help save money [25–27].

Using the experience [28, 29] according to a methodical approach to identifying the influence of seasonal fluctuations on the parameters of the construction object, we will calculate the seasonality indices for the "labor intensity" indicator for various construction objects.

The most common indicator used to analyze seasonal fluctuations of various indicators are seasonality indices I_s, which allow to quantitatively assess the manifestations of winter and summer conditions for any indicator of the activity of enterprises, construction processes or social phenomena [30–32].

Indices of seasonality (or seasonal wave) are calculated as a percentage ratio of the actual levels of the dynamic series to its average annual (adjusted) level [33, 34].

The most optimal method of forecasting seasonality is trend extrapolation, the essence of which is the use of statistical data from previous periods and the use of trend models, which are based on the assumption that the identified trend will persist in the future.

The approach to the study of seasonality, which is based on the decomposition of time series, is widespread. When using the decomposition of time series, it is assumed that the trend and seasonal fluctuations revealed as a result of the decomposition of the series into components will persist in the future, and, therefore, can be used to make a forecast.

3 Results and Discussion

The study of the impact of seasonality on the labor intensity of construction of residential buildings begins with the compilation of monthly data on deviations of the progress of the construction process on various construction sites and the formation of a single numerical series of deviations of the actual construction terms from the design ones. The next step is to determine the impact of seasonal fluctuations on each of the groups according to seasonality indices calculated by the method of simple averages, amplitude, simple and quadratic coefficients of seasonality (Table 1 and Fig. 1).

Table 1 Seasonality indicators for the parameter "labor intensity"

Month	Indicator (ΔLi)								$\overline{x}$	$I_s = \dfrac{\overline{x}}{\overline{X}}$	$I_s - 100$	$(I_s - 100)^2$
	2012	2013	2014	2015	2016	2017	2018	2019				
Labor intensive												
January	40.3	36.7	45.8	42.7	48.8	47.9	41.4	32.1	42.18	1.2	22	484.00
February	58.0	39.9	55.8	52.4	58.4	47.1	46.0	35.5	41.88	1.2	21	441.00
March	59.7	25.7	36.5	35.9	48.0	47.1	32.2	32.8	32.28	0.9	7	49.00
April	48.3	40.1	41.9	32.7	47.1	43.9	42.6	30.7	34.86	1.0	1	1.00
May	38.9	38.0	43.0	40.9	37.7	36.6	31.1	26.7	31.73	0.9	8	64.00
June	31.3	33.8	43.0	38.0	42.2	29.7	25.7	20.8	29.15	0.8	16	256.00
July	45.7	40.7	43.8	40.7	43.5	40.9	30.5	20.7	32.59	0.9	6	36.00
August	40.4	36.2	46.0	40.7	40.3	33.8	29.2	18.8	30.61	0.9	12	144.00
September	41.0	31.5	45.4	38.1	40.7	27.7	24.7	16.1	28.03	0.8	19	361.00
October	45.8	34.1	46.3	40.6	43.4	32.5	26.0	17.1	30.00	0.9	13	169.00
November	43.4	45.0	57.3	46.3	55.5	37.1	34.2	20.6	37.00	1.1	7	49.00
December	53.3	54.6	62.9	52.5	68.5	45.8	39.6	37.0	45.13	1.3	30	900.00
$\overline{X} = \left(\overline{X}_{January} + \overline{X}_{February} + \cdots + \overline{X}_{December} \right) / 12$									34.62			
In total											69.00	1479.00
$R_s = I_{smax} - I_{smin}$											17.1	
$k_s = \dfrac{\sum (I_s - 100)}{12}$											5.75	
$k_{ssq} = \sqrt{\dfrac{\sum (I_s - 100)^2}{12}}$										123.25		

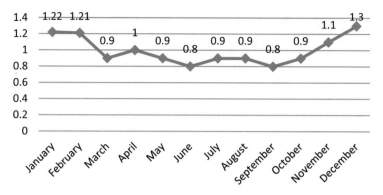

Fig. 1 Indices of seasonality for labor intensity

Analyzing the data in the Table 1, it can be seen that statistical indicators of seasonality assessment indicate the presence of fluctuations. Thus, an increase in the "dust capacity" parameter is observed in January, February, November, and December, and a decrease relative to the average level in March, May, June, July, August, and September. In April, the seasonality index was 1; that is, it coincides with the average annual value. It can be concluded that seasonal fluctuations occur when determining the deviations of labor intensity as well.

The next step is to determine the trend that best determines the direction of changes in labor intensity during the period under study. Linear and polynomial equations were used for the analysis (Fig. 2). Among the three obtained equations, we determine the one that best describes the actual data.

For a linear equation that does not take into account seasonality, the coefficient of determination has a very small value of $R^2 = 0.23$, that is, 23% of the variability of the resulting factor (in our case, deviations of the actual labor intensity from the design one) is explained by this equation.

With a polynomial relationship of the fourth degree, the level of determination $R^2 = 0.468$, i.e. 46.8% of fluctuations in labor intensity during the year can be explained by seasonal factors.

The polynomial of the sixth degree has the coefficient of determination $R^2 = 0.473$, which only slightly exceeds the indicator of the previous equation. That is, 47.3% of labor intensity fluctuations during the year can be explained using the obtained equation.

It can be seen that the polynomial trends approximate the actual data better than the linear equation (Figs. 3 and 4).

Subtracting the values calculated on the basis of the obtained equations from the actual values of labor intensity deviations, we will determine the values of the seasonal component for linear regression and polynomials of the fourth and sixth degrees (Tables 2, 3, 4, 5 and 6).

For polynomial equations of the fourth and sixth degree, the calculations are carried out in the same sequence (Tables 3 and 4).

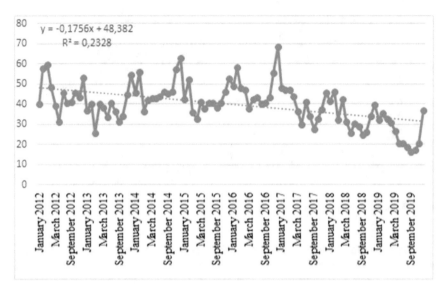

Fig. 2 Determining trends in labor intensity changes using a linear equation

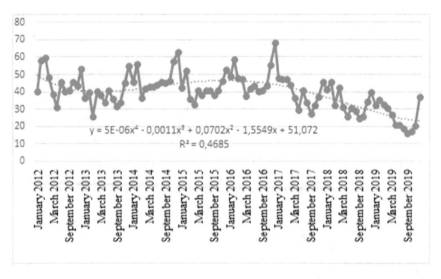

Fig. 3 Determining the trends of changes in labor intensity according to the polynomial equation of the fourth degree

The next stage is the adjustment of the value of the seasonal component in order to obtain the sum of fluctuations, equal to zero, by determining the average value of the sum of averages (average of averages).

For this, the sum of the average values of each month is divided by the number of periods (in this case—12 months). The obtained result is subtracted from the average

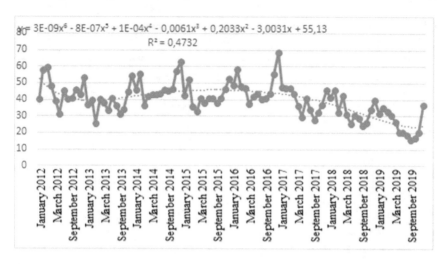

Fig. 4 Determining the trends of changes in labor intensity according to the polynomial equation of the sixth degree

values of each period, as a result of which the sum of all fluctuations is leveled and will be equal to zero (Tables 5 and 6).

According to the Table 5, it can be concluded that the labor-capacity seasonality coefficients increase relative to the average annual level in February, April, May (slightly), November and December. Less than the average annual level of labor intensity deviation in January, March, June, July, August, September, and October. The maximum value of the correction coefficient is in December (5.9), and the minimum value is in September (−4.7). A significant decrease in labor intensity in January ($k_s = -3.7$) is most likely explained by the New Year holidays.

For polynomial models, the correcting coefficients are calculated similarly (Tables 6 and 7).

The correcting coefficients for polynomials of the fourth and sixth degrees change by half-year. Thus, for a polynomial of the fourth degree, the coefficients for January-May exceed the average level, and seasonality coefficients for June-December are lower than the annual average level. According to the data in Table 6, it can be concluded that the maximum value of the correction coefficient is in January (60), and the minimum value is in December (−51.4).

The same trend is observed for polynoma of the sixth degree. k_s exceed the average level in January-May, and the coefficients in June-December are lower than the average annual level.

According to the data in Table 7, it can be concluded that the maximum value of the correction coefficient is in January (36.2), and the minimum value is in December (−27.6).

Table 2 The value of trends and seasonal components for the linear model for the parameter "labor intensity (fragment)"

No.	Month	Labor intensity deviation, (ΔLi)	Linear equation	Seasonal component
1	January 2012	40	48.44	−8.19
2	February 2012	58	48.09	9.92
3	March 2012	60	47.91	11.77
4	April 2012	48	47.73	0.59
5	May 2012	39	47.55	−8.61
6	June 2012	31	47.38	−16.04
7	July 2012	46	47.20	−1.55
8	August 2012	40	47.02	−6.60
9	September 2012	41	46.84	−5.89
10	October 2012	46	46.67	−0.91
11	November 2012	43	46.49	−3.11
...	...	...	...	...
81	September 2018	25	34.06	−9.40
82	October 2018	26	33.89	−7.87
83	November 2018	34	33.71	0.49
84	December 2018	40	33.53	6.10
85	January 2019	32	33.35	−1.28
86	February 2019	35	33.18	2.30
87	March 2019	33	33.00	−0.25
88	April 2019	31	32.82	−2.12
89	May 2019	27	32.64	−5.97
90	June 2019	21	32.47	−11.70
91	July 2019	21	32.29	−11.59
92	August 2019	19	32.11	−13.28
93	September 2019	16	31.93	−15.80
94	October 2019	17	31.76	−14.69
95	November 2019	21	31.58	−10.94
96	December 2019	37	31.40	5.63

Table 3 Values of trends and seasonal components of the polynomial equation of the fourth degree for the "labor intensity" parameter (fragment)

No.	Month	Labor intensity deviation, (ΔLi)	Polynomial equation	Seasonal component
1	January 2012	58	49.59	8.16
2	February 2012	68	48.24	19.71
3	March 2012	60	47.04	12.64
4	April 2012	48	45.98	2.34
5	May 2012	39	45.06	−6.12
6	June 2012	23	44.28	−20.92
7	July 2012	42	43.64	−1.31
8	August 2012	40	43.15	−2.72
9	September 2012	41	42.80	−1.85
10	October 2012	46	42.59	3.16
11	November 2012	43	42.54	0.84
…	…	…	…	…
80	August 2018	29	580.76	−551.60
81	September 2018	25	600.94	−576.28
82	October 2018	26	621.66	−595.64
83	November 2018	34	642.91	−608.71
84	December 2018	32	664.73	−632.60
85	January 2019	36	687.10	−651.53
86	February 2019	35	710.05	−674.58
87	March 2019	33	733.59	−700.84
88	April 2019	31	757.72	−727.02
89	May 2019	29	782.45	−753.28
90	June 2019	27	807.80	−780.37
91	July 2019	27	833.78	−806.41
92	August 2019	29	860.39	−831.56
93	September 2019	13	887.65	−874.85
94	October 2019	14	915.57	−901.84
95	November 2019	21	944.16	−923.53
96	December 2019	27	973.44	−946.40

Table 4 Values of trends and seasonal components of the polynomial equation of the fourth degree for the "labor intensity" parameter (fragment)

No.	Month	Labor intensity deviation, (ΔLi)	Polynomial equation	Seasonal component
1	January 2012	58	52.32	5.43
2	February 2012	68	49.89	18.06
3	March 2012	60	47.79	11.88
4	April 2012	48	46.00	2.32
5	May 2012	39	44.49	−5.55
6	June 2012	23	43.24	−19.88
7	July 2012	42	42.20	0.12
8	August 2012	40	41.38	−0.95
9	September 2012	41	40.73	0.22
10	October 2012	46	40.25	5.50
11	November 2012	43	39.92	3.46
12	December 2012	48	39.71	8.61
…	…	…	…	…
83	November 2018	34	293.93	−259.73
84	December 2018	32	308.77	−276.65
85	January 2019	36	324.42	−288.84
86	February 2019	35	340.90	−305.42
87	March 2019	33	358.26	−325.51
88	April 2019	31	376.54	−345.84
89	May 2019	29	395.80	−366.63
90	June 2019	27	416.08	−388.65
91	July 2019	27	437.44	−410.07
92	August 2019	29	459.92	−431.09
93	September 2019	13	483.59	−470.79
94	October 2019	14	508.50	−494.77
95	November 2019	21	534.72	−514.09
96	December 2019	27	562.31	−535.28

Table 5 Values of trends and seasonal components of the polynomial equation of the fourth degree for the "labor intensity" parameter (fragment)

Month	2012	2013	2014	2015	2016	2017	2018	2019	Average	Seasonal component k_s
1	2	3	4	5	6	7	8	9	10	11
Linear model										
January	−8.2	−9.4	1.8	0.8	9.0	10.3	5.9	−1.3	1.1	−3.7
February	0.0	−6.0	12.0	10.7	18.8	9.6	10.7	2.3	7.3	2.4
March	40.3	−20.1	−7.1	−5.6	8.6	9.9	−3.0	−0.2	2.8	−2.0
April	58.0	−5.5	−1.6	−8.7	7.9	6.8	7.6	−2.1	7.8	2.9
May	59.7	−7.4	−0.3	−0.3	−1.4	−0.3	−3.7	−6.0	5.0	0.2
June	48.3	−11.5	−0.1	−3.0	3.3	−7.0	−8.9	−11.7	1.2	−3.7
July	38.9	−4.4	0.9	−0.2	4.8	4.4	−3.9	−11.6	3.6	−1.2
August	31.3	−8.7	3.2	0.0	1.7	−2.5	−5.1	−13.3	0.8	−4.0
September	45.7	−13.2	2.8	−2.3	2.4	−8.5	−9.4	−15.8	0.2	−4.7
October	40.4	−10.5	3.9	0.3	5.3	−3.5	−7.9	−14.7	1.7	−3.2
November	41.0	0.6	15.1	6.2	17.5	1.3	0.5	−10.9	8.9	4.0
December	45.8	10.5	20.9	12.6	30.7	10.2	6.1	5.6	12.8	5.9
In total	441.1	−85.7	51.4	10.6	108.7	30.5	−11.0	−79.7	4.9	0.0

Table 6 Calculation of the seasonal component for polynoma of the fourth degree

Month	2012	2013	2014	2015	2016	2017	2018	2019		Average	Seasonal component k_s
1	2	3	4	5	6	7	8	9		10	11
Polynomial model of the fourth degree											
January	8.2	−0.1		−2.3	−43.9	−114.7	−225.7	−409.0	−651.5	−179.9	60.0
February	19.7	−3.3		−4.6	−43.9	−116.8	−239.3	−424.4	−674.6	−185.9	54.0
March	12.6	−18.3		−26.4	−66.9	−130.2	−259.4	−455.4	−700.8	−205.6	34.3
April	2.3	−4.4		−23.8	−73.0	−144.5	−275.3	−462.6	−727.0	−213.5	26.4
May	−6.1	−7.3		−25.6	−73.6	−162.6	−294.3	−492.2	−753.3	−226.9	13.0
June	−20.9	−12.6		−28.6	−82.2	−164.6	−314.4	−516.2	−780.4	−240.0	−0.1
July	−1.3	−8.8		−31.1	−88.8	−175.2	−324.4	−530.6	−806.4	−245.8	−5.9
August	−2.7	−12.7		−32.5	−91.7	−185.6	−340.5	−551.6	−831.6	−256.1	−16.2
September	−1.8	−18.9		−36.7	−99.6	−197.6	−360.3	−576.3	−874.9	−270.8	−30.9
October	3.2	−18.0		−39.7	−105.9	−207.7	−371.3	−595.6	−901.8	−279.6	−39.7
November	0.8	−8.9		−32.8	−108.7	−203.8	−382.0	−608.7	−923.5	−283.4	−43.5
December	5.7	−3.2		−38.3	−109.7	−214.3	−391.6	−632.6	−946.4	−291.3	−51.4
In total	19.6	−116.5		−322.4	−987.7	−2017.7	−3778.4	−6255.2	−9572.2	−239.9	0.0

Table 7 Calculation of the seasonal component for polynoma of the sixth degree

Month	2012	2013	2014	2015	2016	2017	2018	2019	Average	Seasonal component k_s
1	2	3	4	5	6	7	8	9	10	11
Polynomial model of the sixth degree										
January	5.4	3.1	12.0	2.2	−12.9	−46.3	−137.0	−288.8	−57.8	36.2
February	18.1	0.3	11.4	5.9	−9.3	−52.7	−144.4	−305.4	−59.5	34.5
March	11.9	−14.2	−8.5	−13.2	−16.9	−65.3	−167.4	−325.5	−74.9	19.1
April	2.3	0.2	−3.8	−15.2	−25.2	−73.7	−166.7	−345.8	−78.5	15.5
May	−5.6	−2.1	−3.4	−11.6	−37.2	−85.2	−188.4	−366.6	−87.5	6.5
June	−19.9	−6.7	−4.0	−15.8	−32.9	−97.6	−204.6	−388.7	−96.3	−2.3
July	0.1	−2.1	−4.0	−17.8	−37.1	−99.9	−211.3	−410.1	−97.8	−3.8
August	−1.0	−5.0	−2.6	−16.0	−40.9	−108.2	−224.6	−431.1	−103.7	−9.7
September	0.2	−10.2	−3.9	−19.0	−46.2	−120.2	−241.8	−470.8	−114.0	−20.0
October	5.5	−8.1	−3.9	−20.2	−49.5	−123.2	−253.8	−494.8	−118.5	−24.5
November	3.5	2.3	6.2	−17.8	−38.7	−125.9	−259.7	−514.1	−118.0	−24.0
December	8.6	9.4	4.1	−13.5	−42.1	−127.6	−276.6	−535.3	−121.6	−27.6
In total	29.2	−33.2	−0.3	−152.0	−388.9	−1125.8	−2476.5	−4877.0	−94.0	0.0

As a result, we get three equations (additive models) that take into account seasonal fluctuations, as they include seasonality coefficients:

$$y = -0.1775t + 48.44 + k_s \qquad (1)$$

$$y = 0.000005t - 0.00111t + 0.0702t - 1.5549 + 51.07 + k_s \qquad (2)$$

$$y = 0.00000004t - 0.000007t + 0.0001t$$
$$- 0.006t + 2.2t - 0.203t + 53.1 + k_s \qquad (3)$$

where y is labor intensity; t is time; k_s is seasonal component.

The level of approximation of the resulting models is determined using the average linear deviation (Table 8).

The linear model has the highest predictive ability (deviation 8.3 times), the polynomial model of the 4th degree has the lowest (deviation 6.8 times). In works [35, 36] it is noted that polynomial models should be used with caution in calculations, as they are capable of significantly deviating from established trends over time. In some cases, the indicators set by the model can take a negative value (which deprives the model of economic meaning). Therefore, the authors recommend recalculating the model every time when new data is received, which should significantly increase the accuracy of forecasting.

4 Conclusion

It has been proven that natural and climatic conditions and seasonality have a moderate influence on the progress of construction, which can be expressed in the form of seasonal fluctuations in indicators of deviations in the progress of construction works and their parameters such as construction terms, labor intensity and estimated cost during the calendar year.

The main stages of forecasting and corrections of certain organizational and technological parameters of construction under the influence of seasonal fluctuations based on the decomposition of numerical series have been determined, and a set of relevant models has been created.

The first stage is the formation of a sample for correlation and regression analysis, the next stage is the identification of the leading trend of the development of the process and the creation of a set of equations that describe this trend, the third stage is the identification of the seasonal component and the creation of equations that include this component in the form of a coefficient seasonality, and the final one is the final selection and setting of special econometric functions for adjusting construction characteristics, with seasonality coefficients, which are determined based on the analysis of the retrospective behavior of data series in previous periods for each

Table 8 Determination of the level of approximation according to each of the models

Month	Actual data	Forecast by linear model	Deviation	Prediction by a polynomial of the 6th degree	Deviation	Prediction by a polynomial of the 4th degree	Deviation
January 2012	40	49.6	−0.23	−5.5	1.09	−130.3	3.26
February 2012	58	55.3	0.05	−9.6	1.14	−137.7	3.03
March 2012	60	50.8	0.15	−27.1	1.45	−158.5	3.66
April 2012	48	55.5	−0.15	−32.5	1.67	−167.6	4.47
May 2012	39	52.6	−0.35	−43.0	2.10	−181.8	5.67
June 2012	31	48.6	−0.55	−53.0	3.27	−195.7	9.38
July 2012	46	50.8	−0.11	−55.6	2.31	−202.2	5.78
August 2012	40	47.9	−0.18	−62.3	2.54	−212.9	6.27
September 2012	41	47.0	−0.15	−73.3	2.79	−228.0	6.57
October 2012	46	48.3	−0.06	−78.2	2.71	−237.0	6.18
November 2012	43	55.4	−0.28	−78.1	2.80	−240.9	6.55
December 2012	53	64.1	−0.20	−81.9	2.69	−248.7	6.15
January 2013	37	47.2	−0.29	−18.2	1.43	−137.0	4.21
February 2013	40	53.2	−0.33	−19.9	1.50	−142.6	4.57
…	…	…	…	…	…	…	…
September 2019	16	32.1	−0.99	369.6	−27.87	616.9	−47.19
October 2019	17	33.4	−0.96	390.0	−27.40	636.0	−45.31
November 2019	21	40.5	−0.96	416.7	−19.19	660.7	−31.02
December 2019	37	49.2	−0.33	440.7	−15.30	682.1	−24.23
			0.083		3.35		6.8

individual object or group of similar objects. The final model, which represents the detected trend and the seasonal component, is selected based on the mean linear error.

The paper measured the deviation of the construction terms from the plan at the objects being built in the cities of Kyiv and Chernivtsi in order to identify the impact on the terms of work, labor intensity and cost of the seasonality factor and justify the correction coefficients to the construction parameters depending on the season, which can to be used by subcontractors for internal company planning and creation of own data indicators based on a methodical approach based on the results of own observations.

On the basis of the introduced approach, a system of prognostic, organizational and technological models for adjusting performance characteristics under the influence of seasonal fluctuations is proposed, which makes it possible to reveal the variable dependence of construction parameters on the average daily temperature of the outside air: (a) labor intensity forecasting model with an accuracy of $\Delta_i = 8.3\%$, it is recommended to use for the stage of work on the construction site and the season; (b) models for forecasting the cost of works and terms with an accuracy of $\Delta_i = 14.1 - 16\%$, it is advisable to use them at the stages of development of PVR, conclusion of subcontracts.

References

1. Stetsenko, S.P., Tytok, V.V., Emelianova, O.M., Bielienkova, O.Y., Tsyfra, T.Y.: Management of adaptation of organizational and economic mechanisms of construction to increasing impact of digital technologies on the national economy. J. Rev. Global Econ. **9**, 149–164 (2020). https://doi.org/10.6000/1929-7092.2020.09.15
2. Pavlichenko, P. V.: Modelyuvannya makroekonomichnogo stanu Ukraïni: tendentsiï rozvitku sektoriv ekonomiki. Molodizhnii ekonomichnii visnik KhNEU im. S. Kuznetsya **3**, 191–197 (2020)
3. Fedun, I., Stetsenko, S., Tsyfra, T, Vershygora, D., Valchuk, B., Andriiv, V.: innovative software tools for effective management of financial and economic activities of the organization. In: Alareeni, B., Hamdan, A. (eds.) Impact of Artificial Intelligence, and the Fourth Industrial Revolution on Business Success. ICBT 2021. Lecture Notes in Networks and Systems, vol. 485. Springer, Cham (2023). https://doi.org/10.1007/978-3-031-08093-7_2
4. Ryzhakova, G., Pokolenko, V., Omirbayev, S., Novykova, I., Bielienkova, O., Kapustian M.: Modern structuring of project financing solutions in construction. In: 2022 International Conference on Smart Information Systems and Technologies (SIST), pp. 1–7 (2022). https://doi.org/10.1109/SIST54437.2022.9945779
5. Terentyev, O., Tsiutsiura, S., Honcharenko, T., Lyashchenko, T.: Multidimensional space structure for adaptable data model. Int. J. Recent Technol. Eng. **8**(3), 7753–7758 (2019). https://doi.org/10.35940/ijrte.C6318.098319
6. Semenova, K.D.: Assessment of seasonal fluctuations in enterprise performance indicators. Soc. Econ. Res. Bull. **25**, 311–315 (2007)
7. Mykhaylov, V.S., Prilypko, Yu.I., Shepel, K.I.: Seasonal fluctuations and calendar effects: separate problems of the theory and practice of statistical evaluation. Stat. Ukraine **4**, 21–26 (2012)

8. Oklander, M.A., Pedko, I.A.: Methods of expert estimates and forecast industry sales volumes. Mech. Econ. Regul. **1**, 69–77 (2016)
9. Bielienkova, O.: The impact of seasonal fluctuations on circulating assets of a construction firm. Investytsiyi: praktyka ta dosvid **10**, 48–53 (2015)
10. Demydova, O., Lytvynenko, O., Moholivets, A., Novak, Y.: Influence of seasonal factors on quality, cost, labor and other parameters of construction. Sci. Heritage **74**, 42–49 (2021)
11. Reznik, N. et al.: Systems Thinking to Investigate the Archetype of Globalization. Springer International Publishing (2022). https://doi.org/10.1007/978-3-031-08087-6_9
12. Meng, G., Liu, J., Feng, R.: Prediction of construction and production safety accidents in china based on time series analysis combination model. Appl. Sci. **12**, 11124 (2022). https://doi.org/10.3390/app122111124
13. Hoła, B., Topolski, M., Szer, I., et al.: Prediction model of seasonality in the construction industry based on the accidentality phenomenon. Archiv. Civ. Mech. Eng **22**, 30 (2022). https://doi.org/10.1007/s43452-021-00348-7
14. Bulakh, V., Kirichenko, L., Radivilova, T.: Time series classification based on fractal properties. In: 2018 IEEE 2nd International Conference on Data Stream Mining and Processing, DSMP 2018, pp. 198–201 (2018). https://doi.org/10.48550/arXiv.1905.03096
15. Pilgrim, I., Taylor, R.P.: Fractal Analysis of Time-Series Data Sets: Methods and Challenges (2019). https://doi.org/10.5772/intechopen.81958
16. Ruiz-Fernández, J., Benlloch, J., López, M., Valverde-Gascueña, N.: Influence of seasonal factors in the earned value of construction. Appl. Math. Nonlinear Sci. **4**(1), 21–34 (2019). https://doi.org/10.2478/AMNS.2019.1.00003
17. Rohatynskyi, R., Harmatiy, N., Fedyshyn, I., Dmytriv, D.: Modeling the development of machine-building industry on the basis of the fuzzy sets theory. Naukovyi Visnyk Natsionalnoho Hirnychoho Universytetuthis Link is Disabled (2), 74–81 (2020). https://doi.org/10.33271/nvngu/2020-2/074
18. Barros, L.C., Bassanezi, R.C., Lodwick, W.A.: Fuzzy sets theory and uncertainty in mathematical modeling. In: A First Course in Fuzzy Logic, Fuzzy Dynamical Systems, and Biomathematics. Studies in Fuzziness and Soft Computing, vol 347. Springer, Berlin (2017). https://doi.org/10.1007/978-3-662-53324-6_1
19. Stetsenko, S., Bielienkova, O., Lytvynenko, O.: Influence of seasonal fluctuations on cost parameters of construction production. Manag. Dev. Complex Syst. **32**, 179–185 (2017)
20. Shalennyi, V.T., Skokova, A.O.: Results of a correlation-regression analysis of the influence of architectural and planning properties and damage to facades on the cost and time-consuming work of restoring their external thermal insulation and equipment. Constr. Man-Made Saf. **42**, 90–97 (2012)
21. Zeltser, R.Y., Bielienkova O.Y., Novak, Y., Dubinin D.V.: Digital transformation of resource logistics and organizational and structural support of construction. Sci. Innovation 15(5), 38–51 (2019). https://doi.org/10.15407/scine15.05.034
22. Honcharenko, T., Chupryna, Y., Ivakhnenko, I., Zinchenco, M., Tsyfra, T.: reengineering of the construction companies based on BIM-technology. Int. J. Emerg. Trends Eng. Res. **8**(8), 4166–4172 (2020). https://doi.org/10.30534/ijeter/2020/22882020
23. Stetsenko, S., Sorokina, L., Goiko, A., Tsyfra, T., Bolila, N.: Cals-model for forming the anti-crisis potential of construction enterprises. Sci. J. Astana IT Univ. **4**, 49–57 (2020). https://doi.org/10.37943/AITU.2020.80.41.006
24. Akselrod, R., Shpakov, A., Ryzhakova, G., Honcharenko, T., Chupryna, I., Shpakova, H.: Integration of data flows of the construction project life cycle to create a digital enterprise based on building information modeling. Int. J. Emerg. Technol. Adv. Eng. **12**(01), 40–50 (2022). https://doi.org/10.46338/ijetae0122_02
25. Mihaylenko, V., Honcharenko, T., Chupryna, K., Andrashko, Y., Budnik, S.: Modeling of spatial data on the construction site based on multidimensional information objects. Int. J. Eng. Adv. Technol. **8**(6), 3934–3940 (2019). https://doi.org/10.35940/ijeat.F9057.088619
26. Myasnikov, A.G.: Information and logical modeling in construction. Int. J. Adv. Trends Comput. Sci. Eng. **9**(1), 304–307 (2020). https://doi.org/10.30534/ijatcse/2020/46912020

27. Honcharenko, T. et al.: Method for representing spatial information of topological relations based on a multidimensional data model ARPN. J. Eng. Appl. Sci. **16**(7), 802–809 (2021)
28. Tytok, V., Bolila, N., Ryzhakov, D., Pokolenko, V., Fedun, I.: CALS–technology as a basis of creating modules for assessment of construction products quality, regulation of organizational, technological and business processes of stakeholders of construction industry under the conditions of cyclical and seasonal variations. Int. J. Adv. Trends Comput. Sci. Eng. **10**(1), 271–276 (2021). https://doi.org/10.30534/ijatcse/2021/381012021
29. Bielienkova, O., Novak, Y., Matsapura, O., Zapiechna,Y., Kalashnikov, D., Dubinin, D.: Improving the organization and financing of construction project by means of digitalization. Int. J. Emerg. Technol. Adv. Eng. **12**(8), 108–115 (2022). https://doi.org/10.46338/ijetae082 2_14
30. Cecilia D., Toffolon M., Woodcock C. E., Fagherazzi S.: Interactions between river stage and wetland vegetation detected with a seasonality index derived from LANDSAT images in the Apalachicola delta, Florida. Adv. Water Resour. **89**, 10–23 (2016). https://doi.org/10.1016/j. advwatres.2015.12.019
31. Imteaz, M.A., Hossain, I.: Climate Change Impacts on 'Seasonality Index' and its Potential Implications on Rainwater Savings. Water Resour. Manag. (2022). https://doi.org/10.1007/s11 269-022-03320-z
32. Tan, X., Wu, Y., Liu, B., et al.: Inconsistent changes in global precipitation seasonality in seven precipitation datasets. Clim. Dyn. **54**, 3091–3108 (2020). https://doi.org/10.1007/s00382-020-05158-w
33. Polaschek, M., Zeppelzauer, W., Kryvinska, N., Strauss, C.: Enterprise 2.0 integrated communication and collaboration platform: a conceptual viewpoint. In: 2012 26th International Conference on Advanced Information Networking and Applications Workshops (2012). https://doi.org/10.1109/WAINA.2012.73
34. Engelhardt-Nowitzki, C., Kryvinska, N., Strauss, C.: Strategic demands on information services in uncertain businesses: a layer-based framework from a value network perspective. In: 2011 International Conference on Emerging Intelligent Data and Web Technologies (2011). https://doi.org/10.1109/EIDWT.2011.28
35. Watson, J., Fitzallen, N., Fielding-Wells, J., Madden, S.: The practice of statistics. In: Ben-Zvi, D., Makar, K., Garfield, J. (eds.) International Handbook of Research in Statistics Education. Springer International Handbooks of Education. Springer, Cham (2018). https://doi.org/10.1007/978-3-319-66195-7_4
36. Goiko, A.F., Mikhels, V.O., Vakhovich, I.V., Pokrovsky, R.L., Hrytsenko, Y.O.: Principles of Planning the Production Program of a Construction Enterprise and Methods of Normalizing its Parameters. KNUCA, Kyiv, Ukraine (2007).

Data-Driven Public Budgeting: Business Management Approach and Analytics Methods Algorithmization

Tetiana Zhyber⑩, Anna Pyslytsya⑩, Hanna Zavystovska⑩, Olena Tymchenko⑩, and Roman Shchur⑩

Abstract Digital data processing tools have enabled progress toward improving solutions for complex systems, such as state budget systems. The creation of general budgeting systems, specifically in regional communities, has shown results for data-driven budgeting using a set of decision analysis methods. Increasing public awareness of digitalizing public services is achieved through participation in the evaluation of expenditure budget results using online tools. A review of scientific works on information asymmetry has shown its significant influence on the quality and consequences of budget decision-making during the budget process. Furthermore, most data analysis methods focus on the scattered problem areas of budgeting. The aim of the publication is the formation of a data-driven budgeting system based on the quality conditions of digitized data collected and the support of the interaction of all budgeting participants, including citizens. The latest data-driven budgeting technology combines transparent data for decision-making, a business approach to achieving goals, and mutual involvement of authorities, executors, and citizens in the formation of the most perfect of the adopted decisions regarding public budgets through adjusting and feedback. For decision-making based on data, MCDM methods have been separated, and simulations of their use have been carried out to predict

T. Zhyber (✉) · A. Pyslytsya · H. Zavystovska · O. Tymchenko
Kyiv National Economic University Named After Vadym Hetman, Kyiv 03057, Ukraine
e-mail: tzhyber@kneu.edu.ua

A. Pyslytsya
e-mail: Pyslytsya.Anna@kneu.edu.ua

H. Zavystovska
e-mail: zavistovskay@ukr.net

O. Tymchenko
e-mail: tymchenko.olena@kneu.edu.ua

R. Shchur
Vasyl Stefanyk Precarpathian National University, Ivano-Frankivsk 76018, Ukraine
e-mail: roman.shchur@pnu.edu.ua

A. Semenov et al. (eds.), *Data-Centric Business and Applications*, Lecture Notes on Data Engineering and Communications Technologies 194,
https://doi.org/10.1007/978-3-031-53984-8_5

results. The findings of the research include both favorable and hindering prerequisites of using information for public budgeting based on gathered data. Limitations and possibilities for the methods of data analysis by algorithmization have been systematized. Based on business data management methods, prospects for data-driven decisions in public budgeting have been highlighted.

Keywords Public budgeting digital system · Data-driven governance · MCDM techniques for public budgeting · Financial planning · Budget transparency · Business management approach in public budgeting

1 Introduction

State-of-the-art public budgeting technologies include the use of software tools to manage and analyze budget data, automated budgeting systems to streamline the budget process, and cloud-based budgeting solutions to improve access and collaboration between budgeting stakeholders—citizens (main taxpayers including) and public units (authorities including). In addition, many public budget systems now include data analysis and predictive modeling for improved decision-making. Successive crises—climate change, the global pandemic of an infectious disease in connection with the massive military aggression of the Russian Federation—exacerbate the issues of effective use of information for making budget decisions. Especially in an environment in which the usual budgeting processes when approving the budget are disrupted by current challenges that create sharply new information for public budgeting needs.

The budget management approach in modern public budgeting is the process of allocating resources to various public expenditure activities within specified budgetary constraints. The main purpose of budget management is to ensure that resources are used in the most efficient and effective manner. A common example of budget management in public budgeting is the use of performance-based budgeting. This approach focuses on ensuring that government resources are allocated in a way that maximizes desired outcomes and results. The business management approach in modern public budgeting typically focuses on improved performance and efficiency in the delivery of public services. This includes practices such as the use of performance-based budgeting, data-driven decision-making, results-oriented management, and strategic planning. For example, a performance-based budgeting approach looks at the outcomes of public programs, which have been executed by public units, rather than traditional inputs such as staffing or personnel. Data-driven decision-making focuses on the use of evidence and data to inform policy decisions. Results-oriented management involves setting measurable goals and objectives to ensure that public programs are meeting their desired outcomes. Lastly, strategic planning focuses on long-term objectives, such as the development of new services or operational cost savings.

As for European countries, growing crisis fronts are exacerbated by long-term challenges associated with low cost-efficiency, limited revenue diversification, and excess inflation involved in segments of the Euro area public budgeting. Accelerated digitalization will help remedy some of these new age-old issues, although it comes at the cost of greater exposure to the threat of cyber risks. To solve the problems of flexibility and prompt response to challenges, the authorities entail the use of the latest information tools when making budgeting decisions on the distribution of budgetary funds, based on the interaction of budgetary units with each other until striving for a common goal, as business management doing. E-Services for the citizens are data reliable and have a high potential for data gathering for public budgeting decisions [1]. Today, the strategic goals for countries' welfare are the Sustainable Development Goals developed by the UN.

Based on the relevance of further research on the use of information for data-driven budgeting, we systematize its conditions and implications in the following parts:

- concern about information formulation for data-driven budgeting to avoid information asymmetry;
- business approach methods algorithmization for making decisions based on data to reach objectives for the public budget.

The findings of the research includes favorable and hindering prerequisites of using information for public budgeting based on gathering data. Limitations and possibilities for the methods of data analysis have been systematised by algorithmisation. Prospectives for data-driven desigions in public budgeting highlited based on business data management methods.

2 Successful Examples of Data-Driven Public Budgeting

At the local level, where the calculation of input data is more controlled due to the smaller amount of input information and less variability in data grouping, local authorities have already created information interfaces for making decisions on city and regional budgets based on data analysis. The Financial Information System for California (FI$Cal) [2], USA, was created as a historic partnership of the Department of Finance, the Department of General Services, the State Controller's Office, and the State Treasurer's Office. Joint cooperation was started in 2005 and completely finished in late 2022. With the purpose of transforming traditional budgeting designs through discussions and paper calculations, separate accounting, separate procurement deals, and old cash management systems to optimize the public management of finance by California's government was created consolidate system. This system has the brand name FI$Cal and allows the region to operate as a single enterprise for the first time in history. Enter the system is forbidden for the members of other countries, so its an official and secure state system.

The example of city Los Angeles in California uses a computerized budgeting system to facilitate budgeting and reporting. Critics had considered the system too costly to create and use 10 years ago, but now the development of digitalization proved the movement of power in the right direction. The system uses web interfaces, data entry and analysis tools, and automated workflows to manage the budget process. Such a system for the public community is based on the business approach to the public budgeting decisions within the public welfare goals frame. The city is also using a cloud system to allow multiple users to access and edit budget data in real-time.

Regional digitization of decision-making regarding the use of public budget funds is developed in South Korea. The region of Seoul in South Korea uses digital data-driven budgeting relying on the "Smart-City" system, and data-driven budgeting is implemented by applying an information platform that has been in operation for a long time. The essence of such budgeting is in adjusting the allocation of budget resources depending on changes in input data and results. Seoul builds like business management system for the budgeting of its local authorities, too.

The consequences of successful informational techniques for budgeting in Seoul, South Korea, nowadays are:

- Online Budgeting allows using online budgeting tools such as artificial intelligence-based big data service platforms (28th stage) and other online budgeting apps. The system in Seoul was intended to have a preventive function to reduce populist policies and bureaucratic processes by heads and assemblymen of local governments leading to the waste or inefficient use of public resources.
- Organizational Planning lets the authorities coordinate information about planning objects with the current states. Using organizational planning to make sure that all projects are within budget and that all goals are achievable.
- Financial Planning means planning ahead for all regional community financial ability, needs, and wants in their linkage based on organizational planning.
- Digital budgeting data increased transparency and accountability. Informational techniques such as budgeting and financial analysis can help to increase transparency and accountability in public budgeting in metropolia. This can improve public trust in the government, as citizens will be able to view detailed information about how and where their tax money is being spent.
- More efficient and effective spending by using algorithms to analyze information and transparency. Through the use of informational techniques, public budgeting can become more efficient and effective in public service providing. Data-driven decision-making can help to ensure that money is being spent in the right areas and that resources are being used in the most effective way.
- Improved coordination and collaboration between public budgeting participators: government, budget units, public units, and citizens. Improved coordination and collaboration between government departments can also be achieved through the use of informational techniques. By having access to the same set of data, different departments can work together in an organized and efficient manner.

– Increased communication in on-line continuance of budgeting to prevent not only past reported mistakes. Increased communication between the public and officials can also be achieved through the use of informational techniques lice "budget windows" and "personalized budgeting participation accounts feedback" [3]. Through the use of online tools, such as online budgeting systems, citizens can access detailed budgeting information and communicate directly with government officials.
– Improved public services by reaching the online prevalent part of the population. Improved public services can also be achieved through the use of informational techniques. Public officials can use data to more accurately assess the needs of the population and develop targeted policies and services by monitoring, questionaries and feedback calculation.
– More efficient decision-making is the core purpose of data-driven public budgeting. By using informational techniques, public officials can make decisions that are based on accurate and up-to-date data. This can help to ensure that policies and services are based on the most current information available. By using informational techniques, public officials can identify areas that are in need of increased funding and direct resources accordingly. This can help to ensure that public money is being used in the most effective way possible.
– Providing improved public safety by immediately connection with the state service and policy creators. Improved public safety can also be achieved through the use of informational techniques. By having access to accurate and up-to-date data, public officials can assess the public safety situation and develop policies and services to address any issues.
– Increased motivation for innovation for public servants because of the need to learn a digital approach to work. Through the use of informational techniques, public officials can identify areas where new technologies or services could be used to improve services or operations. This can help to ensure that public money is being spent in the most effective way possible by the current knowledge and practices.
– Improved access to public services can also be achieved through the use of informational techniques. Improved access to public services by using several devices in rural areas, for an inclusive approach to citizens with stricted physical abilities. By having access to accurate and up-to-date data, public officials can develop policies and services that are tailored to the needs of the population.

Successful informational techniques for data operation during public budgeting in regions of South Korea at Seoul example are [4]:

1. The Seoul government has established a citizen participation approach with Communication and Sharing by administrative portal to strengthen the relationship between citizens and the government.
2. Seoul has implemented a digital instruments to support citizen involvement in the budgeting process like proposing, voting, and monitoring.
3. Citizen Engagement is important in forming digital data and using data-driven budgeting solutions. The city has created an open, collect, and share Seoul

public data: Seoul Open Data Plaza to promote greater public engagement in the budgeting process. The city has created a budget transparency platform to provide the public with easy access to the city's budget information.

4. Feedback from society members is important. The Seoul administration regularly conducts surveys to gain insight into public opinion. South Korean parties have established various advisory boards and activities like forums in governance to provide propositions to the authorities on budget-related issues during new policy implementation.
5. Seoul authorities have opened up its data to the public to improve the transparency of its budgeting process by cutting edge administration infrastructure: IT Superhighway.
6. The city has utilized specially adapted social digital tools to provide information about the budgeting process and to engage citizens in the discussion. Online system integrating and managing various service channels: 31 existing complaints and proposal sites, including one-click e-applications, social media center, 120 citizen discomfort observations, Oasis of 10 million imagination. Korean citizens can use the CCPIS, mobile devices, SNS, and phones to request all types of complaints and proposals for authority designs and services.
7. In addition, the South Korean government holds public hearings to solicit feedback from citizens on budget-related issues, for example, on the wartime forced labor issue in 2023.
8. Seoul has developed programs to educate citizens about the budgeting process and provide them with the tools to become more involved in the decision-making process.

Several data-driven budgeting and management data-driven steps in the Seoul region are platform-based. The rules approved by law apply to improving the collection, transparency, verification, and analysis of data, obtaining a forecast of trends, and making decisions on the needs of society with a promising time lag to prevent problems, and not just react to them. Thus, several interconnected layers for evaluating data on the adoption of budget decisions for the South Korean authorities build a sequence of approval, according to which decisions are divided into areas of public welfare. The processing of incoming information leads to the revision and improvement of decisions on the allocation of budgetary funds for the development of society's welfare [5]. Fixing data in an interconnected system allows all participants to have the same, but centralized regulated if needed, access to information.

In addition, many government bodies (agencies) are now incorporating data analytics into their budgeting communications. Data analytics allows to reduce the asymmetry of information through the availability of data from all participants in the budgeting desigions and budgeting process. Information asymmetries, however, can show up when building a government budgeting data system, so budget data analysis includes incorporating predictive models to identify future trends and potential cost savings. This allows public sector employees to make more informed decisions and improve the overall efficiency of budget processes.

In general, modern government budget technologies are revolutionizing the way governments manage and analyze budget data. Using advanced software tools, automated processes and data analytics, governments can make better decisions and improve the efficiency of budget processes while avoiding information asymmetries at the heart of decision-making.

3   Literature Review

Research by Fisher et al. [6] finds that asymmetric information in public budgeting processes can lead to inefficient resource allocation. This is further discussed by Fanani and Saudale [7], who observe that information asymmetry influences the creation of budgetary slack. Mogues and Olofinbiyi [8] summarize that asymmetric information between government departments and local authorities can result in budgetary decisions that are not in the public interest. Leonard et al. [9] also found that asymmetric information between local authorities and citizens could lead to a breakdown of trust. Finally, Garmann [10] proves that asymmetric information between local communities or governments can cause public budgeting to become politicized, where politically divided governments create much more adequate policies for the citizen's needs. Gathering data, digitization of processes and creating the common digital data for budgeting with the access of public budgeting participators could reduce asymmetric information for budgeting decisions significantly.

There are many studies on the impact of asymmetric information on public budgeting. We structure the most important areas of research, summarizing the following:

1. Significant impact of asymmetric information on public budgeting in developing countries. Publications and studies examine the role of asymmetric information in budgeting and its impact on the fiscal budgeting process in developing countries. In the middle 1990s, these papers considered the asymmetry of information about budgeting through developing country borrowing [11] and the information overload of government officials in the face of the need to respond more frequently to changes in budgeting tasks [12]. Policy on Government spending on goods and services, as a major driver of economic activity, was concentrated on public spending declining as a share of national income for nearly forty years in past and current centuries. This case has been developed in both rich and developing countries. Same time pressures to "rebuild fiscal buffers" were forced by the latest economic crises with increasing public expenditures distribution by budgets in the economy during and after each recession. The idea that discreet but temporary fiscal expansions during economic shocks are sufficient to keep the economy close to its optimal equilibrium level (TDR 2021 [13]) leads to the demand for proper calculations and prognosis in uncertainly.

The influx of foreign and one-off injections of funds has caused distortions in the accuracy of macroeconomic indicators, which are the basis for budgeting decisions at the country or regional level over time. To address this, there has been a critical need to both simplify the methods of processing and deepen the data on financial resources and their purpose in budgetary reallocations in frequently changing circumstances. Currently, studies on the emergence of asymmetry in information problems for budgeting in developing countries and in times of turbulence are focused on the possibilities of subjective influences by its participants in the formation and use of budgets [14]. Identifying inefficient practices of spending society's funds due to low professionalism, establishing interrelationships between the costs of topical priorities for improving the welfare of society and their consequences [15–17]. Modern research also focuses on the qualifications of civil servants and the potential personal interests of government officials who decide on the formation and use of budgets [18].

2. The discussion of the concept of asymmetric information in public budgeting confirms that asymmetric information can lead to adverse outcomes for budgeting and society as a whole, including budget deficits and misallocation of resources [19]. Several decades ago, the agent-principal problem between civil servants and society was considered the basis of information asymmetry [20]. Modern studies detail this problem as an obtained state of a public agent, where a civil servant must show performance in order to further receive budgetary resources, provided that the process of processing information about data showing performance is not properly debugged. Private sector-hired employees in public–private partnerships, for example, can hide information about their expenditures or other results compared to budget units with transparent reporting in order to receive preferences in performance budgeting [21]. It is the case when specially formed information from the contractor forms biased assessments, conclusions and trends in the allocation of budgetary funds. It also considers the readiness for changes and improvement of operating with information and the budget process in key budget units with budgetary powers [22].

3. Direction to improve the quality of information for the state budget with the participation of citizens in setting budgeting goals and accumulating best practices. The consequences of avoiding asymmetric information for the public budget lead to better budgeting when citizens are aware of the details of the budget process [23]. Transparent information about legislation and budget regulations allows independent analysts to evaluate government decisions and provide their recommendations [24]; a survey of civil servants allows drawing conclusions about agency problems [25].

4. Asymmetric information affects government budgeting due to the loss of the fiscal potential of countries and regions. Research suggests ways to reduce the costs of fiscal mismanagement [26]. Ways of influencing the main fiscal players by creating attractiveness in fiscal policy are proposed [27]. The results of such

studies support the advantage of sustained projected inflation and government spending over shocks that make economic players uncertain about government information [28] and potentially reduce the fiscal capacity of developing countries. Accordingly, the constant communication of the government and the consistent fiscal policy of the state are a factor in the proper organization of information in budgeting.

5. As an extension of the previous track, asymmetric information causes fiscal instability as taxpayers, who are important actors in budgeting, lose confidence in the public sphere and their country's authorities. The cited article examines the history of the impact of asymmetric information on fiscal instability and discusses the consequences for the public budget [29]. IMF specialists, after budgeting shocks during the pandemic and intensified hostilities in Europe and Asia, draw attention to the need to define clear rules of the game for both taxpayers and public service providers for budget funds [30].

6. The direction of the search for effective tools to control the avoidance and prevention of information asymmetry when making budget decisions based on data and the creation of modern information systems for budgeting. Again, research comes to the need to pay attention to the conditions for the activities of public agents, in particular in a private–public partnership with a clear prescribing of expectations from professionals of public authorities partners in contracts [31, 32]. A more disciplinary public authority focuses on the control of information on investments to meet public needs [33], including the obligations of such control by responsible executors for budgetary funds [34]. There is also a study that in developed countries budgeting decisions and control over them are guided to a large extent by traditions that have led to success in past years [35].

Returning to data-based decisions in budgeting, next, we highlight the factors affecting the quality of data for budgeting. Qualitative data should be available to all budgeting participants, including citizens and large taxpayers. The negative effects of information asymmetry prevent budget decision-making by removing occasions, which were listed by the researchers previously.

4 Prerequisites for Gathering and Using Data for Public Budgeting

Public budgeting is a vital process for allocating resources in order to meet the needs of society. Adequate and relevant information is essential for this process to ensure efficient and appropriate allocation of resources. The prerequisites for creating data and a pool of appropriate information for decision-making in the state budget system include:

(A) Understanding the needs and priorities of society, and comprehending the potential of financial resources and the possibilities of their use. Social groups divide the needs of society, thus any government cannot completely and accurately rank needs among these groups as a result of limited budgetary funds. Consequently, priority layers rank satisfaction of needs by first, addressing urgent needs over deferred ones, and second, by satisfying the needs of larger groups in society before those of limited groups [36]. It is common for expenditures on cultural events, hobbies, inventions, or space of habitation decorations at the expense of taxpayers' funds to occur after the resolution of problems with the environment, security, and safety through vital public services [37].

   Educational services in the world also ranked from mandatory free primary basic education, to mostly free for recipients of secondary education, to less accessible higher education [38]. However, the world's attitude towards public funding for education of all kinds is evolving, as it is seen as an urgent social need. Information about the financing of development goals that are permanent and reliable leads more visibly to their achievement.

(B) Understanding by the executive and legislative authorities of the country, as well as local authorities, of the potential for available financial resources and the possibilities of their use. This understanding fully depends on relevant, timely, complete, unambiguous, comparable, and high-quality information about the fiscal and financial capabilities of the society in budgeting. The comprehension of the financial and fiscal capabilities of society in budgeting is dependent on the availability and comparability of information not only from the state but also from businesses associated with budgets [39].

(C) Strategic goals and short-term objectives should be clearly defined, accompanied by the necessary actions taken by public units. This is closely linked to the need for an understanding of fiscal and financial potential. Every budget unit should understand its purpose of activity, and not duplicate functions or take conflicting policy stances.

(D) Appropriate methods should be implemented to collect and analyze data related to public budgeting. The data should be divided in terms of financial and management indicators in terms of time and direction. Sorting, grouping, and cleaning the collected data to create satisfactory information for budget design decision-making is a major challenge.

(E) Tools should be created for monitoring and evaluating progress in fulfilling budgeting tasks by responsible executors. This is an integral part of modern performance budgeting. The truthfulness, transparency, and reliability of the budget executor's information must be ensured.

(F) An effective communication strategy should be established between branches of executive governance, legislative units, and citizens. Citizens in modern budgeting are the very important part—common feedback and direct transparent propositions by society form the data for the best decisions in political environment. All participants in the budget execution should use the same methodology for interpreting the data, and communication should be prompt and non-discretionary in terms of time.

(G) An effective decision-making process should be created in budgeting through the use of actual data analysis methods. Further details on this stage of implementation are described later in this publication.

Fundamental principles of quality public data for public budgeting were summarized and adapted with J. White's publication about OECD's principles of good budgetary [40]: timeliness (public data should be current and up-to-date); accuracy (public data should be free from errors, omissions, and irregularities), completeness (public data should include all required components); accessibility (public data should be easily accessed and obtained); consistency (public data should be consistent across different sources); relevance (public data should be relevant to the public's needs); transparency (public data should be transparent in its content and origin).

Fundamental principles of proper use of public data for public budgeting are: data usage to produce high-quality statistics; a standardized way of collection using the latest technologies; accountability is necessary in budgeting; data use should be cost-effective and efficient; cybersecurity to avoid potential misuse of data; data should be used to provide insights and inform decision-making.

Causes of information asymmetry for public units as the generalized subjects of public budgeting costs are assumed by [32].

- Different incentives between decision-makers and information providers. Decision makers have incentives to use the information provided from primary sources of information and use it for their own purposes, such as promoting their own interests or the interests of their followers, while public service providers may have incentives to provide misleading information in order to increase their income.
- Poor communication between decision-makers and suppliers. When communication between decision-makers and suppliers is poor, decision-makers may not be aware of the true degree of information asymmetry or may not be able to use the information effectively.
- Lack of transparency. When decision-makers do not have access to relevant information, they cannot make informed decisions.
- Inadequate incentives for information providers. Information providers may not have adequate incentives to provide accurate or useful information to decision-makers, or may not have sufficient resources to do so. For example, it can be an irrelevant number of hired professionals with proper qualifications in public units.
- Limited access to information. In many cases, decision-makers may not have access to all the information needed to make informed decisions.
- Incompatible IT systems because of unprofessional coordination or the lack of funding during reforms. Incompatible IT systems can lead to information asymmetries between decision-makers and information providers.

- Limited information literacy is the result of previous circumstances. Decision makers may not have the necessary knowledge and skills to effectively interpret and use the information they receive.
- Cultural or language barriers in the international cooperations applying digital technical solutions for the authorities. Cultural or language barriers may prevent decision-makers from understanding the information they receive or may limit their effectiveness in using it.

As a result, misapplication of the budget data by a public servant could be grouped as: using insider information for personal gain; making decisions that are not in the public interest; abusing power or authority to unfairly benefit oneself or others; engaging in conflict of interest; failing to declare interests; misusing public resources; manipulating data or statistics; illegally awarding contracts or favors; disclosing confidential information; accepting gifts, favors or bribes.

5   Methods Algorithmization for Data-Driven Public Budgeting

We have assumed previously that data-driven budgeting is a developed complex approach to public budgeting, that looks at data from a variety of sources to make informed decisions about how to allocate public resources. This method relies heavily on data analysis methods to determine the best use of funds.

Modern tools of data-driven public budgeting methods could based on business management digital tools refer to the process of using data to inform the decisions made by public sector managers when allocating and managing public funds. This approach focuses on efficient and effective resource management based on empirical evidence and available data sources, rather than relying on traditional methods of budgeting. It is especially applicable in areas such as infrastructure, public safety, and public health, where accurate data-driven decisions can have a major impact on the public. This approach also allows for public sector managers to better understand the needs of local communities, resulting in more equitable and effective use of public funds.

The directions for data-driven budgeting that policy creators and funds managers use for decision-making are:

- Cost–benefit analysis involves calculating the cost and benefits of a particular program to determine if it is worth implementing [41].
- Performance Metrics provide the use of quantitative data to measure the performance of a program and make budget decisions based on those metrics [42].
- Benchmarking deals with comparing the performance of a program to similar programs in order to identify areas for improvement [43].

- Program evaluation involves collecting and analyzing data related to the effectiveness of structured (by budget unit's) programs for different directions and making decisions based on the results [44, 45].
- Predictive Modeling by using data-driven models to anticipate future trends and outcomes, and adjusting budget allocations accordingly.

Next, we will focus on the algorithmization of analytic tools (methods) within expenditures evaluation and predictable modeling for decisions in budgeting. These methods of decision-making in budgeting create a logical sequence and take into account the use of such a budgeting outcome management tool as a budget unit decision or government function. Predictive modeling for data-driven public budgeting is the process of using data and analytics to determine the best possible spending decisions for public sector organizations. It is done by analyzing data from past years, such as spending levels, public sector trends, and economic forecasts, and using sophisticated algorithms to identify patterns, forecast outcomes, and provide recommendations for how to best allocate resources. This allows public sector bodies to make decisions, that are more informed, and ensure that their budgeting is as effective and efficient as possible. The application of budget programs is based on data grouping: by the quality and relevance of the input data, by the data of appeals when monitoring, and by the type of expected budgeting result—allocation of sum or time priorities [46, 47].

Accordingly, the following objectives will be addressed with data-driven budgeting:

1. Evaluating the characteristics of input data for budget decision-making;
2. Selecting the most optimal budgeting solution;
3. Predicting the outcome of budget allocation.

The data for budget indicators quality and subsequent grouping thereof determine the capacity of subsequent decisions and results that can be obtained by analyzing the generated database. Pseudocode can be used to describe the potential of methods for assessing the quality of data for budgeting and subsequent decision-making. Pseudocode is written in English, but it is not a set of actual English language codes. Algorithms for assessing the quality of public data can facilitate the improvement of input data for budgeting. The following is an example algorithm for standardizing data and excluding outliers and exceptions, written in pseudocode (which has similarities with programming code but is easier to understand):

```
for each data point
    if outside of range
        remove data point
    else
        keep data point
end for
```

Detailed construction of this algorithm for particular deviations removed is as follows:

```
// Assume data is stored in a two-dimensional array
// named 'data'

// Step 1: Calculate mean values

For each row in data
    meanValue = 0
    For each column in row
        meanValue += row[column]
    meanValue /= row.length
    dataMeanValues[row] = meanValue

// Step 2: Calculate standard deviation

For each row in data
    stdDeviation = 0
    For each column in row
        stdDeviation += (row[column] - dataMeanVal-
ues[row]) ^ 2
    stdDeviation = sqrt(stdDeviation / row.length)
    dataStdDeviations[row] = stdDeviation

// Step 3: Calculate the quality of each row

For each row in data
    quality = 0
    For each column in row
        quality += ((row[column] - dataMeanValues[row]) /
dataStdDeviations[row]) ^ 2
    quality /= row.length
    dataQuality[row] = quality
```

The selection of solutions in public budgeting is distinguished by its orientation towards the satisfaction of social needs, many of which cannot be ignored. At the same time, some groups in society have needs that come into conflict. Consequently, different criteria must be balanced when making budget decisions. The best algorithm for decision-making by budget unit managers for the public good in the face of information asymmetry is the Multi-Criteria Decision Making (MCDM) approach [48, 49]. Modern MCDM proposes a set of state-of-the-art tools that use mathematical optimization techniques to solve complex problems by considering multiple objectives and criteria. It takes into account quantitative and qualitative factors, as well as constraints, to identify the optimal solution.

Multi-criteria Decision Making (MCDM) approach in public budgeting data during uncertainty is especially important. MCDM is an approach to decision-making that takes into account multiple criteria and objectives, as well as various stakeholders' preferences and constraints. The goal of MCDM is to identify a preferred option among a set of alternative options. In public budgeting, the MCDM approach is used to make decisions that involve uncertain data, such as forecasting future budget needs, determining the optimal allocation of resources and balancing the interests of all stakeholders.

Public budgeting data presents a unique challenge for decision-makers due to the uncertainty involved in the data and the need to balance the interests of stakeholder groups. Budget fund managers apply data decision-making when operative unexpected situations occur and examine the MCDM approach to public budgeting data during uncertainty, then provide suggestions for improving decision-making processes.

The MCDM approach has numerous benefits for public budgeting. It allows decision-makers to consider multiple objectives, criteria, and preferences and to weigh the benefits and costs of each option. Furthermore, the MCDM approach enables decision-makers to take into account the preferences of stakeholders and to identify the most preferred option for all stakeholders. Finally, the MCDM approach enables decision-makers to assess the trade-offs between different options and to make decisions that are in line with their objectives.

Despite the numerous benefits of using the MCDM approach in public budgeting data, there are also some challenges associated with it. One of the main challenges is that it can be difficult to identify the optimal solution, due to the complexity of the data and the multiple objectives and criteria that must be taken into consideration. Additionally, the MCDM approach requires an extensive amount of data to be collected and analyzed, which can be time-consuming and costly. Finally, the MCDM approach requires decision-makers to make trade-offs between different objectives and criteria, which can be difficult and contentious. It is important for decision-makers to have a clear understanding of the objectives and criteria that should be used in the decision-making process and the political parties' preferences in a strong democracy. As follows, the information asymmetry should be carefully reviewed and prevented at the criteria-setting stage. Therefore, it is important for decision-makers to have a clear understanding of the circumstances and criteria and the preferences of stakeholders before making any decision after the digital calculation of the data. Finally, the MCDM allows decision-makers to identify the most preferred option and discuss it during approving budgeting decisions. MCDM methods are well-suited for decision-making in situations where there is possible asymmetric information, such as from the private provider's sector of the public good.

Promising examples of MCDM methods for public budgeting include the Analytic Hierarchy Process (AHP) [50], Goal Programming (GP) [51], and Weighted Sum Model (WSM) [52], which we will algorithmize to choose the best public budgeting design. Other algorithms of analytic methods, such as Reinforcement Learning (RL) [53] or Markov Decision Process (MDP) [54], can be applied to analyze the environment and optimize the policy for the public good.

However, let us consider the limitations of the *worst method* for budgeting decision analysis. An inappropriate method for decision-making by public budget fund managers for public goods coverage in the face of information asymmetry using state-of-the-art methods is the Nash Equilibrium algorithm. This algorithm assumes that all agents know the preferences of all other agents and the environment in which the decision is being made is static and deterministic. Nash Equilibrium is a concept from game theory that suggests that the best outcome for a game is the one where no player has an incentive to change their strategy. This means that each player's decision has no effect on the outcome. Thus, Nash Equilibrium does not fit for public budgeting decisions, as these decisions have the potential to change the outcome of the budget allocation in addition.

This additional approach separates the interests of public units as agencies, it creates confrontation instead of necessary in public sector synergy. In a situation of information asymmetry and a dynamic environment, this algorithm fails to accurately capture the consequences of each decision. The algorithm of Nash Equilibrium is constructed as follows:

```python
import numpy as np

# Define the environment
env = np.zeros((5,5))

# Define the rewards for each action
rewards = np.ones((5,5))

# Define the policy
def policy(state):
    actions = ['up', 'down', 'left', 'right']
    return np.random.choice(actions)

# Set the maximum episode length
max_episode_length = 15

# Define the learning rate
alpha = 0.1

# Implement the Nash Equilibrium Algorithm
episode_rewards = []
for episode in range(10):
    state = [0,0] # Start position
    episode_reward = 0
    for step in range(max_episode_length):
        action = policy(state) # Take action
        reward = env[state[0], state[1]] # Get reward
        episode_reward += reward # Update episode reward
        state[0] += action[0] # Update position
        state[1] += action[1] # Update position
        rewards[state[0], state[1]] = rewards[state[0],
state[1]] + alpha * (reward - rewards[state[0],
state[1]]) # Update reward matrix
    episode_rewards.append(episode_reward)

# Output the episode rewards
print(episode_rewards)
```

For example, if a government is deciding how to allocate its budget, it might choose to invest in defense or education. If the government invests in one, it may be leaving out the other, and the outcome of the budget allocation decision will be affected.

To remove the confrontation of the Nash Equilibrium, we can use the Weighted Sum Model method for allocating budget expenditures algorithmization between confronting policies (functions).

The next dominant strategy algorithm simulation for the education and defense expenditures from the state budget (for example, in Ukraine) shows resource allocation for two conflicting interests by imaginary given political parties. This algorithm takes into account the preferences of both parties, their relative bargaining power, and the relative costs of the resources being allocated. The algorithm maximizes the payoffs for both parties and can be used to negotiate a fair and efficient distribution of resources.

The algorithm works by giving each political party a score for each expenditure (e.g. education, defense) based on the preferences of each party, their bargaining power, and the relative costs of the budget expenditures (on functions in our example). The algorithm then totals up the scores to arrive at a total score for each function (or policy). The function (or policy) with the highest total score is then chosen as the dominant strategy for the budget expenditures.

The weighted sum calculation is provided by the formula:

$$\text{Weighted Sum} = E \times w_e + D \times w_d \tag{1}$$

The simulation is: let's assume that government wishes to allocate the state budget expenditures strictly limited sum between budget expenditures on education and defence. The points for preferences of each political party, the relative bargaining power, and the possible maximum budget expenditures are as follows (Table 1).

The algorithm totals up the scores for each function and arrives at the following total scores (Table 2).

Therefore, the dominant strategy is Education, as it has the highest weighing factor. The model is based on the idea that each expenditure should be weighted according to its importance, as the political power of each party in this simulation.

We can use this model to calculate the mid-term public budgeting. It allows to assignment of various weights to different parameters in a budget and the calculation of the total budget.

Table 1 Points by criteria

Preference	Education	Defense
Political party 1	4	3
Political party 2	5	2
Bargaining power	0.4	0.6
Max. budget expenditures	4	4

Table 2 Results of weighing

Function	Total score of political pover	Weighing factor (w)	Budget expenditures distribution
Education	3.6	0.54	4.32
Defence	3.0	0.46	3.68

An example of the weighted sum model by Python code is given below:

```python
def weighted_sum(parameters, weights):
    total = 0
    for i in range(len(parameters)):
        total += parameters[i] * weights[i]
    return total
```

Another example of strict choice could be a government deciding between investing in healthcare or improving public transportation. If the government chooses to invest in one, it will be limiting the amount of resources available to invest in the other. This means that the outcome of the decision will be affected by the choice that is made, and one of the public needs will not be met. Nevertheless, the Weighted Sum Model method is a simple but effective tool for allocating budget expenditures. It allows decision-makers to consider the relative importance of each expenditure and optimize the budget accordingly.

After considering the Nash Equilibrium, we moved to the prediction of output in budgeting allocations, which was supported by the proposed before MCDM Methods. The Analytic Hierarchy Process (AHP) is a structured technique for organizing and analyzing complex decisions. It is used to determine the relative importance of a number of factors that are influencing an overall decision.

The analytic Hierarchy Process (AHP) [48] creates a decision-making algorithm that could be used in budget allocation. Here is an example of how it could be used to allocate funds between educational expenditures, healthcare expenditures, and defense expenditures from the public budget:

1. Establish the criteria: Educational expenditures, Healthcare expenditures, and Defense expenditures.
2. Assign weights to each criterion based on importance:

   Educational Expenditures (0.5).
   Healthcare Expenditures (0.2).
   Defense Expenditures (0.3).

3. Establish the relative importance of each criterion with respect to one another:

Educational Expenditures > Healthcare Expenditures (4:1)
Educational Expenditures > Defense Expenditures (2:1)
Healthcare Expenditures > Defense Expenditures (3:1)

4. Calculate the resultant weightings:

Educational Expenditures $(0.5 \times 4/5) = 0.4$
Healthcare Expenditures $(0.2 \times 4/5) = 0.16$
Defense Expenditures $(0.3 \times 4/5) = 0.24$

5. Allocate the budget accordingly:

Educational Expenditures $(0.4 \times 100\%) = 40\%$
Healthcare Expenditures $(0.16 \times 100\%) = 16\%$
Defense Expenditures $(0.24 \times 100\%) = 24\%$

Therefore, the budget should be allocated as 40% for educational expenditures, 16% for healthcare expenditures, and 24% for defense expenditures.

Next, we construct the algorithm of the Analytic Hierarchy Process for budget expenditures by functions for example of the government in Ukraine. We took the actual budget expenditures from the state budget of Ukraine in 2021, before the active war intervention by Russia had started. Education (256), Defence (122), and Healthcare (30.5) in mln hryvnias are three main functions of government that have been chosen to consideration. To begin, a matrix with criteria and subcriteria should be created. Each criterion will be given a weight that will be calculated to determine the relative importance of each criterion. The weights will be used to compare different combinations of expenses. The algorithm of decision-making is built as follows.

1. Create a matrix with the criteria and subcriteria
2. Determine the relative importance of each criterion
3. Calculate the weights of each criterion
4. Compare the different combinations of expenses
5. Use the weights to make a decision on the expenditures.

```python
# Step 1
# Create a matrix with the criteria and subcriteria
# criteria - Education, Defence, Healthcare
# subcriteria - expenditure amount

criteria = ["Education", "Defence", "Healthcare"]
subcriteria = [256, 122, 30.5]

matrix = [[criteria[i], subcriteria[i]] for i in range
(3)]

# Step 2
# Determine the relative importance of each criterion

rel_importance = [0.3, 0.2, 0.5]

# Step 3
# Calculate the weights of each criterion

weights = []

for i in range(len(rel_importance)):
    weights.append(subcriteria[i] * rel_importance[i])


# Step 4
# Compare the different combinations of expenses

# Step 5
# Use the weights to make a decision on the expenditures

# Sum the weights
total_weight = 0
for weight in weights:
    total_weight += weight

# Determine the percentage of each weight
percentages = []
for weight in weights:
    percentages.append(weight/total_weight * 100)

# Make a decision based on the percentages

# Education has the highest percentage, so should have
the highest priority
```

It should be noted that in 2023, education expenditures lost their priority in fact in Ukraine. Budget expenditures on higher education, which are funded from the central and state budgets, have been cut in 2022–2023 due to the damages caused by Russian war activities.

As we mentioned before, budgeting objectives designed by management instruments like programs for responsible executors are the precondition of modern budgeting. The expected objectives of programming activities by budget units in public budgeting can be divided into linear and non-linear layers. Linear programming is a method of optimizing a linear objective function subject to linear constraints. Non-linear programming, on the other hand, is a method of optimizing a non-linear objective function subject to non-linear constraints. The main difference between the two is that linear programming is able to find a global optimal solution, while non-linear programming is only able to find a local optimal solution. Linear programming is used in budgeting activities such as cost analysis, resource allocation, and project scheduling, while non-linear programming is used for applications such as mathematical modeling, decision-making, and engineering design.

Goal programming is a mathematical optimization technique, that could be used in analyzing public budget expenditures by directions or sets of actions. In fiscal constraints, goal programming at the same time is an important component of public budgeting and can be used to maximize or minimize a certain variable in the budgeting allocation, like public service costs [44]. To create an example of goal programming, the sums of expenditures on functions of government could be considered. It is based on the concept of goal-seeking and uses linear programming techniques to optimize a set of goals subject to a set of constraints. It could use a multi-criteria decision-making approach to calculate the best solution based on predetermined objectives and limitations.

Goal programming could used to determine optimal allocations of resources in areas such as health care, public works, education, and other public services. This Goal Programming simulation creates the circumstances for budget expenditures in total by several components, for example budget unit's expenditures by separate directions. Algorithm with the Goal Programming expenditures by three previous government functions to predict budget expenditures allocation constructed as follows:

```python
# Import Libraries
import numpy as np

# Set Up the Problem
# Set Variables
objectives = np.array([[1], [1], [1]])
m = 3
n = 3

# Set Coefficients
A = np.array([[1, 1, 0],
              [1, 0, 1],
              [0, 1, 1]])
b = np.array([[200], [150], [100]])

# Set Target
c = np.array([[0.4], [0.3], [0.3]])

# Set Up the Optimization
# Solve the Problem
x = np.linalg.inv(A).dot(b)

# Calculate the Target Function
z = np.transpose(objectives).dot(x)

# Calculate the Error Vectors
e_plus = c - x/z
e_minus = x/z - c

# Calculate the Penalty Vectors
d_plus = np.maximum(e_plus, 0)
d_minus = np.maximum(e_minus, 0)

# Calculate the Penalty Function
penalty = np.transpose(d_plus).dot(d_plus) + np.trans-
pose(d_minus).dot(d_minus)
# Print Results
print("Target Function: {:.2f}".format(z[0][0]))
print("Error Vectors: " + str(np.around(e_plus, 2)) +
str(np.around(e_minus, 2)))
print("Penalty Function: {:.2f}".format(penalty[0][0]))
print("Optimized Solution: " + str(np.around(x, 2)))
```

The method of analysis helps guide public budgeting decisions by analyzing expected cost distribution. To specify the expected result, we propose to use the Goal Programming method in the different directions of expenditures on education, healthcare, and defense with the desired constraints on the result of budget expenditures allocation. For this purpose, the Goal Programming Algorithm was created for state budget expenditures in three directions based on the mentioned data from 2021.

Given the following constraints for the algorithm by the data of expenditures by functions from 2021 in mln UAH:

- Education expenditure: 256
- Defence expenditure: 122
- Healthcare expenditure: 30.5

The following algorithm can be used to allocate the budget expenditures:

```
def goalProgramming(expenditures):
    # Initialize variables
    x_1 = 0
    x_2 = 0
    x_3 = 0

    # Set up the goal equation
    goal = x_1 + x_2 + x_3

    # Set up the constraints
    education_constraint = x_1 == expenditures[0]
    defence_constraint = x_2 == expenditures[1]
    healthcare_constraint = x_3 == expenditures[2]

    # Solve the goal programming problem
    goal = minimize(goal, [education_constraint, de-
fence_constraint, healthcare_constraint])

    # Return the solution
    return goal.x

# Call the goal programming algorithm
expenditures = (256, 122, 30.5)
solution = goalProgramming(expenditures)

# Print the solution
print("Allocation of budget expenditures:")
print("Education:", solution[0])
print("Defence:", solution[1])
print("Healthcare:", solution[2])
```

To optimize decision implementations in data-driven budgeting Reinforcement Learning (RL) can be used as to model base for discussion of serial public budgeting decisions. A standard example of how to use RL for this purpose is outlined below:

1. Define the goals for the budgeting decision.
2. Identify the environment which the budgeting decision will take place.
3. Identify the available actions and states of the environment.
4. Define a reward function that captures the goal of the budgeting decision.
5. Create a policy using RL to determine the best actions as the reward to take in order to optimize the impact of referred to policy government function realization.
6. Evaluate the policy using simulations to assess its performance.
7. Implement the policy in practice and measure its performance.

Here is an example of RL code that can be used to create a policy for the discussing and completing of budgeting decisions:

```
import numpy as np

# define states
states = [0, 1, 2, 3, 4]

# define actions
actions = [0, 1]

# define rewards
rewards = [[0, 0, 0, 0, 10],
           [0, 0, 0, 0, 10],
           [0, 0, 0, 0, 10],
           [0, 0, 0, 0, 10],
           [0, 0, 0, 0, 10]]

# initialize Q-values
Q = np.zeros((len(states), len(actions)))

# define learning rate
alpha = 0.8

# define discount factor
gamma = 0.95

# define initial state
```

```
initial_state = 0

# create the Q-learning algorithm loop
for i in range(1000):

    # set current state to the initial state
    current_state = initial_state

    # loop until the terminal state is reached
    while current_state != 4:

        # choose an action randomly
        action = np.random.choice(actions)

        # select the next state using the action
        next_state = np.random.choice(states)

        # compute the reward
        reward = rewards[current_state][action]

        # compute the temporal difference
        td = reward + gamma * Q[next_state,
np.argmax(Q[next_state, :])] - Q[current_state, action]

        # update the Q-value
        Q[current_state, action] += alpha * td

        # set the current state to the next state
        current_state = next_state

# print the Q-table
print(Q)
```

Markov Decision Process (MDP) is a tool for making public budgeting decisions in defined strictly constraints, which could be macroeconomic indicators. In this approach, the decision maker has to specify the states of budget expenditures directions, actions due to executors of public services choosing, transition probabilities, rewards, discount factor, and cost associated with each state and action. The goal is to determine an optimal policy that maximizes the expected total discounted reward. This method is very applicable in terms of using the budget savings approach, but since we cannot talk about saving on the public good—using the Markov solution, we focus on such relationships between budgeting performers in budget expenditures that will lead to the best output for the same money (value). For modern public budgeting, these actions are constructed in program directions and have a projective basis.

For example, a public budgeting decision process involving MDP can consider a situation where the decision maker needs to allocate a budget over multiple years. The states can include variables such as inflation rate, unemployment rate, public infrastructure development, public welfare programs, public health programs, public education programs, and so on. The decision maker can then specify the available actions, such as increasing the budget for a certain program or decreasing it. The transition probabilities and rewards can be determined based on the impact of the action on the states. The cost associated with each state and action can be determined based on the expected return on investment. Finally, the discount factor can be used to account for the time value of money.

Once the MDP is set up, the decision maker can use optimization algorithms such as value iteration or policy iteration to find the optimal policy. This policy will help the decision maker to allocate the budget in a way that maximizes the expected total discounted reward. The algorithm is built as follow:

```python
import numpy as np

# states
states = ["inflation_rate", "unemployment_rate", "in-
fra_dev", "welfare_prog",
          "health_prog", "edu_prog"]

# actions
actions = ["increase_budget", "decrease_budget"]

# transition probabilities
transition_prob = {
    "inflation_rate": {
        "increase_budget": 0.3,
        "decrease_budget": 0.7
    },
    "unemployment_rate": {
        "increase_budget": 0.2,
        "decrease_budget": 0.8
    },
    "infra_dev": {
        "increase_budget": 0.9,
        "decrease_budget": 0.1
    },
    "welfare_prog": {
        "increase_budget": 0.5,
        "decrease_budget": 0.5
    },
    "health_prog": {
        "increase_budget": 0.8,
```

```
            "decrease_budget": 0.2
        },
        "edu_prog": {
            "increase_budget": 0.7,
            "decrease_budget": 0.3
        }
}

# rewards
rewards = [2, -1, 3, 4, 6, 5]

# discount factor
discount = 0.9

# cost associated with each state and action
costs = {
    "inflation_rate": [10, -5],
    "unemployment_rate": [20, -10],
    "infra_dev": [15, -7],
    "welfare_prog": [30, -15],
    "health_prog": [25, -12],
    "edu_prog": [20, -10]
}

# setting up the MDP
states_len = len(states)
actions_len = len(actions)

# transition probability matrix
transition_matrix = np.zeros((states_len, actions_len,
states_len))
for i, s in enumerate(states):
    for j, a in enumerate(actions):
        for k, s_n in enumerate(states):
            transition_matrix[i, j, k] = transi-
tion_prob[s][a]

# reward matrix
reward_matrix = np.zeros((states_len, actions_len))
for i, s in enumerate(states):
    for j, a in enumerate(actions):
        reward_matrix[i, j] = rewards[i] + costs[s][j]

# value iteration
```

```
def value_iteration(transition_matrix, reward_matrix,
discount, tolerance=1e-5):
    # initializing the values
    values = np.zeros(transition_matrix.shape[0])

    while True:
        # calculating expected values
        expected_values = np.sum(transition_matrix * val-
ues, axis=2)

        # updating the values
        new_values = reward_matrix + discount * ex-
pected_values

        # calculating the maximum difference
        diff = np.max(np
```

Budgeting on the macro level differs from the company's budgeting by the total complexity of objectives first. So when we propose the goal programming with multiple objectives, we should understand the limitations of information quality, mentioned in the previous chapter. The goal programming simplex method is a mathematical optimization technique that can be used to solve a linear programming problem with more than one objective function. It is used to solve problems that involve both constraints and objectives and is based on linear programming.

The Goal Programming Simplex Method is used to determine the optimal solution to a goal programming problem by finding a collection of feasible allocations that satisfy the constraints and maximize the value of the objectives. The algorithm works by converting the problem into a linear programming problem and then solving the linear programming problem using the simplex method.

The simplex method is an iterative algorithm that starts from an initial feasible solution and then cycles through a series of improved solutions, eventually arriving at the optimal solution. The simplex method is a powerful tool for solving goal programming problems, but it can be difficult to use in some cases, for example, if the actual inflation is much more differ than planned.

```
// Initialize the problem data

 SET number_of_constraints
 SET number_of_objectives
 SET constraint_coefficients
 SET objective_coefficients

// Convert to linear programming problem

 SET new_constraint_coefficients
 SET new_objective_coefficients

// Solve using simplex method

 DO
    SET i = 0
    WHILE i < number_of_constraints
       Calculate the new constraints
       i = i + 1
    END WHILE
    Calculate the new objective value
    IF new_objective_value > current_objective_value
       SET current_objective_value = new_objective_value
    END IF
    IF current_objective_value > best_objective_value
       SET best_objective_value = current_objective_value
    END IF
    Repeat until best_objective_value is reached
 END DO
```

The following is an example of Goal Programming Simplex Method algorithmization for defense expenditures in public budgeting in Ukraine for next year:

```
# Import Packages
from pulp import *

# Set Variables
x = LpVariable.dicts('x', [1, 2, 3,4], lowBound=0)

# Set Objective
prob = LpProblem("Goal Programming - Defence Expendi-
tures", LpMaximize)
prob += 4 * x[1] + 2 * x[2] + 6 * x[3] + 8 * x[4], "Total
Expenditure"

# Set Constraints
prob += x[1] + x[2] + x[3] + x[4] <= 40, "Total Expendi-
ture Limit"
prob += x[1] + 2*x[2] + 0.5*x[3] + 0.25*x[4] >= 20, "Min-
imum Expenditure in Research and Development"
prob += 0.75*x[1] + 0.5*x[2] + 0.25*x[3] + 0.5*x[4] >=
30, "Minimum Expenditure in Environmental Protection"

# Solve
prob.solve()

# Print Results
  print ("Total Defence Expenditure:", value(prob.objec-
  tive))
  print ("\nThe Optimal Allocation")
  for v in prob.variables():
      print (v.name, "=", v.varValue)
```

In the active phase of the war, increasing uncertainty and disproportionate threaten effective decision-making in the preparation of the state budget. At the same time, the accelerated development of artificial intelligence tools, in parallel with a well-formed database, can form prospects for machine learning. Such machine-learning tools for budget decision-making are possible thanks to built-in feedback algorithms and assessment of the prospects for the development of current societies' needs. Due to direct communication of the government with the performers and recipients of public services, the political discussion during budgeting discussions in parliament could lose the leading role in budget formation. Despite that, the parliament's discussions were significant performances of the previous two centuries of public budgeting.

Half of the countries in the lowest quartile of the latest Transparency International Corruption Perception Index are in conflict. There is a strong link between corruption as inefficient budget resources distribution, fiscal resource ineficial allocation in total, and war [55]. In war, the public accountability of political actors becomes increasingly difficult, which creates ideal conditions for corruption, giving politicians the freedom to use their power to advance their own interests. Algoritmisation of budget base allocation for public good's sake during unstable times could help with efficient public budget usage in the lack of transparency of budgeting during war, crisis, and disasters. Such direction of research we consider as prospective for a new approach to budgeting transparency.

6 Conclusions and Limitations

The conclusion to the algorithmization in data-driven public budgeting prerequisites and methods can be drawn in the frame of data quality and the development of a suitable algorithm. The prerequisites for data-driven public budgeting include the availability of reliable data, the availability of suitable tools for data analysis, and the development of suitable algorithms for decision-making.

The business management approach in public budgeting includes managing public funds that focus on improving the efficiency, effectiveness, and accountability of every public services executor and beneficiary in the budgeting process. Therefore, methods for analyzing data on the results obtained as an output of the budget funds use can be a significant help in reducing the asymmetry of information at the stage of making budget decisions. To avoid information asymmetry, a strong system of gathered quality data provisioning and citizen feedback about public services is needed, which currently exists in regional-level data-driven budgeting examples. It relies on the principles of openness, inclusion, and new digital tools applying to ensure that public resources are allocated to the most prospective needs and trends. Therefore, a unified methodology for applying methods for processing budget information, negotiating, and making budget decisions is important.

For relevant and quality data for a data-driven budgeting approach strategic orientation and goals should provide the framework for decision-making, prioritizing resources, and developing a balanced budget that meets both financial and operational objectives. No less important is communication and transparency in the application of the analysis methodology between different branches of government, performers, and citizens, a clear demonstration of the grounds for arguments in budgeting for performers, taxpayers, and citizens.

The methods of algorithmization for data-driven public budgeting include the use of predictive models, a business management approach to public budgeting efficiency, and data mining techniques. These techniques help to identify the necessary data and develop an algorithm that can be used to make decisions. Moreover, the use of these techniques can also help to improve the accuracy of the decisions taken, as well as the speed at which they are taken. Finally, it is important to note that these techniques should be used responsibly, so that the decisions taken are based on reliable data and are in line with the relevant guidelines. Methods of analysis that are limited to alternatives in optimizing choice should be used with caution in public budgeting. Since the public sector cannot use savings or stop funding due to lack of funds or the transfer of resources to other areas. The system of the public good presupposes, first of all, the security of citizens. Therefore, MCDM methods have a good prospect for making budget decisions in the face of increasing opportunities and capacities in the formation and processing of information for budgeting. Modern data-driven budgeting oriented on budget units and private executors in public–private partnerships includes all budgeting subjects acting with a business management approach to the public welfare.

The publication does not provide the data arrays necessary for the formation of managerial decisions on budgeting using business-oriented methods of data analysis. The main limitations for preparing the publication were the actual secrecy of data on the budgets of the country and regions, due to the martial law in Ukraine, and the unprecedented disproportion in budgets revenues and expenditures associated with active hostilities and the migration of people and businesses as a result of the war, which started by russians. Therefore, to construct the evidence, methods of algorithmization and systematization of evidence and conclusions obtained from previous studies were used. Strengthening the fight against corruption during the war should be just as important a task as ensuring defense and preserving the economy. This is because war not only does not reduce corruption but also actually encourages it.

References

1. Hübl, B.F., Šepeľová, L.: Design criteria of public e-services. In: Kryvinska, N., Greguš, M. (eds.) Developments in information and knowledge management for business applications. In: Studies in Systems, Decision and Control, vol 421. Springer, Cham (2022). https://doi.org/10.1007/978-3-030-97008-6_5
2. FI$Cal—State of California. One state. One system. Homepage. https://fiscal.ca.gov/. Last accessed 21 Jan 2022
3. Choi, I.: What explains the success of participatory budgeting? Evidence from Seoul autonomous districts. J. Deliberative Democracy 10(2), Article 9 (2014)
4. Seoul Solutions Homepage. E-Government: Seoul Ranked 1st in the UN's E-Government Survey Five Consecutive Times in the Last Decade (2020). https://www.seoulsolution.kr/en/egov. Last accessed 21 Jan 2023
5. Advisory Services Program PEMNA Homepage. The Effective Medium-Term Budgetary Framework for the Royal Government of Cambodia. https://www.pemna.org/bbs/Publications_AdvisoryServicesProgram/view.do. Last accessed 21 Jan 2023
6. Fisher, J.G., Maines, L.A., Peffer, S.A., Sprinkle, G.B.: Using budgets for performance evaluation: Effects of resource allocation and horizontal information asymmetry on budget proposals, budget slack, and performance. Account. Rev. 77(4), 847–865 (2002)
7. Fanani, Z., Saudale, G.E.K.: Influence of information asymmetry and self-efficacy on budgetary slack: an experimental study. Jurnal Akuntansi dan Keuangan 20(2), 62–72 (2018)
8. Mogues, T., Olofinbiyi, T.: Budgetary influence under information asymmetries: evidence from Nigeria's subnational agricultural investments. World Dev. 129, 104902 (2020)
9. Leonard, D.K., Bloom, G., Hanson, K., O'Farrell, J., Spicer, N.: Institutional solutions to the asymmetric information problem in health and development services for the poor. World Dev. 48, 71–87 (2013)
10. Garmann, S.: Political budget cycles and divided government. Reg. Stud. 52(3), 444–456 (2018)
11. Anayiotos, G.C.: Information asymmetries in developing country financing. IMF Working Papers (079), A001 (1994)

12. Kearns, P.S.: State budget periodicity: an analysis of the determinants and the effect on state spending. J. Policy Anal. Manag. **13**(2), 331–362 (1994)
13. UNCTAD Trade and Development Report 2021.: From Recovery to Resilience—The Development Dimension. United Nations Publication. Sales No. E.22.II.D.1. Geneva (2021)
14. Jalali Aliabadi, F., Gal, G., Mashyekhi, B.: Public budgetary roles in Iran: perceptions and consequences. Qual. Res. Account. Manag. **18**(1), 148–168 (2021). https://doi.org/10.1108/QRAM-11-2018-0084
15. Ahmed, Z., Cary, M., Shahbaz, M., Vo, X.V.: Asymmetric nexus between economic policy uncertainty, renewable energy technology budgets, and environmental sustainability: evidence from the United States. J. Clean. Prod. **313**, 127723 (2021)
16. Nugroho, A., Takahashi, M., Masaya, I.: Village fund asymmetric information in disaster management: evidence from village level in Banda Aceh City. In: IOP Conference Series: Earth and Environmental Science, vol. 630(1), pp. 012011, IOP Publishing (2021)
17. Ullah, S., Ozturk, I., Sohail, S.: The asymmetric effects of fiscal and monetary policy instruments on Pakistan's environmental pollution. Environ. Sci. Pollut. Res.Pollut. Res. **28**, 7450–7461 (2021)
18. Sun, S., Andrews, R.: The determinants of fiscal transparency in Chinese city-level governments. Local Gov. Stud. **46**(1), 44–67 (2020)
19. Levaggi, R.: Asymmetry of information in public finance. In: Jackson, P.M. (eds.) Current Issues in Public Sector Economics. Current Issues in Economics. Palgrave, London (1992). https://doi.org/10.1007/978-1-349-22409-8_10
20. Hung, F.-S., Lee, C.-C.: Asymmetric information, government fiscal policies, and financial development. Econ. Dev. Q. **24**(1), 60–73 (2010). https://doi.org/10.1177/0891242409333548
21. Paarporn, K., Chandan, R., Alizadeh, M., Marden, J.R.: A General Lotto game with asymmetric budget uncertainty. arXiv:2106.12133 (2021)
22. Mauro, S.G., Cinquini, L., Sinervo, L.-M.: Actors' dynamics towards performance-based budgeting: A mix of change and stability. J. Public Budg. Account. Financ. Manag.Budg. Account. Financ. Manag. **31**(2), 158–177 (2019)
23. Setyoko, P.I., Kurniasih, D.: SMEs performance during Covid-19 pandemic and VUCA Era: how the role of organizational citizenship behavior, budgetary, participation and information asymmetry? Int. J. Soc. Manag. Stud. **3**(4), 105–116 (2022)
24. Saliterer, I., Korac, S., Moser, B., Rondo-Brovetto, P.: How politicians use per-formance information in a budgetary context: new insights from the central government level. Public Adm. **97**(4), 829–844 (2019)
25. НАДС Homepage.: Звіт «Державна служба: Ваша точка зору» 2021 рік, https://nads.gov.ua/storage/app/sites/5/DIYALNIST/UPRAVLINJA%20PERSONALOM/Analitika%20ta%20doslidgenja/zvit-2021-dlya-druku-merged.pdf. Last accessed 22 Jan 2023 (in Ukrainian)
26. Park, S.: Organizational Performance and Government Resource Allocation: Panel Evidence from Washington State's Public Programs. Public Perform. Manag. Rev. 1–26. (2022)
27. Ehigiamusoe, K.U., Samsurijan, M.S.: What matters for finance-growth nexus? A critical survey of macroeconomic stability, institutions, financial and economic development. Int. J. Financ. Econ.Financ. Econ. **26**(4), 5302–5320 (2021)
28. Chen, H., Hongo, D.O., Ssali, M.W., Nyaranga, M.S., Nderitu, C.W.: The asymmetric influence of financial development on economic growth in Kenya: evidence from NARDL. SAGE Open **10**(1), 2158244019894071 (2020)
29. Mishkin, F.: Asymmetric Information and Financial Crises: A Historical Perspective, Financial Markets and Financial Crises, pp. 69–108. National Bureau of Economic Research, Inc. (1991)
30. Goncalves, G.H.H., Lagerborg, A., Medas, P., Nguyen, A.D.M., Yoo, J.: The Return to Fiscal Rules. IMF Staff Discussion Notes, Fiscal Affairs Department (2022)
31. Nguyen, D.A., Garvin, M.J.: Life-cycle contract management strategies in US highway public–private partnerships: Public control or concessionaire empowerment? J. Manag. Eng. 04019011 (2019)

32. Aben, T.A., van der Valk, W., Roehrich, J.K., Selviaridis, K.: Managing in-formation asymmetry in public–private relationships undergoing a digital transformation: the role of contractual and relational governance. Int. J. Oper. Prod. Manag. **41**(7), 1145–1191 (2021)
33. Gu, B., Chen, F., Zhang, K.: The policy effect of green finance in promoting industrial transformation and upgrading efficiency in China: analysis from the perspective of government regulation and public environmental demands. Environ. Sci. Pollut. Res.Pollut. Res. **28**(34), 47474–47491 (2021)
34. Yang, L., Qin, H., Gan, Q., Su, J.: Internal control quality, enterprise environmental protection investment and finance performance: an empirical study of China's a-share heavy pollution industry. Int. J. Environ. Res. Public Health **17**(17), 6082 (2020)
35. Biolsi, C., Kim, H.Y.: Analyzing state government spending: balanced budget rules or forward-looking decisions? Int. Tax Public Financ.Financ. **28**, 1035–1079 (2021)
36. World Bank.: World Development Report 1988. Chapter 5: Improving the Allocation of Public Spending. https://doi.org/10.1596/9780195206500
37. Dewan, S., Ettlinger, M.: Comparing Public Spending and Priorities Across OECD Countries. Center for American Progress (2009)
38. World Conference in Barcelona to shape the future of higher education 2022| UNESCO. Homepage. https://www.unesco.org/en/articles/unesco-world-conference-barcelona-shape-fut ure-higher-education. Last accessed 23 Jan 2023
39. Azarenkova, G.M. Buriachenko, A., Zhyber, T.: Anticorruption efficacy in public procurement. Financ. Credit Activity Probl. Theory Pract. **2**(33), 66–73 (2021). https://doi.org/10.18371/fca ptp.v2i33.206412
40. White, J.: What are budgeting's purposes?: Comments on OECD's principles of good budgetary governance. OECD J. Budg.Budg. **14**(3), 1–18 (2015)
41. Mishan, E.J., Quah, E.: Cost-Benefit Analysis. Routledge (2020). https://doi.org/10.4324/978 1351029780
42. Argento, D., Kaarbøe, K., Vakkuri, J.: Constructing certainty through public budgeting: budgetary responses to the COVID-19 pandemic in Finland, Norway and Sweden. J. Public Budgeting, Account. Financ. Manag. **32**(5), 875–887 (2020). https://doi.org/10.1108/JPB AFM-07-2020-0093
43. Buriachenko, A., Zhyber, T., Paientko, T.: Deliverology Implementation at the local government level of Ukraine. In: ICTERI Workshops, pp. 338–350 (2020)
44. Zhyber, T., Solopenko, T.: Substantive and legal aspects of determining the components of the budget program efficiency indicator on the example of financing the secondary health care. In: Economic Bulletin. Series: Finance, Accounting, Taxation, vol. 5, pp. 88–97. University of the State Fiscal Service of Ukraine (2020) (in Ukrainian)
45. Robinson, M.: Connecting evaluation and budgeting. In: ECD Working Paper Series;No. 30. Independent Evaluation Group, World Bank Group. Washington, DC (2014)
46. Stoshikj, M., Kryvinska, N., Strauss, C.: Efficient managing of complex programs with project management services. Glob. J. Flex. Syst. Manag. **15**, 25–38 (2013). https://doi.org/10.1007/ s40171-013-0051-8
47. Engelhardt-Nowitzki, C., Kryvinska, N., Strauss, C.: Strategic demands on information services in uncertain businesses: a layer-based framework from a value network perspective. In: 2011 International Conference on Emerging Intelligent Data and Web Technologies (2011). https:// doi.org/10.1109/EIDWT.2011.28
48. Ren, J., Ren, X., Liu, Y., Man, Y., Toniolo, S.: Sustainability assessment framework for the prioritization of urban sewage treatment technologies. In: Waste-to-Energy, pp. 153–176. Academic Press (2020)
49. Taherdoost, H., Madanchian, M.: Multi-Criteria Decision Making (MCDM) methods and concepts. Encyclopedia **3**(1), 77–87 (2023)
50. Mosadeghi, R., Warnken, J., Tomlinson, R., Mirfenderesk, H.: Comparison of Fuzzy-AHP and AHP in a spatial multi-criteria decision making model for urban land-use planning. Comput. Environ. Urban Syst.. Environ. Urban Syst. **49**, 54–65 (2015)

51. Keown, A.J., Martin, J.D.: Capital budgeting in the public sector: a zero-one goal programming approach. Financ. Manag. 21–27 (1978)
52. Perwira, Y., Apriani, W.: Application of Weighted Sum Model (WSM) for determining development priorities in rural. Jurnal Teknik Informatika CIT Medi-com **12**(2), 72–87 (2020)
53. Zheng, Y., Xie, X., Su, T., Ma, L., Hao, J., Meng, Z., et al.: Wuji: Automatic online combat game testing using evolutionary deep reinforcement learning. In: 2019 34th IEEE/ACM International Conference on Automated Software Engineering (ASE), pp. 772–784. IEEE (2019)
54. Boutilier, C., Lu, T.: Budget allocation using weakly coupled, constrained Markov decision processes. In: Proceedings of the Thirty-Second Conference on Uncer-tainty in Artificial Intelligence (UAI'16), pp. 52–61. AUAI Press, Arlington, Virginia, USA (2016)
55. Kos, D.: War and corruption in Ukraine. Eucrim: the European Criminal Law Associations' fórum (2), 152–157 (2022)

Research of Information Platforms and Digital Transformation Algorithms for Post-war Recovery of Ukrainian Business

Oleksandra Mandych⬤, Jacek Skudlarski⬤, Tetiana Staverska⬤,
Oleksandr Nakisko⬤, Oksana Blyzniuk⬤, Halyna Lysak⬤,
and Hanna Morozova⬤

Abstract The article is devoted to the theoretical justification and elaboration of applied possibilities scenarios of creating a software product with modeling digital adaptation strategies for the recovery of Ukrainian business in the post-war period. The paper presents the research methodology and principles of designing information management systems. Determined that the methodology of designing a digitized platform should include the systematization of existing potential and components, the development of mechanisms, tools, strategies, algorithms, and scenarios for the adaptation and development of management systems, the analysis of the possibilities of organizational changes in digitalization, and reengineering of business processes, the involvement of risk management and anti-crisis management strategies. Working out the scientific-methodical and practical principles of the functioning information platform, strategies for digital adaptation, transformation, and innovative information technologies transfer involves substantiating the advantages of engineering technologies in business, the possibility of using the transfer of information technologies for recovery, creating a digital platform for the integration business entities into the external environment with the aim of attracting innovative solutions through the information technologies transfer, substantiating approaches to the creation of an information platform, adapting existing theoretical provisions to the applied use of models by business entities, developing a scientific and methodical approach to the digital business transformation on the proposed information platform for a gradual recovery

---

O. Mandych (✉) · T. Staverska · O. Nakisko · O. Blyzniuk · H. Lysak · H. Morozova
State Biotechnological University, Kharkiv, Ukraine
e-mail: ol.mandych@biotechuniv.edu.ua

H. Morozova
e-mail: g.morozova@btu.kharkov.ua

J. Skudlarski
Warsaw University of Life Sciences, Warsaw, Poland
e-mail: jacek_skudlarski@sggw.pl

in the post-war period. Based on the research results, algorithms and procedures of the scenario approach to digital business transformation are proposed.

Keywords Digital technologies · Digital adaptation strategies · Information platform · Business process reengineering · Transfer of innovative information technologies · Scenario approach · Digital transformation

1 Introduction

Digital transformation requires the whole world to gradually transition to cloud technologies, as well as the creation of innovative databases to store the entire array of information platforms for various fields. Accentuation of the Internet with data all over the world occurs according to a geometric progression. The actualization of issues of the implementation of digital transformation is caused by the proven economic effect of the use of information resources based on the results of analytics of the last 3–5 years. The existing results of developed countries based on NRI models are an example of the need to create a modern reboot, first of all, of economic systems [1–4].

The growing demand for digital technologies also creates the prerequisites for the formation of advanced systems within different countries. The creation of a single, effectively functioning platform is not possible due to a number of reasons, in particular, centralized management, existing private commercial activity, and technical development of systems. It is known that information platforms and the corresponding access to networks are models whose development scenarios vary widely. The studied NRI indices allow us to draw conclusions about the rapid growth and dynamics of the introduction of the digital world to different countries and their economic systems, about the lack of a single trajectory of electronic modification of existing data management tools, about the existing disparities between the opportunities and needs of achieving the general goals of digitalization, as one of the prerequisites for ensuring sustainable development of countries, economies, spheres, business subjects. If some countries can manage data using the latest technologies, others can only use technologies from earlier stages of digitalization development. Thus, digital transformation should be reflected in the area of involvement of various spheres of functioning and various possible scenarios [5–7].

According to the project of the Recovery Plan of Ukraine, proposed by the Ministry of Digital Transformation of Ukraine for the period 2022–2032, according to the materials of the working group "Digitalization" (Lviv 2022), among the current problematic issues regarding the provision of maximum digitalization, the involvement of institutional support tools, as well as the restoration of destroyed information and communication systems, the issue of restoring domestic business is gaining particular importance. Thus, for 2023–2025, the stage of recovery, restarting the economy and institutions, is already planned [8]. At this stage, the predicted results regarding the development of digitization, information network systems, optimization and centralization of information support, integration with European platforms and joining global information resources, digitalization of the country's economy, etc., have been formed. It should be noted that the post-war plan for developing the national economy includes the maximum involvement of modern technologies in business development based on creating opportunities to use open data technologies, cloud technologies, digital technologies, etc. The creation of information management systems that will allow effective strategic management models to be applied in practice will provide wider access to business digitization procedures (especially for "small" and "medium" sized companies), which is a necessary result of the development of a modern digital platform for business.

2 Business Process Management on Digital Transformation Framework

The concept of "Business Process Management" outlines the existence of connections between the components of the company's management system, taking into account the possibilities of using information systems, and also proves the multi-level management information system and represents the step-by-step structuring and introduction into business processes [9]. Emphasizing the need for applied consolidation of management subsystems into a single information system, the involvement of modern tools for digitalization and visualization of processes, the use of cloud technologies, and the synthesis of the components of the management and financial accounting system of companies in order to obtain competitive advantages for business through the involvement of digitalization procedures is the basic rationale for the feasibility of creating a corporate business -systems with software modules for solving operational and strategic level tasks [10]. Existing approaches in modeling digitized management processes with the involvement of information technologies, a certain level of influence on management decision-making procedures, and formalized main areas of restructuring of management systems when introducing progressive technologies represent the basic level of creating a digital transformation in the management of business projects [11].

In project management, the understanding of digital transformation due to the transition from an industrial economy puts forward new requirements for business entities—a new concept of revolutionary changes in the system of globalization of economic transformations and institutional support, paradigms of business conduct, criterion functionality, criteria for innovation and the involvement of innovative models, infrastructural opportunities and social responsibility, etc. [12]. Digitization for Ukrainian businesses is built on the principles of entrepreneurial initiative with the simultaneous development of European integration processes. The results of market activity in recent years prove the maximization of the integrative components of Ukrainian business; in addition, there are initiatives and characteristics of the constant growth of intellectual capital in the field of IT technologies. The growth of such intellectual capital is a platform for the creation, moderation, and implementation of digital transformation models for business entities. The massive development of digital transformation tools is represented by such technologies as Big Data and Business Analytics, artificial intelligence, data transmission platforms for the creation of cloud web services, blockchain technology, and quantum cryptocurrency technology, the infrastructure segment for data collection in cyber systems, technologies for creating smart infrastructures, multi-agent and multi-tasking technologies, technologies for creating digital communication systems and channels, etc. However, applied use in business project management has not only its own perspectives but also opportunities that are limited by the existing potential for digital transformation in the realities of business conditions.

The dynamics of digital transformations create the need to develop a corporate integrated information system, which will include infrastructure support for digitalized business, information and communication systems, and digitalization systems for managing business processes [13]. At the same time, the creation of an information platform requires the search for software and the implementation of the transfer of innovative information technologies. Features of information technologies for business have their differences depending on the management and project workload.

Studies confirm the ideas of the need to create integrated information technologies for business digitalization, the need for maximum involvement in modern digital technologies of Industry 4.0, and the focus in management systems on key IT strategies as a means of attracting innovative changes [14]. The main economic methodology of digital changes should be revealed by questions and new topics, such as the future of business in the conditions of digital transformation, the change of business boundaries of companies, the involvement of modern information technologies in the field of Internet auctions and online behavior of buyers, IT research as a stimulator for the activation of digital management systems (finance, marketing, operations management, and even human resources management) [15]. It is assumed that the development of the transfer of innovative digital technologies for business is the basis for the involvement of modern technologies to ensure uninterrupted work in the field of digital transactions, electronic commerce, the internal management system of companies, and the digitalization of business in general [16].

Globalization affects the adoption of digital technologies, as indicated by country-level data on the Globalization Index (KOF), Digital Evolution Index (DEI), Digital Adoption or Adoption Index (DAI), Digital Economy and Society Index (DESI), Economy Digitalization Index (Boston Consulting Group, e-Intensity), global innovation index (GII), world digital competitiveness index (WDCI), as well as total factor productivity (TFP), network readiness index (NRI), information and communication technology development index (IDI), etc. In the study of indices of this type, the effective indicator is the used advanced modeling of panel data. Examples of basic indices are presented (Fig. 1).

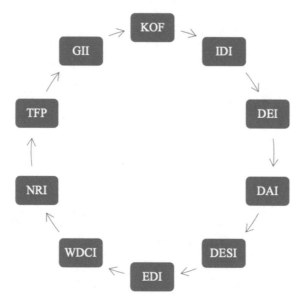

Fig. 1 Globalization indices for adapting digital transformation business processes [13–19]

Conducted research on the globalization index (KOF) of Ukraine, as one of the important coefficients for the digital transformation of business, allows us to assert significant shifts in the directions of economic, trade, and financial globalization over the past 30 years or so, as well as a full correlation with the global growth trend [18, 19] (Figs. 2, 3 and 4).

Of course, the presence of a wide scale of correlation between actual indicators applies to a lesser extent to the levers of digital transformation [20]. For the introduction of modern digital technologies, it is important to obtain a growing curve that demonstrates the development of the country's economic and financial systems and the ability to accept new transformational changes [21].

3 Innovative Information Technologies Transfer

The research presents that globalizing digital transformations have a positive effect on information technology transfer; new innovative products are created with the help of digital technologies, and countries that introduce digital technological changes achieve an increase in the convergence of the implementation of digital technologies [22].

Digital transformation creates a separate impact on the socio-economic component and business responsibility, affects the intensification of the use of information technologies on the management system, business productivity, labor productivity, and population employment, creates the need to involve modern technologies

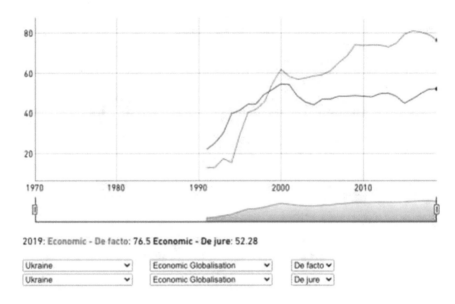

2019: Economic - De facto: **76.5 Economic - De jure**: 52.28

| Ukraine | ∨ | Economic Globalisation | ∨ | De facto ∨ |
| Ukraine | ∨ | Economic Globalisation | ∨ | De jure ∨ |

Fig. 2 KOF globalization index—economic index Ukraine [17]

(electronic freelancing, electronic outsourcing, start-employment, etc.). The result of digital transformation in this direction can be models of digitalization according to various scenarios of their implementation; in terms of evolutionary and forced modeling, proposals have been made for state support for digital business transformation [23]. One of the directions of digital transformation in a certain area is the search for opportunities to involve SMART technologies. The use of SMART information technologies for management systems and project management for the purpose of building goals and management tasks demonstrates that these models occupy important places in the process of choosing possible variations of business development [24].

Attracting the tools of digital transformation for the business of any country requires, first of all, in addition to the available opportunities and the latest information technologies, an actual platform for effective functioning. Ukrainian business in the conditions of the pre-war state during the crisis due to the COVID-19 pandemic was characterized to a greater extent by the insufficient level of use of the available potential, non-compliance with the requirements of the external environment, insufficient consideration of the occurrence of risks, which required a change in the approach to management, updating the methodological and methodical tools for improving the configuration organizational structure, technologies, processes, procedures for the development of strategic management, which, in turn, should ensure organizational changes. Gradual processes of digital transformation have changed the market environment for business and strengthened the role of the latest information technologies. The conditions of the war period nullified the results of economic, innovative, and technological growth of recent years and led to a new, more powerful

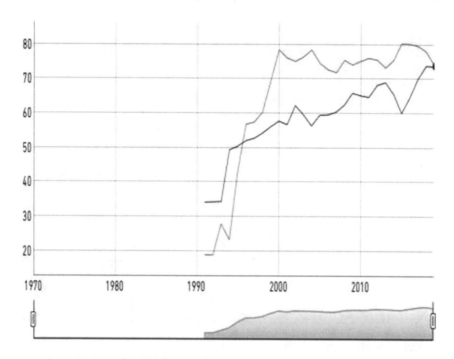

2019: Trade - De facto: 74.2 **Trade - De jure:** 73.72

| Ukraine | ⌄ | Trade Globalisation | ⌄ | De facto ⌄ |
| Ukraine | ⌄ | Trade Globalisation | ⌄ | De jure ⌄ |

Fig. 3 KOF globalization index—trade index Ukraine [17]

crisis. The process of post-war recovery of Ukrainian business is already starting, but under the existing realities, it is precisely the mechanisms of the complete transformation of the information space that require more attention. The working hypothesis of the project is the need to involve information systems in the fields of management, marketing, finance, accounting, audit, HR, and others in order to reengineer business processes and adapt the business environment to the realities of the market situation (Fig. 5).

Transformation in the digital space aims to transfer business to an information platform, the creation of which is the idea of the project. The created platform will have the form of a software product (multifactor matrix model), which will take into account all areas of business (management, economic, operational, etc.) and, at the same time, will have a complete analytical unit for determining the effectiveness of the company's activities and, based on the received data, algorithms for modeling development prospects. Thus, companies using this platform can independently solve practical tasks in their activities, fully relying on scientifically based models. A

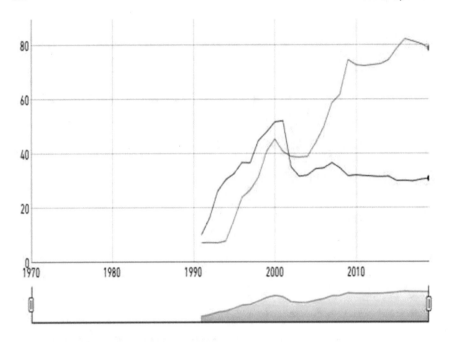

2019: Financial - De facto: **78.8 Financial - De jure: 30.84**

| Ukraine ∨ | Financial Globalisation ∨ | De facto ∨ |
| Ukraine ∨ | Financial Globalisation ∨ | De jure ∨ |

Fig. 4 KOF globalization index—financial index Ukraine [17]

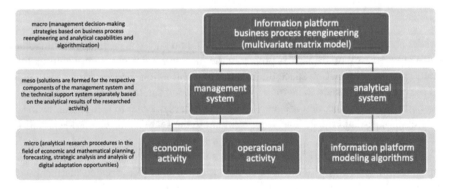

Fig. 5 The level procedure for the introduction of the information platform [own development]

feature of the construction of an information product is its complexity, i.e., models are formed starting from the analysis of current data, go through modeling stages using economic and mathematical programming packages, provide algorithms for choosing strategies with a description of the methodological base and opportunities in accordance with the SWOT analysis of the results obtained at the previous stages, as well as simulate possible scenarios of market behavior for business.

The determined trends of the digital transformation of the Ukrainian pre-war economic platform make it possible to single out the main components and components, influencing factors by industry and industrial direction, the general dimensions and scaling of the involvement of digital technologies [25]

The main conceptual provisions and possibilities of previous research in the areas of sectoral support of information and communication technologies, meaningful content and digitization models when digital technologies are introduced, the impact on the human and intellectual capital of Ukrainian business, the peculiarities of the formation of the digital labor market, competitive opportunities, weaknesses and risks when the global introduction of digital transformations create a basis for the formation of models of digital development of Ukrainian business for the period of post-war recovery. The developed models for today's realities have only theoretical content due to the fact that the organizational and strategic changes that the country underwent in 2022 caused a new format for finding solutions. Opportunities of economic, technical, technological, etc. nature at the moment are only potentially predictable; in addition, Ukrainian business is forced to work in conditions of complete uncertainty. The national digital transformation program and digitalization strategy, which provided for the consolidation of the digital strategy of the G20 countries by 2020, was designed on the basis of appropriate institutional, legal, and infrastructural support [26–34].

The created model of the development of the digital transformation of the Ukrainian system must completely change due to the crisis situation of 2022–2023. The involvement of digital technologies and, in general, digital transformation in business processes is currently not an easy task for most companies, especially from the territories of "gray zones" (zones of active hostilities, where the material and technical base was completely or partially destroyed, other losses for business). Existing information technologies for the digitalization of business or management of business projects require separate capabilities: financial, technological, and, most importantly, HR. It should be understood that the digitalization of business is a complex, multifunctional process that can be fully involved from the experience of world practice. Digital transformation requires a business to find resources for its implementation, and, first of all, it is human capital. The HR potential of most of the affected companies and the overall migration of HR potential due to the situation in 2022 tend to decrease sharply compared to the pre-war period. It should be understood that modern tools, for example, for project management and business digitalization (Jira, Trello, Redmine, etc.) are also difficult for HR or a project manager without previous work skills, and in the realities of today's labor market, it is difficult to find HR with relevant knowledge within Ukraine is more complicated. Thus, for the rapid adaptation and interpretation of digital data for business projects, it is necessary to look for new approaches to digital transformation.

To solve the problem, it is necessary to predict the mechanisms of scientific research in the field of creating digital tools, adapting the theoretical provisions of digital models to a more simplified form in the form of cases with the possibility of applied use. One of the directions for solving the outlined problems should also be the involvement of the transfer of innovative information technologies for the spread of digitalization (to a greater extent, for leveling the financial, technical, and technological components of Ukrainian business).

The digitization of companies' business processes is the basis for integration in the information space with other counterparties, opportunities to attract not only experience but also the transfer of innovative information technologies, which will play a decisive role in business recovery. In addition, the transfer of information technologies is a necessary adaptation to modern market requirements. Currently, Ukrainian businesses almost do not use the transfer of innovative information technologies. To a greater extent, innovative development is justified only in theoretical research. That is why, in our opinion, the introduction of models and the possibility of adding innovations through the transfer of information technologies to the developing information product is an important block (Fig. 6).

The effective assessment of the transfer of innovative technologies, unfortunately, has a declining trend. At the same time, the features of regression and the crisis situation of martial law allow us to forecast the results of 2023–2023 at the minimum

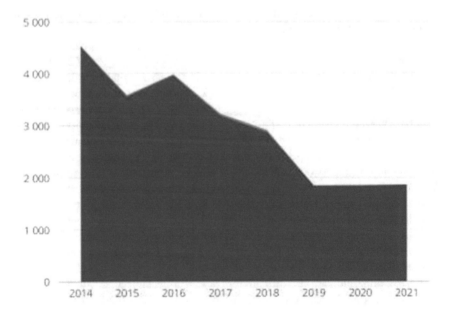

Fig. 6 Transfer of innovative technologies in Ukraine [35]

possible level of obtaining performance indicators. Of course, first of all, this will be caused by a low level of financial support. However, it should be noted that the real consolidation of commercial efforts for post-war recovery will take place precisely through the involvement of information technologies and the transformation of the commercial component of the business. One should also understand the general level and likely inequalities between individual business entities in the middle of spheres and industry affiliation. Moreover, "large" enterprises have more attention and opportunities, while the situation is lower in comparison with "small" businesses. The implementation of the policy of transfer of innovative technologies is proportional to the involvement of information and digital technologies.

The peculiarities of the current state of development of "small" and "medium" entrepreneurship in the country allow us to draw conclusions about the low level of digitalization of business processes, the low interest in digitalization, as well as about the low ability to join the transfer of innovations through the involvement of the information space. The post-war recovery of Ukrainian businesses requires not only the search for new ideas but also the creation of powerful platforms that will help companies implement the management strategies needed today.

4 Digital Transformation Methodology

Digital business transformation processes and reengineering management systems companies create the need to involve tools and comprehensive study of issues. At the same time, it is necessary to take into account the existing results and prospects for the design of innovative models in this field. The research methodology is presented through the integration of the authors' scientific results and the involvement of world experience.

In terms of the development of business informatization, research is aimed at the globalization of information systems and network space, digital transformation, algorithmization, and modeling of management information systems with the involvement of management, marketing, finance, and other tools. The research is based on the principles of involving an interdisciplinary approach, which requires the application of complexity, systematicity, and adaptability between the direction of information systems and technologies (technical component) and business process management systems (managerial and economic component) (Fig. 7).

The process approach is based on the development of mechanisms for adapting the strategies of digital business transformation entities. The synergistic approach is outlined by the integration of the business environment in the process of digitization, digitization, information visualization, and the creation of modern management systems. The functional approach involves the creation of basic strategic management systems with theoretical, methodological, and practical substantiation of competitive business recovery and development models, elaboration of the toolkit, its components, and components to ensure the search for probable optimal models of management decision-making. The structural approach is based on the procedural provision

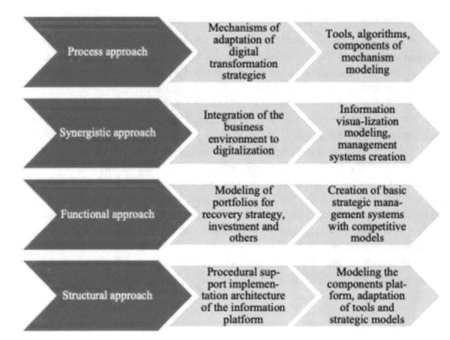

Fig. 7 Complex modeling of digital transformation adoption procedures [own development]

of generation, implementation, and implementation of business recovery and development strategies through the creation of the appropriate architecture and architecture of the information platform (software product), modeling, forecasting, and algorithmization of the company's strategic management system, structuring of all components and components into developed technological solutions. New approaches and prospects for further research will be based on the creation of a software product, substantiation of updated methods of integrated digital data arrays according to the components of business processes and management systems, development of new strategic development models with the involvement of economic and mathematical programming methods, game theory and probability of choice, analytical data, etc.

The economic interpretation of information support is aimed at researching the mechanisms of competitive development of companies based on strategic management, analytical models, concepts of accounting and analytical, financial and management support, risk management technologies, and anti-crisis management systems. Reengineering of digital systems of companies should be based on models of active and proactive behavior, market technologies of business conduct and adaptive mechanisms of strategic development. When designing digital transformation strategies, it is necessary to involve the experience of ensuring the viability of companies in the conditions of an economic crisis due to the COVID pandemic, as well as

Fig. 8 Algorithms of the first stage of digital business transformation and strategizing procedures [own development]

Digital transformation of business and strategizing procedures

State of development of information systems and technologies

The existing potential for engaging in digital transformation

Components of the information platform, mechanisms and tools

Strategies, algorithms and scenarios of adaptation and development of management systems

working out the modeling patterns of anti-crisis situational projects, with the possibility of simultaneously taking into account the complete toolkit of the management system and chain industry specialization.

The main goal of the research in the direction of developing opportunities for digital business transformation entities is to develop theoretical, methodological, and applied recommendations for the creation of an information software product for the company's management system, which will provide analytical research, help in modeling strategic management scenarios, and also contribute to the development innovative directions of development for the recovery of Ukrainian business in the post-war period. It is appropriate to distinguish two staged processes, taking into account the distribution of the functional load of reengineering systems (Figs. 8 and 9).

Solving the set goal has the following basic tasks for the company:

- to investigate the current state of development of information systems and technologies in the field of business, to systematize the existing potential for attracting digital transformation;
- systematize the components of the information platform (management, marketing, finance, accounting, audit, etc.), develop mechanisms and appropriate tools;
- to develop strategies, algorithms, and scenarios for the adaptation and development of management systems;
- analyze the possibilities of introducing organizational changes by business entities taking into account digitalization in the system of strategic management, digitalization, and reengineering of business processes, as well as the involvement of risk management and anti-crisis management strategies;

Fig. 9 Algorithms of the
second stage of digital
business transformation and
strategizing procedures [own
development]

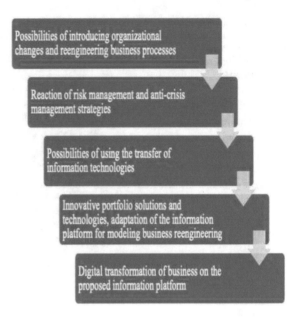

Fig. 9 Algorithms of the second stage of digital business transformation and strategizing procedures [own development]

– to substantiate the modern advantages of using engineering technologies in business, as well as the possibilities of using the transfer of information technologies in the conditions of post-war recovery (reinvestment strategy);
– create a digital platform for the integration of business entities in the external environment with the aim of attracting innovative solutions through the transfer of information technologies and portfolio technologies;
– justify approaches to creating an information platform, adapt existing theoretical provisions to the applied use of strategic models by business entities;
– to develop a scientific and methodological approach to the digital business transformation on the proposed information platform for a gradual recovery in the post-war period.

The constituent studies are distributed in such a way as to fully create a methodology on an economic basis (full description of the management system) for applied use on an information platform (software product).

5 Business Re-Engineering Based on Applied Use Digitalization

The adaptation of Ukrainian businesses to the global digital transformation solves one of the socio-economic problems of the country—there is a profitable business, there are jobs, there is the filling of the budget with taxes, and there is the development of the country. It is impossible to restore the country's economy without restoring its main operators. The current state of operation of "small" and "medium" businesses, especially in the territories of "gray zones", where active hostilities are being or were previously being conducted, allows us to talk about a low level, first of all, of production activity due to existing losses of the material and technical base, financial activity and the level of digitalization of business processes, about the low interest in digitalization, as well as the low level of ability to attract the transfer of innovative technologies in the information space. The post-war restoration of domestic business will require not only the search for production ideas but also the design of a powerful system of management, financial, and information support, which should help businesses in the implementation of recovery strategies. The final result of the research is the formation of the architecture of a new information platform for management systems with a formed strategic management methodology, which will create a new software product and form a new target audience for its use—the Ukrainian business as a real sector of the national economy. A feature should be the ability to use it in different industries, which is taken into account in the modeling due to the multi-industry architecture of filling the database of data arrays. The adaptation of Ukrainian businesses to the global digital transformation solves one of the socio-economic problems of the country—there is a profitable business, there are jobs, there is the filling of the budget with taxes, and there is the development of the country.

The following methodological recommendations will be provided for the practical use of the developed product with business process analytics algorithms and modeling of strategic business ideas for various areas:

- theoretical and practical cases, description of the possibilities of implementation and adaptation of digital technologies for business;
- methodological basis of opportunities, mechanisms, and prospects of digital transformation for management systems;
- business process reengineering models, key competence formation structures, and market success factors (competitive strategies);
- analytical matrices for diagnostics of the effectiveness of the functioning of companies (all necessary economic, financial, and management indicators with a full description of values, calculations, and dependencies);
- algorithms for conducting SWOT analysis, vector analysis, modeling based on game theory and selection of probabilities, expert assessments, etc., in specialization and industry specifics;
- scenarios and forecast models of risk management in conditions of uncertainty;
- models of proposed strategies separately in the areas of management, marketing, finance, accounting, and auditing of companies;
- elaborated industry formats of information systems (according to the algorithms of the full cycle of project management);
- theoretical and applied basis in the field of transfer of innovative information technologies for the recovery of Ukrainian business through inclusion in the integration programs of digitalization;

– procedures and tools for the use of information systems and technologies in company management, educational materials for using the program, for filling in blocks, analyzing data in matrices, and working out the effectiveness of proposed solutions.

The proposed information platform model should become a finished product that any company can use for its competitive development. The practical implementation of the idea consists in the fact that companies will be able to independently carry out a full diagnosis of the effectiveness of their activities, receive their indicators and relevant comparisons of their values, will be able to conduct a strategic analysis of their activities, will be able to identify risks that relate specifically to their business, and will also be able to choose by their results, anti-crisis or competitive strategies—models and scenarios of market behavior.

The identified opportunities will enable companies to obtain an economic effect—to increase their performance, increase the profitability and level of business profitability, reduce costs due to the involvement of modern effective management tools and increase capital investment based on the created investment portfolios and opportunities to attract the transfer of information technologies, will increase business productivity, in general. Social effect—the implementation of the program can be fully performed by existing employees. The software product will have full methodical and educational support, and all proposed activities will have a scientific, theoretical, and practical justification. The importance of the obtained results for various industries lies in the long-term nature and maintenance of positive market positions by consolidating the effect of digital transformation through the implementation of appropriate strategic organizational changes within companies. Attracting state financial support will allow Ukrainian companies to achieve the effect of saving resources (financial, human, informational, etc.) and help to go through the digital transformation at a faster pace, which today is extremely important for the post-war recovery of Ukrainian business.

Algorithms for digital business transformation should be reflected in the research and development of many directions, which should be combined with each other in terms of meaningful content. According to the scenario approach to the algorithmization of digital transformation, the following steps are proposed, which are formed according to two main blocks. The first block provides for the design of research procedures for the theoretical and methodological foundations of designing information management systems in the field of digitalization of strategic management of business entities. Algorithmization should focus on the following tasks: research the current state of development of information systems and technologies in the field of business, systematize the existing potential for attracting digital transformation, systematize the components of the information platform (field of management, marketing, finance, accounting, audit, etc.), work out mechanisms and appropriate tools, develop strategies, algorithms and scenarios for the adaptation and development of management systems, analyze the possibilities of introducing organizational changes by business entities in digitization in the strategic management system, digitalization and reengineering of business processes, as well as the features of involving risk management and anti-crisis management strategies, etc. The second block involves working out the scientific, methodical, and practical principles of the information platform (software product) functioning, working out digital adaptation strategies and prospects for the development of digital transformation and the transfer of innovative information technologies for post-war business recovery. Procedures for substantiating the advantages of using engineering technologies in business, as well as the possibility of using the transfer of information technologies in the conditions of recovery (reinvestment strategies) to attract innovative solutions and portfolio technologies, to adapt existing theoretical provisions to applied use, to develop a scientific and methodological approach to the digital transformation business are planned on the proposed information platform for a gradual recovery in the post-war period.

Digital transformation aims at gradual involvement in business processes, while any changes and interventions will cause a reaction in the management system. When involving the tools of digital transformation, it is expedient to create models of scientific substantiation of the relevant procedures and processes. Performance indicators and, in particular, scientific results are planned as the expected results of the procedural approach and modeling of behavior scenarios in the direction of digital transformation:

- the methodological basis of the possibilities of digital transformation with the effectiveness of the functioning of information systems in the economy;
- a digitalized model of the company's management system in terms of economic mechanisms, models, tools, algorithms, and procedural support of an activity, which aims to take into account all components and components, their relationships and interdependencies;
- models of market behavior of business entities in the war and post-war periods, as well as a methodology for proactive formation of management strategies for the recovery and development of companies was developed;
- the methodological basis of reengineering business processes with the involvement of digitalization innovations and the use of the proposed information platform;
- the methodological toolkit of risk management and anti-crisis strategic management in the context of introducing organizational changes;
- procedures for the formation and use of business process engineering models and the involvement of the transfer of innovative information technologies in ensuring the creation of unique competitive advantages for the implementation of the strategy of ensuring the competitiveness of business development;
- information platform modeling mechanisms, software product components, procedural support and algorithms for creating business portfolios for recovery (reinvestment), taking into account the predicted impact of external and internal environmental factors;
- scenarios of reaction and adaptation of business entities to the introduction of digital models of market behavior, which will contribute to the fullest use of the existing digital potential of Ukrainian companies;

– models for creating an information platform and software products for applied use in the company's management system with algorithmization of economic-mathematical programming procedures and modeling of development strategies.

The difference between the existing results and possibilities of the digital transformation of Ukrainian businesses in the war and post-war periods lies in the need to create an information product that will allow forming an updated model of combining information technologies with management systems. At the same time, the idea of developing the concept of digital transformation of Ukrainian business for the period of post-war recovery, as well as modeling mechanisms and algorithms for the applied implementation of digitalization technologies, becomes important. The management of business projects should involve the use of existing developments for the improvement and acquisition of further development of strategic management models in various industries for companies of various specializations and individual reengineering business models. The issues of theoretical systematization of the directions of digital transformation of Ukrainian business, methodological substantiation of the concept of creating an information platform for management systems, applied use of digital technologies, involving the transfer of innovative information systems and reengineering of business processes in accordance with the proposed organizational and strategic changes require further resolution.

The proposed modeling of financial architecture and architecture, an integrated complex of management mechanisms, is a scientific product that enterprises will be able to fully use in their activities. The practical implementation of the study is aimed at the scientific and methodical provision of business for the independent conduct of a full diagnosis of the performance of the activity, obtaining all the necessary economic, managerial, and financial indicators and relevant comparisons of their values. Enterprises will be able to independently conduct strategic analysis based on scientifically based approaches, which will be supported by practical cases. Businesses will be able to identify risks that concern them, and will be able to choose market behavior scenarios, anti-crisis models or competitive recovery strategies based on the results of their own diagnostics. The identified opportunities will enable the business to obtain an economic effect by increasing its effectiveness, increasing profitability indicators and profitability levels, reducing costs involving modern management, innovation, and information tools, and using cloud technologies in conditions of the complexity of doing business in territories or relocation. An increase in capital investments based on the development of investment portfolios and the possibility of attracting tools for transferring information and cloud technologies will increase the level of business recovery in general.

The digital adaptation and design of the information platform for the reengineering of business processes involve the elaboration of the theoretical and methodological basis for the formation of strategies for recovery and competitive development and the elaboration of all components of the financial architecture of business entities. It is necessary to develop a methodological basis for management strategies for the development of business entities based on the combination of the fields of management, finance, economics, accounting, auditing, insurance, marketing, social responsibility of business, etc.

The economic interpretation of the results according to the research idea is demonstrated through the prism of the mechanisms of competitive development of enterprises on the basis of strategic management, analytical models, concepts of accounting and analytical, financial and management support, risk management technologies, and anti-crisis management systems. The basis for the creation of a digital platform involves the formation of an applied model for determining effective performance indicators and market activity of enterprises using a wide range of methods of financial analysis and diagnostics, economic-mathematical modeling, procedural programming models, software control, game theory matrices, equilibrium, situational and vector analysis, competence and expert assessment, etc. All defined calculation models are part of the digital platform and digital adaptation strategy (adaptation of financial diagnostics to the digital dimension).

Development of methodological approaches to the justification of types of organizational changes will allow in further research to ensure correspondence between the existing potential of the existing situation to the requirements of the constantly changing external environment and uncertainty, as well as in accordance with the goals and priorities of reengineering business processes in the financial activities of business entities. Developed models for assessing the potential of agribusiness entities will allow identifying unused reserves to ensure the creation of new and consolidation of existing competitive advantages, which is the basis of the recovery strategy. It is important to pay attention to the scenarios of reaction and adaptation of business entities to the introduction of digital models of market behavior and marketing strategies to increase the effectiveness of functioning in selected object markets. The planned competitive development of business entities will be based on the implementation of recovery strategies aimed at creating competitive advantages, the social source of which should be human capital. When human capital is involved in business activities, intellectual capital will be formed or preserved. The latter, with effective distribution, will provide a rapid increase in capital return. The growth of capital return should become the basis for provoking the need for further investments in the formation, capitalization, development, and use of synthesized capital as a phenomenon of economic science. Synthesized capital, which combines human, intellectual, and social capital, is one of the manifestations of business socialization. Social effect—the implementation of the program is suitable for existing employees. The software product (scientific development of the information e-platform) will have scientific and methodological support; all proposals will have theoretical and practical justification. The importance of the result obtained from the involvement of this financial e-platform for various industries will be the long-term nature of maintaining market activity and positions through the introduction of digital transformation with the simultaneous implementation of relevant strategic and organizational changes in the internal environment. Involvement of donation mechanisms, foreign investments, and state funding will allow Ukrainian business to undergo digital adaptation and transformation at a faster pace. Today, digitalization is extremely important for the post-war recovery of Ukrainian business, but it requires, first of all, the search for resources and justification for the possibility of use.

6 Conclusions

The relevance of the research tasks is confirmed by the existing plans of the Ministry of Digital Transformation of Ukraine regarding the restoration of the Ukrainian economy on the basis of informatization, digital transformation, and digitization, and is also carried out in accordance with the National Economic Strategy developed until 2030 with clearly defined vectors of digital development. Modern scientific research is aimed at strengthening the role of digitalization and its maximum possible implementation in the real sector of the economy, which today represents Ukrainian business. The tasks of digital transformation fully confirm the working hypothesis of digitization and correlate with all procedural stages of the modern development of business entities. The main idea is revealed through the presented tasks and is aimed at creating a large-scale information platform that will allow Ukrainian businesses not only to adapt to the digital dimension in the existing conditions of martial law but also to involve modern tools of digital integration (engineering, technology transfer, etc.) for business recovery.

The conditions of the war period nullified the results of recent years and led to a new, more powerful crisis. The process of post-war recovery of Ukrainian business is already starting, but under the existing realities, financial mechanisms and digital development need more attention. There is a need to involve information systems in the field of management, marketing, finance, accounting, auditing, etc., in order to reengineer business processes and adapt the business environment to the realities of the market situation. Transformation in the digital space is aimed at transferring business to an information platform, the creation of which is the working idea of the project. The developed platform will take the form of a software product (multifactor matrix model) that will take into account all areas of business (management, economic, operational, etc.) and, at the same time, will have a complete analytical unit for determining the efficiency of the enterprise and algorithms for modeling development prospects. A feature of the construction of an information product is its complexity, i.e., models are formed starting from the analysis of current data, go through the stages of modeling with the help of economic and mathematical programming packages, provide algorithms for choosing strategies with a description of the methodological base and opportunities in accordance with the SWOT analysis of the results obtained at the previous stages, as well as simulate possible scenarios of market behavior of businesses. Since digital transformation is currently a difficult task for most enterprises, the project also involves the adaptation of theoretical provisions of economic science to a more simplified form in the form of cases with the possibility of applied use.

The main idea of the research is revealed through the presented tasks and is aimed at designing a financial architecture and a large-scale information platform that will allow Ukrainian businesses not only to adapt to the digital dimension in the existing conditions of martial law but also to involve modern tools of digital integration (engineering, technology transfer, cloud technologies, etc.) for post-war recovery.

The paper presents theoretical developments in the field of substantiating the importance, necessity, and possibilities of implementing modern information systems, develops theoretical models, substantiates the stages and procedures of business digitalization, and outlines models and mechanisms separately for the management and accounting systems of companies. The presented scientific results of analytics only partially highlight the outlined directions for developing information technologies for business processes. The available developments are of a theoretical nature, proving the need to create corporate information systems and introduce modern, innovative information technologies. The results of the study demonstrate the importance of globalization trends in the development of information systems and technologies, digital transformation, digitalization of business processes, and the transfer of innovative information technologies.

The final result of the project is the formation of financial architecture and architecture for management systems with a formed strategic management methodology, which will create a new software product—an information resource and form a new target audience for its use—Ukrainian business as a real sector of the national economy. This is a completely new product that has no analogs. A feature is an ability to use in various industries, which is taken into account in the modeling due to the multi-industry architecture of filling the database of data arrays. Modeling recovery strategies for business with the help of an information software product will allow to take into account the full toolkit of the latest market tools, which were previously not used to their full extent due to the usual complexity of combining theory and practice.

References

1. IDC.: Worldwide Global DataSphere Forecast, 2021–2025: The World Keeps Creating More Data—Now, What Do We Do with It All? (2020)
2. Forbes.: Zettabytes By 2025 (2022). [online] Available at: <https://www.forbes.com/sites/tom coughlin/2018/11/27/175-zettabytes-by-2025/?sh=758dfff05459>
3. IDC.: Worldwide Global Storage Sphere Forecast, 2021–2025: To Save or Not to Save Data, That Is the Question (2021)
4. ITU (2021) [online] Available at: https://www.itu.int/hub/2021/11/facts-and-figures-2021-2-9-billion-people-still-offline/
5. The Network Readiness Index 2022 Stepping into the new digital era. How and why digital natives will change the world (2022). Editors Soumitra Dutta and Bruno Lanvin. Portulans Institute, p. 262. ISBN: 979-8-88862-905-5
6. Network Readiness Index Database.: Portulans Institute, p. 110 (2022)
7. The Economist.: What Gen-Z graduates want from their employers. [online] Available at: <The Economist, 2022. What Gen-Z graduates want from their employers (2022). [online] Available at: <https://www.economist.com/business/2022/07/21/what-gen-z-graduates-want-from-their-employers>

8. Project of the Recovery Plan of Ukraine. Materials of the "Digitalization" working group. National Council for the Recovery of Ukraine from the Consequences of War (2022) [online] Available at: https://uploads-ssl.webflow.com/625d81ec8313622a52e2f031/62c4577defe5 bf7afedc5b4a_%D0%94%D1%96%D0%B4%D0%B6%D0%B8%D1%82%D0%B0%D0% BB%D1%96%D0%B7%D0%B0%D1%86%D1%96%D1%8F.pdf

9. Yurchuk, N.P.: Information systems and technologies as an innovation in the business process management system. Efficient Econ. (5) (2018). [online] Available at: http://www.economy. nayka.com.ua/?op=1&z=6323

10. Afanasyeva, I.I.: The information system of management accounting in the conditions of digitalization of the economy. Econ. Manag. 1(49), 32–40 (2021)

11. Tereshchenko, L.: Modeling technologies of management information systems in the enterprise management system. Econ. Soc. (27) (2021). https://doi.org/10.32782/2524-0072/2021-27-55

12. Grynko, A., Grynko, P.: Digital transformation of business: theories, problems, mechanisms. Sci. Collect. "InterConf" (123), 41–49 (2022)

13. Lingur, L.: Integrated approaches to the formation of a CSR information system for small and medium-sized enterprises. Econ. Soc. (22) (2020). https://doi.org/10.32782/2524-0072/2020-22-38

14. Tajudeen, F.P., Nadarajah, D., Jaafar, N.I., Sulaiman, A.: The impact of digitalisation vision and information technology on organisations' innovation. Eur. J. Innov. Manag.Innov. Manag. 25(2), 607–629 (2022)

15. Brynjolfsson, E., Wang, C., Zhang, X.: The economics of IT and digitization: eight questions for research. MIS Q. 45(1), 473–477 (2021)

16. Nandal, N., Nandal, N., Mankotia, K., Jora, N.: Investigating digital transactions in the interest of a sustainable economy. Int. J. Modern Agric. 10(1), 1150–1162 (2021)

17. KOF Globalisation Index (2022) [online] Available at: https://kof.ethz.ch/en/forecasts-and-ind icators/indicators/kof-globalisation-index.html

18. Gygli, S., Haelg, F., Potrafke, N., Sturm, J.-E.: The KOF globalisation index—revisited. Rev. Int. Organ. 14(3), 543–574 external page (2019). https://doi.org/10.1007/s11558-019-09344-2call

19. Dreher, A.: External pagedoes globalization affect growth? evidence from a new index of globalizationcall_made. Appl. Econ. 38(10), 1091–1110 (2006)

20. Stoshikj, M., Kryvinska, N., Strauss, C.: Efficient managing of complex programs with project management services. Glob. J. Flex. Syst. Manag. 15, 25–38 (2013). https://doi.org/10.1007/s40171-013-0051-8

21. Engelhardt-Nowitzki, C., Kryvinska, N., Strauss, C.: Strategic demands on information services in uncertain businesses: a layer-based framework from a value network perspective. In: 2011 International Conference on Emerging Intelligent Data and Web Technologies (2011). https://doi.org/10.1109/EIDWT.2011.28

22. Skare, M., Soriano, D.R.: How globalization is changing digital technology adoption: An international perspective. J. Innovation Knowl. 6(4), 222–233 (2021)

23. Popelo, O., Kychko, I., Tulchynska, S., Zhygalkevych, Z., Treitiak, O.: The impact of digitalization on the forms change of employment and the labor market in the context of the information economy development. Int. J. Comput. Sci. Netw. Secur. 21(5), 160–167 (2021)

24. Ivanov S.M.: Analysis of the advantages of using smart technologies in the economy. Econ. State (7), 35–38 (2018)

25. Pyshchulina, O.: Digital economy: trends, risks and social determinants Razumkov Center, p. 125. Zapovit Publishing House (2020). ISBN 978-966-2050-07-3

26. Digital Economy Board of Advisors—National Telecommunications and Information Administration United States Department of Commerce [online] Available at: https://www.ntia.doc.gov/category/digital-economy-board-advisors

27. A Digital Single Market Strategy for Europe—EUR-Lex [online] Available at: https://eur-lex.europa.eu/legal-content/EN/TXT/?uri=celex%3A52015DC0192

28. Universal Service.: Shaping Europe's digital future—European Commission [online] Available at: https://ec.europa.eu/digital-single-market/en/universal-service

29. Directive on universal service and users' rights relating to electronic communications networks and services (Universal Service Directive)—European Commission [online] Available at: https://ec.europa.eu/digital-single-market/en/news/directive-universal-service-and-users-rights-relating-electronic-communicationsnetworks-and

30. Chivot, E.: A Roadmap for Europe to Succeed in the Digital Economy—Center for data innovation (2019). [online] Available at: https://s3.amazonaws.com/www2.datainnovation.org/2019-roadmap-for-europe-digital-economy.pdf

31. Commerce Department Digital Economy Agenda [online] Available at: https://www.nist.gov/system/files/documents/director/vcat/Davidson_VCAT-2-2016_post.pdf

32. Order of the CMU.: On the approval of the Concept for the development of the digital economy and society of Ukraine for 2018–2020 and the approval of the plan of measures for its implementation No. 67 of January 17 2018 (2018)

33. Zhekalo, G.I.: Digital economy of Ukraine: problems and development prospects. Sci. Bull. Uzhhorod Natl. Univ. (26), part 1 (2019). [online] Available at: http://www.visnyk-econom. uzhnu.uz.ua/archive/26_1_2019ua/12.pdf

34. Digital transformation (digitalization) of regions of Ukraine. Analytical note [online] Available at: http://academy.gov.ua/pages/dop/198/files/4ba4c1b4-cefe-4f27-b58b-3aee7c8cf152.pdf

35. Pisarenko, T.V., Kvasha, T.K., Paladchenko, O.F., Molchanova, I.V., Kochetkova, O.P.: Implementation of medium-term priority areas of innovative activity at the national level in 2021: analytical report. UkrINTEI, p. 95 [online] Available at: https://mon.gov.ua/storage/app/media/nauka/2022/08/08/Analit.dov.Real.seredn.pr.napr.2021-08.08.2022.pdf

Diagnostics as a Tool for Managing Behavior and Economic Activity of Retailers in the Conditions of Digital Business Transformation

Nataliia Kashchena⑩, Hanna Chmil⑩, Iryna Nesterenko⑩, Olena Lutsenko⑩, and Nadiia Kovalevska⑩

Abstract The research is devoted to the conceptualization of methodological principles and methodological tools development for diagnosing retailers' digital behavior and economic activity in the context of digital business transformation. The theoretical basis of the methodological platform for the diagnosis of the retailers' digital behavior and economic activity has been formed as the basis for the implementation of the diagnostic process, focused on the special information formation used to make management decisions. A set of management actions related to the diagnosis of digital behavior and its changes in order to increase the efficiency of economic activity and the retailers' activities in the conditions of digital transformation of business processes are determined. Methodical approaches have been developed to assess the adaptive retailers' digital behavior through the identification of the level of mastering digital technologies and readiness for digital transformation based on data on the determinants of the formation of such behavior. The methodical tools development has been defined as promising research directions.

Keywords Digitization · Digital behavior · Retail · Economic activity · Management · Diagnostics · Assessment methods

N. Kashchena (✉) · H. Chmil · I. Nesterenko · O. Lutsenko
State Biotechnological University, 44 Alchevskikh Str., Kharkiv 61002, Ukraine
e-mail: n.kashena@btu.kharkov.ua

H. Chmil
e-mail: hannachmil@btu.kharkov.ua

I. Nesterenko
e-mail: i.nesterenko@btu.kharkov.ua

O. Lutsenko
e-mail: l.a.lytsenko@btu.kharkov.ua

N. Kovalevska
V. N. Karazin Kharkiv National University, 4 Svobody Sq., Kharkiv 61022, Ukraine
e-mail: n.kovalevska@karazin.ua

© The Author(s), under exclusive license to Springer Nature Switzerland AG 2024
A. Semenov et al. (eds.), *Data-Centric Business and Applications*, Lecture Notes on Data Engineering and Communications Technologies 194,
https://doi.org/10.1007/978-3-031-53984-8_7

1 Introduction

1.1 Relevance

Digitalization of business has changed all aspects of life and created new opportunities for stimulating economic activity and development. Information communication technologies (ICT) and artificial intelligence have become drivers of socio-economic growth and the formation of a new quality of life. According to the World Bank, the introduction of digital technologies leads to the blurring of geographical and physical borders and opens new perspectives for the economic, social, and cultural development of countries, as well as for the growth of regional and global competitiveness [1].

In today's global challenges, exacerbated by geopolitical and geoeconomic shifts, digital technologies, as noted in the European Commission's message, are indispensable and necessary for work, learning, entertainment, communication, shopping, and access to everything from health services to culture [2]. Despite the apprehension of most Europeans because of the perceived threat to their economic security, the digital transformation of the economy provides all people with new sources of prosperity. It also enables entrepreneurs to innovate, build, and grow their businesses in European and global markets. Along with this, there was a need for clear and agreed rules and tools for measuring the level of mastering of digital technologies, the existing digital potential of the enterprise, the possibilities of transforming digital behavior, and, on this basis, improving the efficiency of business management in various sectors of the economy. The retail sphere is not an exception since entrepreneurial activity in the field of bringing goods from the producer to the consumer has, without a doubt, been one of the most important sectors of the economy throughout time.

In the pre-war period, the retail sector was one of the most profitable branches of the Ukrainian economy. In total, retailers earned more than UAH 430 billion [3]. According to [4], despite the loss of a third of trading points at the beginning of the war, the trade industry, along with agriculture and the processing industry, kept Ukraine's gross domestic product from falling even further in 2022. At the same time, the results of surveys of business representatives at the end of 2022 indicate pessimistic expectations regarding economic activity and prospects for sustainable development against the background of active hostilities and terrorist attacks. The sectoral index of expectations of business activity of trade enterprises in January 2023 decreased to 33.9 from 42.1 in December 2022 due to a reduction in the supply of goods [5].

Under certain circumstances, it is vital for every retailer to assess the state and prospects of their activity in the context of ensuring effective functioning and socio-ecological-economic standards of the quality of life of current and future generations. As a result, there is growing scientific and practical interest in diagnosing the digital behavior of retailers as a scientific basis for making balanced management decisions to increase their economic activity, strengthen competitive advantages, and ensure sustainable development in the future.

Interpretation, prerequisites for formation, assessment, identification of factors of change and means of correction of digital behavior, stimulation of economic activity, and development of activities of retail trade enterprises in the conditions of digital transformation of the economy are the subject of close attention of both domestic and foreign scientists.

Digital transformation, as rightly noted [6], is fundamentally changing consumer expectations and behavior, putting enormous pressure on traditional companies and disrupting numerous markets. New innovative business models are being formed, the functioning of which requires the adjustment of economic behavior adapted to digital challenges and the availability of assets and opportunities necessary for successful digital transformation and increasing the economic activity of retailers. In this context, the results of previous studies concerning:

- understanding the constant changes in consumer behavior due to digital adaptation, divergence, and noise, and the far-reaching ability of digital technologies to reconfigure and reconfigure every aspect of business and consumer behavior, taking into account the micro and macro aspects of the business environment, presented in works [7, 8];
- factors that influence consumer behavior in conditions of technological changes and accompanying social and economic external effects, which made it possible to develop a theory of consumer behavior and, based on it, a conceptual model of digital consumer behavior during digital transformation to assess the relationship between the use of digital technologies and social—the economic development of countries, in particular, their impact on the retail ecosystem and attracting people to online consumption [9, 10];
- online consumer behavior patterns, the erosion of institutional retail as the main interface for the client, the benefits and values of many digital technologies in the retail sector, and the favorable factors of digital transformation in the context of small retailers through the lens of stakeholder theory [11–14];
- theoretical basis of economic activity of business entities, applied aspects of measuring economic expectations and economic activity of enterprises, factors of stimulation, and forecasting of economic activity of retailers [15–21].

1.2　Goals and Objectives

Paying tribute to the achievements of leading modern scientists, it should be noted that in the conditions of digital transformation of business, the problem of diagnosing digital behavior and the economic activity of retailers requires further scientific developments. The priority issues, based on the specifics of their activities and the latest digital trends in retail, are a systematic approach, consistency of diagnostic procedures and methods for assessing digital capabilities, digital potential, economic activity, and further development.

This work aims to scientifically substantiate and develop methodological provisions and apply recommendations for the formation of conceptual foundations and development of a methodological toolkit for diagnosing retailers' digital behavior and economic activity in the context of digital business transformation.

The direction of scientific research involves the step-by-step implementation of a set of tasks for two interrelated objects: adaptive digital behavior and economic activity [22]. This section provides a rationale for the concept and methodological tools for assessing adaptive digital behavior. It is proposed to solve the following scientific and practical tasks:

– conceptualization of the methodological platform for diagnostics of digital behavior and economic activity of retailers;
– development of a methodical approach to assessing the level of mastering of digital technologies and adaptive digital behavior of the enterprise;
– development of a methodical approach to assessing the level of the enterprise's readiness for digital transformation and the formation of adaptive digital behavior.

The chapter ends with a summary and a perspective for further work related to the methodological toolkit for assessing the economic activity of retailers.

2   Theoretical and Methodological Basis and Methods of Diagnosis of Digital Behavior and Economic Activity of Retail Enterprises

2.1   Conceptualization of the Methodological Platform for the Diagnosis of Digital Behavior and Economic Activity of Retailers

In the digital economy, diagnosis contributes to solving the problems of retailers related to the modification of digital behavior and ensuring the desired level of their economic activity. As a component of managerial activity, diagnostics is a complex of actions aimed at providing informational support for management [23] that requires management personnel to have systematic theoretical knowledge. It is the skill that applies in the use of instrumental methods for assessing digital behavior, identifying and forecasting the level of economic activity, monitoring the implementation of measures to improve the company's competitive position, and taking into account modern global challenges and consumer preferences.

The effectiveness of diagnostics is ensured by the appropriate methodology. In modern scientific discourse, methodology (Greek methods: method and logos— science and knowledge) is most often interpreted as teaching about scientific methods of knowledge and transformation of the world; philosophical and theoretical basis, a set of research methods that are used in any field of science in accordance with the specifics of the object of its knowledge; a system of knowledge about the theory of science or a system of research methods; theory of research methods; teaching about

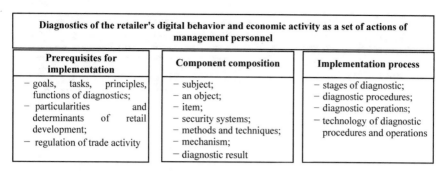

Diagnostics of the retailer's digital behavior and economic activity as a set of actions of management personnel		
Prerequisites for implementation	**Component composition**	**Implementation process**
− goals, tasks, principles, functions of diagnostics; − particularities and determinants of retail development; − regulation of trade activity	− subject; − an object; − item; − security systems; − methods and techniques; − mechanism; − diagnostic result	− stages of diagnostic; − diagnostic procedures; − diagnostic operations; − technology of diagnostic procedures and operations

Fig. 1 The system-forming basis of the methodology of diagnosing digital behavior and economic activity of the retailer

scientific methods of cognition or as a system of scientific principles of cognition, on the basis of which scientific research activity is planned, its concept is developed, and a rational and conscious choice of a set of methods, means, techniques of its study is carried out [24].

Supporting the opinion of respected scientists that «methodology is a teaching about the organization of activity» [25], we believe that organizing work on the diagnosis of digital behavior and economic activity of a retailer means organizing it into a complete system with clearly defined prerequisites, component composition and the process of its implementation (Fig. 1).

The given approach to understanding the architecture of the methodology for diagnosing digital behavior and the retailer's economic activity ensures the development of an appropriate methodological platform—the theoretical and methodological basis for the implementation of the diagnostic process, procedures, and operations. The objective basis for its justification is the development of a tree of problem analysis (Fig. 2), according to which the concept of a diagnostic process should be defined, aimed at ensuring the desired level of digital capabilities, economic activity, and sustainable development of the retailer in the face of rapid changes in the business environment and preferences consumers.

The architectural construction of the system for diagnosing the retailer's digital behavior and economic activity involves the allocation of functional and supply components. The functional component is focused on the performance of diagnostic functions (monitoring, assessment, causal analysis, diagnosis, and prognosis) and is supported by the supply component. In its structure, the supply component integrates organizational, regulatory, informational, methodical, and technical software. Additionally, it is decisive for the productive functioning of the mechanism and effectiveness of the process of diagnosing digital behavior and economic activity in the information system of managing a trade enterprise.

The process of diagnosing digital behavior and economic activity of retailers essentially involves a continuous sequence of stages, procedures, and operations for obtaining and processing information about digital capabilities, potential and level of

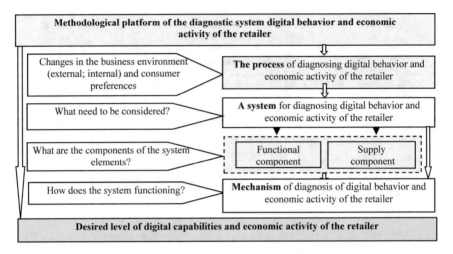

Fig. 2 Analysis tree of digital behavior diagnosis problem and economic activity of trade enterprises

activity, their change under the influence of the business environment and consumer preferences, namely:

(1) preparatory stage: determining the purpose of diagnosis and setting tasks; selection of a set of enterprises; appointment of people responsible for diagnostics, distribution of work, determination of deadlines; identification of information sources for diagnostics;

(2) information stage: use of tables, diagrams, development of expert assessment questionnaires; collection and measurement of criterion signs (indicators) of digital readiness, capabilities, and economic activity of the enterprise; processing information, systematizing it by components, determining indicators;

(3) analytical stage: selection and justification of methods of multidimensional analysis; selection of homogeneous clusters and clustering structure; determination and assessment of factors influencing digital behavior and economic activity;

(4) creative stage: construction of general indicators, determination of the standard; comparison and identification of results; development of a model of digital transformation and economic activity of the enterprise;

(5) experimental stage: analysis of forecast values and influence of formation factors; creation of a classification system for signs of identified problems; determination of the causes of identified problems and ways of their localization;

(6) recommendation-executive stage: justification and adoption of a management decision, coordination of measures for its implementation; planning and organizational support for decision implementation; implementation of management decisions.

The specified sequence of stages of the diagnostic process is regulated by the appropriate mechanism. The mechanism for diagnosing the retailer's digital behavior and economic activity «is a complex system category, which in a broad sense represents a system of interconnected and interdependent organizational, economic, technological and methodological elements (principles, approaches, norms, methods, methods, tools, indicators), which ensure the performance of targeted diagnostic functions. In a narrow sense, the diagnostic mechanism is a set of rules, special measures, and procedures for conducting a diagnostic study to identify factors of the internal and external environment that affect the activity of the enterprise, with the aim of providing recommendations on reducing negative impacts and improving (restoring) the state of the enterprise and monitoring their effectiveness» [26].

The result of the diagnosis of the retailer's digital behavior and economic activity is the formation of special information that is used to make management decisions. The creation of such information is determined by the informational needs of the diagnostic subject and is aimed at their maximum satisfaction (regardless of the volume, complexity of the structure, and movement of information flows in the management information system of the enterprise) with an emphasis on such characteristics as: relevance, probability, truthfulness, neutrality, comprehensibility, and comparability.

The combination of appropriate methods, techniques, means, and techniques in the diagnostic process, which, through the implementation of the research objectives, leads to the transformation of the original information into information for making management decisions, forms a methodology for the diagnosis of digital behavior and economic activity of the retailer. Its effectiveness is ensured by the appropriate mechanism that starts and supports the diagnostic process, thereby ensuring the integrity of the system of digital behavior and economic activity diagnostics in the retailer's management information system.

On the basis of the above and the results of previous studies [27, 28], a conceptual model of the methodological platform for the diagnosis of digital behavior and economic activity of retailers was developed (Fig. 3), based on the principles of systematicity in accordance with the defined tasks identifies the prerequisites and component composition of the methodology and through the involvement of the mechanism of implementation of diagnostic procedures ensures the effectiveness of the process of obtaining information for management.

The developed model allows you to significantly deepen and expand the understanding of the essence of the system of diagnostics of digital behavior and economic activity of a retail enterprise, its structure, regularities, goals and tasks of operation, to provide high-quality instrumental support for the implementation of diagnostic procedures and operations, to streamline the process of obtaining information for making management decisions.

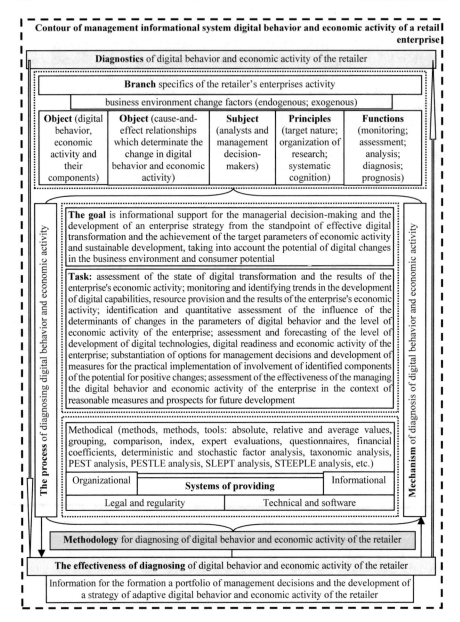

Fig. 3 A conceptual model of a methodological platform for diagnosing digital behavior and economic activity of retailers

2.2 Methodical Tools for Evaluating the Digital Behavior of Retailers

The instrumental component of the proposed conceptual model of the methodological diagnostic platform is the methodical support for evaluating the digital behavior and economic activity of retailers. Earlier [28], it was proved that economic activity is determined by the economic activity of enterprises, is a form of manifestation of their economic behavior, and is formed under the influence of the economic thinking of people who take an active part in all business processes, in particular, digitalization of management and delivery of goods from the producer to the direct consumer with involving the latest information and communication technologies.

Digitalization significantly changes traditional business processes, including management processes, and during digital transformation a new format of economic thinking of the manager is formed—digital thinking. It is focused on the introduction of innovations, modernization of the business model, digitization of assets, and wider use of the latest technologies to improve the experience of employees, customers, suppliers, partners, and stakeholders [29]. New economic thinking, strengthened by digital hard and soft skills, forms digital behavior which determines the activity of retailers in the business environment.

From the position of a holistic vision of multi-level interactions and interdependencies, the digital behavior of retail enterprises is interpreted as an external manifestation of their purposeful activity, which, through a combination of internal and external digital capabilities by means of the adaptation mechanism, reproduces the integrity of the socio-economic system and allows to act to strengthen the digital capacity and activate economic activity retailer. The degree of adaptability of personnel to digital transformations, perception and attitude to new technologies, the level of their mastery, and the range of use from different types of digital behavior (proactive, active, transitional, passive). It determines the technological capacity, the ability to quickly receive, process, and analyze information about the business environment for making strategic and operational management decisions [30, 31]. Therefore, it is extremely important for retailers to diagnose digital behavior that is adaptive to modern challenges, and to develop, based on its results, management decisions regarding further digital transformation and stimulation of economic activity.

The complex management actions related to the diagnosis of digital behavior and its changes in order to increase the efficiency of the economic activity of retailers is presented as a part of the developed reference model of economic activity management in the conditions of digital transformation of business processes (Fig. 4).

Its advantage is the definition of interrelated functional components, the interaction of which, through the mechanism involving the appropriate methodological toolkit, allows obtaining the necessary information to determine the type of adaptive behavior and digital capabilities, as a basis for the development and adoption of adaptive management decisions to increase, accelerate or ensure digital transformation in order to stimulate and increasing the retailer's economic activity.

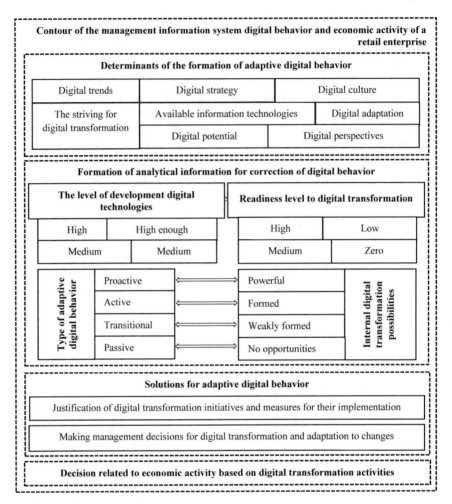

Fig. 4 A reference model for managing the economic activity of retailers in the context of digital transformation of business processes

Within the framework of the proposed model, the key component of the complex management actions of the management apparatus is the formation of analytical information for adaptation to digital changes and correction of digital behavior. It involves evaluating the adaptive digital behavior of retail enterprises through the identification of the level of mastering of digital technologies and readiness for digital transformation based on data on the determinants of the formation of such behavior.

To conclude, this allows the following: to determine the current state of digital behavior and the degree of readiness for further digital transformation; identify priority areas and digital opportunities of new digital development; and identify limitations, barriers, and risks that inhibit the acceleration of digital transformation at the enterprise.

2.3 Methodical Approach to Assessing the Level of Development of Digital Technologies and Adaptive Digital Enterprise Management

In order to determine the level of mastering of digital technologies, a methodical approach has been developed which, through prioritization and actual use of digital technologies, allows to provide a quantitative and qualitative characteristic of the type of adaptive digital behavior (proactive, active, transitional, or passive). This approach involves the use of expert survey tools, ranking analysis, and weighted assessments. It is implemented in the following sequence:

1. Formation of a package of digital technologies used in the field of retail trade

   It was established that in the process of digital business transformation, retailers use: artificial intelligence and cognitive technologies, BigData, mobile applications, electronic logistics, electronic trade, additive technologies and 3D printing, Internet of Things (IoT), cloud computing, drones, social media and platforms, chatbots and intelligent assistants, website, blockchain, robotics, virtual/augmented reality;

2. Expert survey and verification of consistency of experts' opinions

   Presupposes: the choice of a linguistic rating scale; carrying out a survey of experts; processing survey results by building the appropriate matrices (results, linked ranks, etc.) and checking the correctness of their compilation; verification of the consistency of experts' opinions regarding the importance of types of digital technologies for retail trade enterprises.

   In order to establish the significance of digital technologies for retail, experts (managers and/or employees of retail enterprises, information and communication technologies, scientists, etc.) are suggested to use a 5-point rating scale with the following interpretation: 1 point—very low, 2 points—low, 3 points—average, 4 points—high, 5 points—very high. After the survey, a matrix of the results of an expert assessment of the significance of types of digital technologies for retail trade enterprises is formed. Next, experts' opinions are normalized, a matrix of connected ranks is built, and its correctness is verified.

   The following formula is used to normalize point estimates:

$$x_{ij} = l + \frac{t+1}{2}, \tag{1}$$

where l is the quantity of objects more important than the group; t is the number of objects in the group (link length).

The correctness of the matrix is checked based on the determination of the checksum according to the formula:

$$\sum_{j=1}^{n} x_{ij} = \frac{n \times (n+1)}{2},$$
(2)

where n is the quantity of objects (digital technologies).

To find out the consistency of experts' opinions, the concordance coefficient is calculated considering the connected ranks according to the formula:

$$W = \frac{12 \sum_{j=1}^{n} d_i^2}{m^2 (n^3 - n) - m \sum_{i=1}^{m} T_i'}$$
(3)

where m is the number of experts; n is the number of alternatives; d_i^2 is the square of the deviation of the sums of the rank estimates of experts for the ith digital technology (x_i) from the arithmetic mean of the sums of ranks for all digital technologies $(\overline{x} = \frac{\sum x_i}{n})$; T_i is an indicator of connected ranks.

$$T_i = \sum_{q_i=1}^{Q_i} \left(t_{q_i}^3 - t_{q_i} \right),$$
(4)

where Q_i is the number of groups of the same ranks; t_{q_i} is the number of ranks in each group.

The maximum value of the concordance coefficient is 1 (all opinions of experts are maximally agreed), and the minimum value is 0. The threshold value is 0.6: the value of the concordance coefficient above this value means that the experts are quite unanimous in their assessments, and their opinions can be used to choose the best alternatives If the concordance coefficient is less than 0.6, the opinions of experts differ significantly and require a second round of the survey.

The statistical significance of the concordance coefficient is checked based on the Pearson test, which is calculated according to the formula:

$$\chi^2 = mW(n-1).$$
(5)

The calculated value of criterion χ^2 is compared with the critical value obtained from the table of critical points of distribution χ^2 at the specified level of significance $\alpha = 0,05$ and the number of degrees of freedom $k = n - 1$. If the calculated value χ^2 is greater than the tabular value $\chi^2 > \chi^2 kp$, then the

W—value is not random and really characterizes the presence of a fairly high degree of agreement of experts' opinions.

3. Establishing the significance of digital technologies for digitization and further digital transformation of retail trade enterprises

   It provides for the calculation of the weighting factors of digital technologies, their ranking, and grouping according to the priority of development at retail trade enterprises. The weighting coefficients of digital technologies are calculated by determining the ratio between the sum of the rank estimates for the *jth* digital technology (x_i) and the arithmetic mean of the sum of the ranks for all digital technologies ($\overline{x}$). The highest rank (first) is assigned to the digital technology with the largest value of the weighting factor. Based on the obtained results, digital technologies are divided into primary, secondary and promising.

4. Determining the level of development of digital technologies by a retail enterprise

   It involves modeling the level of mastering of digital technologies (R_{DT}) by retailers and its calculation based on the obtained economic-mathematical model. The model is built considering the determined significance of digital technologies based on the formula:

$$R_{DT} = \frac{\sum_{i=1}^{n} DT_i}{5},$$ (6)

where DT_i is a weighted assessment of the level of development of the *i*th digital technology.

$$DT_i = B_{dt_i} \times \omega_{dt_i},$$ (7)

where B_{dt_i} is a point estimate of the level of mastering the *i*th digital technology; ω_{dt_i} is an assessment of the significance of the *i*th digital technology for the retail trade enterprise.

5. Identification of the type of adaptive digital behavior of the retailer

   It provides for the determination of the type of adaptive digital behavior of retail enterprises depending on the level of their mastery of digital technologies using the following identification scale: $0.75 < R_{DT} < 1$—proactive; $0.5 < R_{DT} < 0.75$—active; $0.25 < R_{DT} < 0.5$—transitional; $0.01 < R_{DT} < 0.25$—passive.

6. Development of solutions regarding the priorities of the formation of adaptive digital behavior in order to accelerate the processes of digital transformation and increase the economic activity of retailers

It involves the development of alternative options for management solutions for further digital transformation and the selection of the most optimal of them depending on the level of mastering of digital technologies and the established type of adaptive digital behavior of retail enterprises.

Based on the proposed methodical approach, with the aim of determining the importance of digital technology for retail enterprises in the FMCG sector, an expert survey was conducted, in which 15 experts participated, 10 of whom are representatives of the teaching staff of leading educational institutions specializing in research on the development of retail trade; 5 people are specialists in the field of digital marketing. The survey was conducted by the questionnaire method. Based on the results of the expert survey, a matrix of results and a matrix of normalized ranks were compiled, the correctness of the matrix was checked, the concordance coefficient was calculated considering the connected ranks, and its significance was determined based on the Pearson consistency criterion, the priority of the adoption of digital technologies by retailers was substantiated. The generalized results of the calculations carried out within the framework of the assessment of the importance of digital technologies for retail enterprises in the FMCG sector are presented in Table 1.

Based on the obtained coefficients of the significance of types of digital technologies for retail trade enterprises (concordance coefficient $W = 0.65$, which indicates a fairly high degree of consistency of experts' opinions), they were ranked and divided into primary, secondary, and promising in the future. According to the obtained results, such digital tools as websites, e-commerce, e-logistics, artificial intelligence, and social media turned out to be of primary importance for retail enterprises of the FMCG sector at this stage; secondary—chatbots, BigData, cloud computing, Internet of Things (IoT), mobile application; promising—virtual/augmented reality, drones, blockchain, robotics, adaptive technologies and 3D printing.

According to the results of the expert survey, a mathematical model of the level of development of digital technologies by retail enterprises in the FMCG sector was built:

$$P_{DT} = \left(\begin{aligned} &0,089 \times X1 + 0,080 \times X_2 + 0,061 \times X_3 + 0,091 \times X_4 + 0,093 \times X_5 + \\ &0,024 \times X_6 + 0,064 \times X_7 + 0,074 \times X_8 + 0,02 \times X_9 + 0,008 \times X_{10} + \\ &0,086 \times X_{11} + 0,102 \times X_{12} + 0,052 \times X_{13} + 0,036 \times X_{14} + 0,040 \times X_{15} \end{aligned} \right) \Big/ 5 \qquad (8)$$

where P_{DT} is the level of mastering digital technologies; X_1 is the artificial intelligence and cognitive technologies; X_2 is the BigData; X_3 is the mobile application; X_4 is the electronic logistics; X_5 is the electronic trade; X_6 is the additive technologies and 3D printing; X_7 is the Internet of Things (IoT); X_8 is the cloud computing; X_9 is the drones; X_{10} is the social media and platforms; X_{11} is the chat bots and intelligent assistants; X_{12} is the site; X_{13} is the blockchain; X_{14} is the robotics; X_{15} is the virtual/augmented reality.

Approbation of the proposed model and the developed methodical approach was carried out during the study of the adaptive digital behavior of retail enterprises in the FMCG sector in Kharkiv. The studied sample included 175 retail outlets of leaders of the national market and regional chains in the FMCG sector of Kharkiv: «Silpo», «ATB», «Chudo Market», «Rost», Klass», «Posad», «Tavria V», «Vostorg», «SPAR», «WelMart», which made up about 3% of the general population (which

Table 1 Generalized parameters for assessing the importance of digital technologies for the sector's retail trade enterprises Experts FMCG

Types of digital technologies and their conventional designation		Experts															x_i	d_i	d_i^2	ω_{dt_i}	Rank
		1	2	3	4	5	6	7	8	9	10	11	12	13	14	15					
Artificial intelligence and cognitive technologies	X_1	9	11	11.5	9.5	11.5	11,5	10.5	11	7.5	11	12.5	14.5	5	11	14	161	41	1681	0.089	4
BigData	X_2	9	11	9	9.5	6.5	6,5	10.5	11	13.5	11	12.5	9	5	11	9	144	24	576	0.080	7
Mobile application	X_3	9	3.5	5	7	6.5	6,5	4	11	7.5	3,5	12.5	9	5	11	9	110	−10	100	0.061	10
Electronic logistics	X_4	9	11	14	13	11.5	11,5	10.5	11	7.5	11	12.5	9	11.5	11	9	163	43	1849	0.091	3
E-commerce	X_5	9	11	14	13	11.5	11,5	10.5	11	7.5	11	12.5	9	11.5	11	14	168	48	2304	0.093	2
Additive technologies and 3D printing	X_6	9	3.5	1,5	2	4	4	1	3	2	3,5	2	3	1.5	2.5	1	43.5	−76.5	5852.25	0.024	14
Internet of things (IoT)	X_7	9	3.5	5	4.5	11.5	11,5	10.5	5,5	2	11	6	9	11.5	5.5	9	115	−5	25	0.064	9
Cloud computing	X_8	9	11	5	4.5	11.5	11,5	10.5	11	7.5	11	6	9	11.5	5.5	9	133.5	13.5	182.25	0.074	8
Drones	X_9	1	3.5	1,5	2	4	4	4	1	2	3,5	1	1	1.5	2.5	3.5	36	−84	7056	0.020	15
Social media and platforms	X_{10}	9	11	9	13	11.5	11,5	10.5	11	13.5	11	6	9	11.5	11	9	157.5	37.5	1406.25	0.088	5

(continued)

Table 1 (continued)

Types of digital technologies and their conventional designation		Experts															x_i	d_i	d_i^2	ω_{dt_i}	Rank
		1	2	3	4	5	6	7	8	9	10	11	12	13	14	15					
Chat-bots and intelligent assistants	X_{11}	9	11	11,5	7	11.5	11,5	10.5	11	13.5	11	6	9	11.5	11	9	154	34	1156	0.086	6
Site	X_{12}	9	11	14	13	11.5	11,5	15	11	13.5	11	12.5	14.5	11.5	11	14	184	64	4096	0.102	1
Blockchain	X_{13}	2	11	9	2	4	4	4	5,5	7.5	3,5	6	9	11.5	11	3.5	93.5	-26.5	702.25	0.052	11
Robotization	X_{14}	9	3.5	5	7	1.5	1,5	4	3	7.5	3,5	6	3	5	2.5	3.5	65.5	-54.5	2970.25	0,036	13
Virtual/augmented reality	X_{15}	9	3.5	5	13	1.5	1,5	4	3	7.5	3,5	6	3	5	2.5	3.5	71.5	-48.5	2352.25	0.040	12
$\sum$		120	120	120	120	120	120	120	120	120	120	120	120	120	120	120	1800		32,308.5	1	

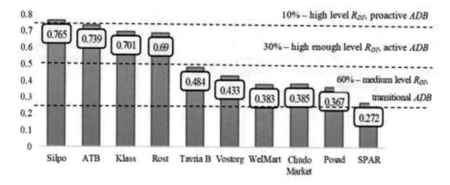

Fig. 5 The results of the evaluation of the level of development of digital technologies in supermarket chains in Kharkiv

includes almost 6,900 objects) and is sufficient to ensure the representativeness of the results. The results of the assessment of the level of development of digital technologies in supermarket chains in Kharkiv and the type of adaptive digital behavior (*ADB*) are presented in Fig. 5.

As shown above, adaptive digital behavior can be characterized as proactive in only 10% of the studied enterprises, in 30% as active, and in the majority as transitional, for which the syndrome of "outdated technologies" remains characteristic. Significant gaps were identified between the sufficiently low level of mastering of digital technologies at the retail enterprises of the sample population and the available digital capabilities for primary and secondary digital technologies, which indicate a lag in the processes of digital transformation in the field of retail trade in the FMCG sector.

In order to accelerate the digital transformation of retail enterprises, a package of measures has been developed, which consists of directing efforts to develop existing digital capabilities and search for new/create breakthrough digital technologies. Their implementation requires the formation of appropriate internal digital capabilities and determines the need to develop a methodical approach to assess the level of readiness for digital transformation (Table 2).

Implementation of the recommended measures requires a preliminary assessment of internal digital capabilities (digital maturity) and prioritization of digital transformation using all available tools, such as: budgetary, low-budget, and non-budgetary, as well as initiating the necessary changes for further mastering digital technologies and adjusting adaptive digital behavior to ensure economic activity of the retailer. The indicator of the level of readiness for digital transformation serves as a methodological tool for assessing digital maturity.

Table 2 Characteristics of the enterprises of the selective universe by the level of mastering of digital technologies and solutions for adaptive digital behavior

Supermarket chains in Kharkov	Indicator of the level of development of digital technologies	Characteristics of the level of development of digital technologies	A type of adaptive digital behavior	Recommended management decision
Silpo	0.765	High	Proactive	Search for breakthrough digital opportunities, proactivity in decision-making regarding the implementation of innovative digital technologies
ATB	0.739	Sufficiently high	Active	Focusing efforts on the development of new external digital opportunities
Klass	0.701			
Rost	0.69			
Tavria B	0.484	Average	Transitional	Increasing efforts to develop unused external digital opportunities
Vostorg	0.433			
Walmart	0.383			
Chudo market	0.385			
Posad	0.367			
SPAR	0.272			

2.4 Methodical Approach to Assessing the Level of Readiness of the Enterprise for Digital Transformation and the Formation of Adaptive Digital Behavior

The relevance and importance of the consequences of the processes of digital transformation and their priority in the strategy of economic activity and development have led to significant scientific interest in researching the problems of assessing the digital readiness of enterprises for transformation. A comparative analysis of available methodical approaches proves that the methodical toolkit developed so far is mostly aimed at characterizing the current state of digitalization at the enterprise and determining priorities for ensuring its adaptability and/or improvement in accordance with the changes caused by the consequences and impact of the digital revolution.

In order to further develop the existing methodological toolkit, based on the results of the content analysis of the methods of assessing the digital maturity of the enterprise, four key parameters and their corresponding attributes of the internal digital capabilities of retailers regarding the introduction/mastery of digital technologies have been determined. A methodical approach has been developed to assess the enterprise's desire for digital transformation, which is based on the application of the index method and the method of expert assessments and involves the calculation of the integral index of digital readiness according to the geometric mean formula based on the partial indices of the determined key parameters and attributes of internal digital capabilities of retailers (Table 3). Its advantages are the possibility of quantitative and qualitative characterization of the enterprise's readiness for digital transformation, determination of available internal digital capabilities, and adoption of adaptive management decisions regarding directions of digital development.

The calculation of the partial and integral index of the digital readiness of retailers for transformation is carried out on the basis of the following formulas:

$$I_{LA} = \frac{\sum_{i=1}^{n} B_{LA_i}}{5},$$ (9)

where I_{LA} is the index of aspirations of the governing office; B_{LA_i} is the grade in points ith attribute of the parameter of the digital vision of managers; n is the quantity of parameter attributes LA.

$$I_{DP} = \frac{\sum_{i=1}^{n} B_{DP_i}}{5},$$ (10)

where I_{DP} is the index of digital potential; B_{DP_i} is the grade in points ith attribute of digital potential parameters; n is the quantity of parameter attributes DP.

$$I_{DC} = \frac{\sum_{i=1}^{n} B_{DC_i}}{5},$$ (11)

where I_{DC} is the index of digital culture; B_{DC_i} is the grade in points ith attribute of the parameter of digital values at the enterprise; n is the quantity of parameter attributes DC.

$$I_{DS} = \frac{\sum_{i=1}^{n} B_{DS_i}}{5},$$ (12)

where I_{DS} is the digital strategy index; B_{DS_i} is the grade in points ith attribute of the desire parameter of the steering apparatus; n is the quantity of parameter attributes DS.

$$I_{DR} = \sqrt[4]{I_{LA} + I_{DP} + I_{DC} + I_{DS}},$$ (13)

Table 3 Parameters and attributes of the enterprise's digital readiness for digital transformation

Partial indices	Parameters	Attributes	Conventional designation
Aspiration of the steering apparatus (LA)	The digital vision of managers	Digital thinking and initiatives of the management apparatus	LA_1
		Satisfaction with the level of development of digital technologies at the enterprise	LA_2
		Recognition of the priority/perspective/ necessity of digitization	LA_3
Digital potential (DP)	Personnel	Availability of modern ICT specialists	DP_1
		Sufficient modern ICT specialists	DP_2
		Qualification of modern ICT specialists	DP_3
		Skills of using digital technologies of employees of other areas/ professions	DP_4
	Finances	Sufficient funding for digital transformation	DP_5
		Financial opportunities for financing digital transformation	DP_6
	Digital infrastructure	The level of material and technical support for digital transformation	DP_7
		Possibilities of digital transformation with the existing material and technical base	DP_8
	Digital awareness	Knowledge of trends and modern digital technologies	DP_9
		Conducting monitoring and analysis of digitization trends	DP_{10}
Digital culture (DC)	Digital values in the enterprise	Training and promotion of digital transformation at the enterprise	DC_1
		Digital thinking of employees	DC_2

(continued)

Table 3 (continued)

Partial indices	Parameters	Attributes	Conventional designation
		Involvement of employees in digital transformation at the enterprise	DC_3
		Motivating and encouraging the adoption/ use of digital technologies by employees	DC_4
Digital strategy (DS)	Digital Leadership (place) on the market	Availability of a digital transformation plan	DS_1
		Place in the market by the level of digital transformation	DS_2
		Adequacy of digital transformation processes to modern trends/ tendencies	DS_3

where I_{DR} is the integral index of the digital readiness of the enterprise for digital transformation.

Based on the values of the integral index, a qualitative characteristic of the enterprise's readiness for digital transformation is provided. The following identification scale is used: $0.75 < R_{DT} < 1.0$—high; $0.5 < R_{DT} < 0.75$—average; $0.25 < R_{DT} < 0.5$—low; $0.01 < R_{DT} < 0.25$—zero.

Using of the developed methodical approach, the level of readiness for digital transformation (*LRDT*) of the enterprises of the sample population of the FMCG sector of Kharkiv was assessed (Fig. 6).

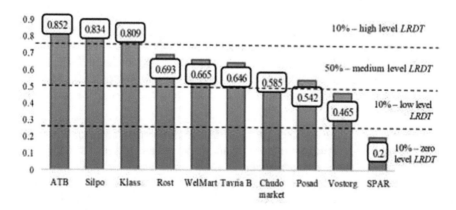

Fig. 6 The results of assessing the level of readiness for digital transformation in supermarket chains in Kharkiv

It was established that, on average, the sample population of enterprises has an average ($R_{DT} = 0.632$) level of digital readiness for transformational changes. This means that the digital capabilities for adaptive digital behavior are formed, but further efforts are needed to bring them to a level adequate to the needs for further digital transformation.

Among the investigated supermarket chains, the group with a high level of digital readiness included three chains—«ATB», «Silpo», and «Klass», which made up 30% of the sample population. According to the values of the integral index, the level of readiness for digital transformation is recognized as high. A high level of digital readiness contributes to the acceleration of digital transformation processes, an increase in the level of mastering of digital technologies, the introduction of innovative digital technologies and, overall, the strengthening of digital leadership in the regional market.

The internal digital capabilities of these networks for adaptive digital behavior in the context of digital transformation can be characterized as powerful. Enterprises of these networks have the highest partial index of digital potential, which exceeds 0.9. They are provided at the appropriate level with IT specialists in the required number, with the necessary qualifications, have the appropriate material and technical base, mostly sufficient financial capabilities and the necessary funding for the implementation and mastering of digital technologies and ensuring adaptive digital behavior. The presence of a digital strategy significantly strengthens the digital readiness of the «ATB», «Silpo», «Klass» supermarket chains. However, assessments of the level of mastering of digital technologies testify to problems regarding its implementation, as a result of which the need to develop and implement a new strategy of digital adaptive behavior, taking into account the available internal digital and external digital capabilities, is actualized. In general, for further digital transformation with a focus on innovative digital technologies, it is necessary to strengthen the digital potential with the necessary resources in all components, especially financial and human.

3 Conclusion

The study of the theoretical and methodological basis and methods of diagnosing digital behavior in managing the economic activity of retail enterprises in the conditions of the digital transformation of the economy allows us to draw the following conclusions.

1. Diagnostics play a significant role in the management information system of retailers. It allows you to carry out a detailed "survey" and substantiate the options for achieving the desired parameters of digital behavior and economic activity of the enterprise based on the available financial, resource, and digital capabilities and their changes under the influence of factors of the external and internal business environment, as well as monitor and evaluate the effectiveness of the management decisions made.

2. Diagnostic methodology combines relevant methods and techniques and determines the methods of obtaining information to achieve previously formulated research goals. The proposed conceptual model of the methodological platform for diagnostics of digital behavior and economic activity of trade enterprises integrates the purpose and tasks of diagnostics, which reflect the object and subject of research, with the principles and functions that act as its basis, determine the format of supply systems and the sequence of stages of the diagnostic process, and optimizes the choice methodical toolkit of diagnostics, which as a whole provides high-quality information support for the development, adoption, and implementation of management decisions, aimed at achieving the target parameters of adaptive economic behavior and economic activity, taking into account dynamic changes in the business environment and consumer preferences.

3. A set of management actions related to the diagnosis of adaptive digital behavior and its adjustment to increase retailers' efficiency and economic activity aimed at assessing their ability and opportunities to perceive digital transformation. Methodical approaches for evaluating the level of mastering of digital technologies and the level of enterprise digital readiness for digital transformation are proposed. The advantages of the proposed methodological approaches are the possibility of quantitative and qualitative characterization of the type of adaptive digital behavior and digital maturity of the enterprise to the perception of digital transformations, identification of priority areas of digital development, and justification of measures to accelerate digital transformation on this basis.

Suggestive directions for further scientific research are the development of methodological tools for evaluating the economic activity of retail enterprises in the conditions of digital transformation and the development of adaptive strategies for the retailers' digital activity.

References

1. The EAEU 2025 DIgital Agenda: Prospects and Recommendations. Overview Report. https://documents1.worldbank.org/curated/en/850581522435806724/pdf/EAEU-Overview-Full-ENG-Final.pdf. Accessed 31 Jan 2023
2. 2030 Digital Compass: the European way for the Digital Decade. Communication from the commission to the European parliament, the council, the European economic and social committee and the committee of the regions https://eur--lex-europa-eu.translate.goog/legal-content/EN/TXT/HTML/?uri=CELEX:52021DC0118&rid=4&_x_tr_sl=en&_x_tr_tl=uk&_x_tr_hl=uk&_x_tr_pto=sc. Accessed 31 Jan 2023
3. Retail lost a third of its outlets. How one of the most profitable industries works during wartime. https://forbes.ua/news/riteyl-vtrativ-tretinu-svoikh-torgovikh-tochok-yak-pratsyue-odna-z-naypributkovishikh-galuzey-v-umovakh-viyni-25032022-5017
4. Ukraine 2022–2024. Macroeconomic research. Capital Times. https://www.capital-times.com/dolya-it-u-vvp-ukraini-2030/. Accessed 31 Jan 2023
5. Business expectations of enterprises, IV quarter of 2022 https://bank.gov.ua/ua/news/all/dilovi-ochikuvannya-pidpriyemstv-iv-kvartal-2022-roku. Accessed 31 Jan 2023

6. Verhoef, P., Broekhuizen, T., Bart, Y., Bhattacharya, A., Dong, J., Fabian, N., Haenlein, M.: Digital transformation: a multidisciplinary reflection and research agenda. J. Bus. Res. **122**, 889–901 (2021)
7. Rajagopal.: Consumer dynamics. In: Agile Marketing Strategies, pp. 25–53. Palgrave Macmillan, Cham (2022). https://doi.org/10.1007/978-3-031-04212-6_2
8. George, B., Paul, J.: Digital Transformation in Business and Society. Springer International Publishing, New York, NY (2020). ISBN 978-3-030-08276-5 ISBN 978-3-030-08277-2 (eBook). https://doi.org/10.1007/978-3-030-08277-2
9. Yegina, N., Zemskova, E., Anikina, N., Gorin, V.: Model of consumer behavior during the digital transformation of the economy. Ind. Eng. Manag. Syst. **19**(3), 576–588 (2020)
10. Jiang, Y., Stylos, N.: Triggers of consumers' enhanced digital engagement and the role of digital technologies in transforming the retail ecosystem during COVID-19 pandemic. Technol. Forecast. Soc. Chang. **172**, 121029 (2021)
11. Richard, M., Chebat, J.: Modeling online consumer behavior: preeminence of emotions and moderating influences of need for cognition and optimal stimulation level. J. Bus. Res. **69**(2), 541–553 (2016)
12. Ferreira, M., Moreira, F., Pereira, C., Durão, N.: The digital transformation at organizations–the case of retail sector. In: Trends and Innovations in Information Systems and Technologies, vol. 18, pp. 560–567. Springer International Publishing (2020)
13. Ziaie, A., ShamiZanjani, M., Manian, A.: Systematic review of digital value propositions in the retail sector: new approach for digital experience study. Electron. Commer. Res. Appl. **47**, 101053 (2021)
14. Candelo, E., Casalegno, C., Civera, C.: Digital transformation or analogic relationships? A dilemma for small retailer entrepreneurs and its resolution. J. Strateg. Manag. **15**(3), 397–415 (2022)
15. Kulinyak, I.: Theoretical aspects of the interpretation of the concept of «economic activity of the enterprise». Bull. Natl. Univ. Technol. Des. Kiev. Econ. Sci. **4**, 69–79 (2017)
16. Turilo, A., Vcherashnia, I.: Theoretical and methodological approaches to the definition of the concept of «economic activity of the enterprise». Financ. Ukraine **10**, 79–84 (2011)
17. Kilian, L.: Measuring global real economic activity: do recent critiques hold up to scrutiny? Econ. Lett. **178**, 106–110 (2019)
18. Claveria, O., Monte, E., Torra, S.: A new approach for the quantification of qualitative measures of economic expectations. Qual. Quant. **51**(6), 2685–2706 (2017)
19. Caballero, J., Fernández, A., Park, J.: On corporate borrowing, credit spreads and economic activity in emerging economies: an empirical investigation. J. Int. Econ. **118**, 160–178 (2019)
20. Maio, P., Philip, D.: Economic activity and momentum profits: further evidence. J. Bank. Financ. **88**, 466–482 (2018)
21. Krutova, A., Kashchena, N., Chmil, H.: Enterprises' economic activity stimulation as a driver of national economy sustainable development. Econ. Strategy Perspect. Dev. Trade Serv. Sphere **1**(31), 162–173 (2020)
22. Engelhardt-Nowitzki, C., Kryvinska, N., Strauss, C.: Strategic demands on information services in uncertain businesses: a layer-based framework from a value network perspective. In: 2011 International Conference on Emerging Intelligent Data and Web Technologies (2011). https://doi.org/10.1109/EIDWT.2011.28
23. Kashchena, N.: Scientific and applied platform of trade enterprises economic activity digital management transformation. In: Sustainable Development: Modern Theories and Best Practices: Materials of the Monthly International Scientific and Practical Conference/Gen. Edit. Olha Prokopenko, pp. 17–18. Teadmus OU, Tallinn (2021)
24. Kalinina, L.: Scientifi c discourse of modern methodologies of organizational mechanism of management in the education area. Native Sch. **1**, 8–17 (2017)
25. Kothari, C.: Research Methodology: Methods and Techniques. New Age International, New Delhi, India (2004). https://books-google-com-ph.translate.goog/books?id=hZ9wSHysQDYC&printsec=frontcover&_x_tr_sl=en&_x_tr_tl=uk&_x_tr_hl=uk&_x_tr_pto=sc#v=onepage&q&f=false. Accessed 31 Jan 2023

26. Kryvoviaziuk, I., Kost, Y.: Diagnosis of financial and economic activity of an industrial enterprise: monograph, p. 200. LNTU, Donetsk-Lutsk, Ukraine (2012)
27. Chmil, H.: Adaptive behavior of consumer market subjects in conditions of digital transformation of the economy: theory, methodology and practice: monograph, p. 377. I.S. Ivanchenko Publisher, Kharkiv (2021)
28. Kashchena, N.: Accounting and analytical management of the economic activity of trade enterprises: theory, methodology, practice: monograph, p. 389. I.S. Ivanchenko Publishing House, Kharkiv, Ukraine (2021)
29. What is Digital Transformation? Theagileelephant.com. website. http://www.theagileelephant.com/what-is-digital-transformation. Accessed 31 Jan 2023
30. Volontyr, L., Potapova, N., Ushkalenko, I., Chikov, I.: Optimization methods and models in business activity: Training manual, p. 404. Vinnitsa, Ukraine. VNAU (2020)
31. Youssef, A., Boubaker, S., Dedaj, B., Carabregu-Vokshi, M.: Digitalization of the economy and entrepreneurship intention. Technol. Forecast. Soc. Change **164** (2022). https://doi.org/10.1016/j.techfore.2020.120043

Methodical Tools for Identification and Quality Control of Design Products

Anatoly Goiko, **Lesya Sorokina**, **Ljudmila Shumak**,
Oleksandr Filippov, and **Artem Strakhov**

Abstract The article presents the results of the analysis of the literature on domestic and foreign experience in the field of design and construction. The work is devoted to topical issues of design and construction products, in particular, the issues of rationality and quality control of design products, the intensity of influence on the coefficient of rationality of design products of the material consumption factor, and other factors that are considered. Linear combinations of the outputs of the empirical model of the relationship between material consumption (x1j) and the coefficient of rationality of design products (yj) are analyzed. The problems of the activity of construction enterprises are clarified. The main mistakes that the Customer makes during the repair and construction season are analyzed and ways to eliminate such omissions are considered. A possible variant of the instrumentation is proposed, which is based on dimensionless relative values. The initial value of this type of model is a weighted average linear combination of the values of input variables or certain constants. The issue of fuzzy inference algorithms is considered. The general structure of a microcontroller using fuzzy logic is shown. An analysis of the formulas was made, which showed that the influence of material consumption and other factors on the quality and price of design products is multidirectional. A graph of the membership function of the material consumption of project products is presented (Author's development). The work done testifies to the periodic consideration of the

A. Goiko · L. Sorokina · L. Shumak (✉) · O. Filippov · A. Strakhov
Kyiv National University of Construction and Architecture, Povitroflotsky Avenue, 31,
Kyiv 03037, Ukraine
e-mail: shumak_lv@knuba.edu.ua

A. Goiko
e-mail: goiko.af@knuba.edu.ua

L. Sorokina
e-mail: sorokina.lv@knuba.edu.ua

O. Filippov
e-mail: filippov_jv@knuba.edu.ua

A. Strakhov
e-mail: strakhov_ao@knuba.edu.ua

economic component in design and construction. And these questions remain open and relevant to this day.

Keywords Methodological tools · Design products · Pricing · Membership functions · Linguistic term · Fuzzy logic · Function parameters · Material consumption of design products

1 Introduction

1.1 Problem Statement

The customer defines the main requirements for the project results. It provides financing of the project at the expense of its own or borrowed funds. And can also enter into contracts with the main executors of the project. The customer is the future owner of the quality results of the project. And he should not overpay for a project where there may be mistakes. No Contractor shall work on a project to the detriment of its project enterprise. The contractor can stop his work if the declared cost is enough for only half of the project. Then, the fate of the project will become uncertain. Due to the high-quality organization of project processes, the project enterprise can offer a good price, using standards and methods of effective work. However, the cost of the project can be reduced either with the help of the quality of the project work or with the help of the volume of work. As they say: measure seven times, cut once! This directly concerns pre-project and project work with the Customer, who is interested in the implementation of the project and the achievement of its goals [1–3].

Taking into account the wishes of the Customer, the calculation of the approximate price of project works is formed, which includes the collection of initial documentation, consulting, and the formation of a technical task. The customer wants the project to be properly evaluated, as usually the project is voluminous. It takes 1–2 weeks of work to specify all the requirements, write out the tasks, and give them to an expert for evaluation. However, at the same time, it is almost impossible to avoid economic conflicts associated with the unlimited desires of the Customers to minimize financial and time costs for the development and implementation of the project. The customer is in a hurry and will not wait. And then, there is only one way out: set an approximate price, that is, a price "from the ceiling". As a result, the Customer not only demands a discount for the duration of the work but also demands that the price of the project be fixed in the contract (fixed price). The contractor agrees to all conditions because he wants to get the Customer. The contractor starts the work, and it becomes clear:—the requirements meant something else;—technical nuances of solutions have arisen;—additional circumstances emerged;—you still need to refine something otherwise, you won't get what you want. This whole process can be called the "approximation trap". Trusting the specialists, the Customer receives a detailed project estimate calculation, but with errors. In order not to overpay project companies for a project in which there may be errors, in the process of developing concepts and gathering information, it is important that the Customer expresses his requirements and is present in the discussion of the most important points. One of the most important factors of successful cooperation between the Customer and the Designer is the issue of trust in the partner (both external and internal). However, complete confidence in the designer's professional competence, honesty, and responsibility does not exempt the Customer from the need to carefully check the correctness of the design and estimate documentation. To increase the efficiency of this stage of the investment and construction process, scientists offer a lot of interesting things. development [4–7]. However, there are still no effective methods for substantiating the contractual cost of design works that would take into account the quality and timeliness of completed design tasks; that is, a significant part of the problem of pricing design products remains undefined.

1.2 The Purpose of the Article

Development of a methodological toolkit for checking the quality of project documentation for the purpose of matching the Designer's remuneration with the results of his work.

2 The Main Results of the Study

Due to the variety of construction products, as well as design solutions, which is further complicated by the time gap between the creations of similar options, at the stage of pricing of design products, the problem of the ratio of value indicators arises. Its consequence is the impossibility of selecting similar projects with the characteristics that are given in the Guidelines for determining the cost of design, scientific design, prospecting works, and examination of project documentation for construction [8]. At the same time, we believe that the list of parameters of the similarity of design documentation in the above-mentioned Guidelines for Design [8] is not exhaustive, since it covers only the technical characteristics of the objects under construction, but in no way takes into account the economic efficiency of their construction processes. However, omissions, mistakes, or even small miscalculations in projects lead to considerable cost overruns during the construction of buildings of varying complexity. First of all, the need for resources for additional work or correction of already completed work is increasing, which is necessarily accompanied by additional costs for the organization, management, and safety of construction production. This exacerbates the need for an effective scientifically based methodological toolkit for identification and quality control of project products, the results of which will have a decisive impact on the formation of the price and cost of project works.

We offer a possible version of such a toolkit, which is based on dimensionless relative values. In this way, it becomes possible to avoid errors in calculations due to devaluation over time of monetary indicators of construction, the causes of which can be inflationary processes, fluctuations in the national currency exchange rate, and finally, macroeconomic and geopolitical factors. At the same time, the provisions of the Guidelines on determining the cost of construction [9], which will enter into force on January 1, 2023, are taken into account.

The choice of the above indicator coefficients is due to the following considerations: firstly, construction is a material-intensive branch of material production. In addition, both the cost of construction and the future value of real estate in the local primary and secondary markets mainly depend on the quality and cost of the building materials used. As a result, the high cost of material resources used in construction causes higher requirements for the organization of their production consumption, namely methods of delivery and storage control over use, and quality of lying in the process of construction of structural elements of buildings. Secondly, a lot of construction work is still performed manually; instead, the operation of construction machines and mechanisms due to the permanent rapid increase in the cost of energy carriers and components becomes a source of additional increase in the cost of construction. Thirdly, the organization of material and technical support of construction processes is also accompanied by costs, which are called general production costs (hereinafter, general production costs). Moreover, ZVV is related not only to the cost of personnel work but also to the cost of other material and technical resources services of third-party organizations. Fourthly, all errors in project documentation, which, in fact, reduce the quality of project products, are immediately detected due to an increase in the cost of material and technical resources, taking into account the practical experience of managing construction production personnel. Instead, the funds for the elimination of errors in the projects in terms of construction organization are accumulated in the ZVV, in particular in their 3rd block, regarding which the following should be noted. The set of costs for the organization of construction production includes such economic elements as "Material costs", "Depreciation", and "Other operating costs", which, as specified in the Guidelines on pricing in construction [9], are taken into account in the 3rd block of the ZVV. This amount can be established according to the data of the sections of the local estimates, since their summaries indicate not only direct costs but also the entire amount of the ZVV, excluding the costs of wages for the general production staff. The latter, according to [9], is the 1st block of social security contributions; the 2nd block is a deduction for social activities, which is calculated as a percentage of the 1st block, usually at the level of the Single Social Contribution rate, disregarding additional costs for payment of temporary disability. Then, the 3rd block is the difference between the total amount of the ZVV and the size of the 1st and 2nd blocks.

Thus, the quality of the project products can be established on the basis of the amount of additional management costs (3rd block of the ZVV) in comparison with the cost of the necessary material resources.

To build the model, the data of the design and estimate documentation of 23 objects of various purposes, industrial, commercial, and administrative real estate, which were developed for different regions of Ukraine after 2014, were used. Structural elements, in which the material capacity exceeded 80%, deserved special attention in these estimates. After all, these are high-cost sections of the estimate, which require significantly greater responsibility of the executors not only during construction but also at all stages of pre-project work, research work, and design.

The output value of a model of this type is a weighted average of linear combinations of the values of input variables or certain constants.

The linear models contained only one variable factor—the indicator of the material intensity of the project products. Actually, it is the input of the algorithm of fuzzy logical conclusion. According to the theory of fuzzy logical inference, the weighting factors by which linear combinations are multiplied reflect the extent to which each specific indicator of material intensity belongs to a certain fuzzy linguistic term. A fuzzy linguistic term is a concept that can be expressed in ordinary human language. So, there are 5 terms in the proposed algorithm:

- $x11 =$ "acceptable material capacity";
- $x12 =$ "average material capacity";
- $x13 =$ "high material intensity";
- $x14 =$ "very high material intensity";
- $x15 =$ "critical material capacity".

The acceptance rate shows the level of confidence in fractions of a unit about how well the actual value of a variable, such as material density, corresponds to an acceptable, average, high, or some other level. This measure of confidence is calculated using special membership functions, which are linear or non-linear, bounded on one or both sides. It is worth noting that successful scientific and applied developments, carried out on the basis of algorithms of fuzzy logical inference, find more and more applications when solving the problems of management of construction economics; this is evidenced by publications [10–12]. In this regard, the works [11, 12], where the parameters of fuzzy logical inference algorithms are determined using hybrid neuro-fuzzy models, deserve special attention. For this purpose, scientists use the Anfisedit editor of the MATLAB software environment. This method of forming the empirical model was used in this study. That is, on the basis of the training sample of data, which contained both indicator indicators, both the material intensity and the rationality coefficient of the design products, in the MATLAV software environment, the parameters of the membership functions of the 5 terms of the input variable were calculated in the MATLAV software environment through 300 learning epochs— the value of the slope and the free constant for the 5 terms of the output variable. The training ended when the sum of the squares of the algorithm's error reached its minimum value. The marginal error of the algorithm was 8.18 ± 2.76 basis points, i.e., in 95 cases out of 100, the application of this algorithm would have resulted in a value of the coefficient of rationality of the design product, which would not have coincided with the actual value, and the deviation would have been in the range from 5.42 to 10, 94 basis points. Of course, the model is slightly biased, it over-estimates the actual result, however, and traditional linear and non-linear models obtained using correlation-regression analyses were characterized by significantly larger values of approximation errors. At the same time, as numerous experiments showed, the error values were so high that the models should be defined as statistically insignificant and unsuitable for practical use. In addition, this algorithm is intended for a pessimistic assessment of the quality and cost of project products; the result of the fuzzy algorithm ensures the determination of the maximum acceptable indicator of the rationality of project management for construction customers under a certain budget for the services of Designers. In view of the above, we consider it expedient to examine the obtained empirical model in more detail. The work of the fuzzy logical inference algorithm begins with fuzzification—the introduction of fuzziness. In other words, on the basis of clear values of material intensity, the degree of acceptance is calculated, whether its value will be acceptable, average, high, very high, or critical. For this purpose, during the design of the empirical model in the Anfisedit editor, the type of membership function of the Gaussian type, 2-sided, was selected. Hybrid neuro-fuzzy modeling provides the minimum possible model error precisely in the case of using the membership function of this type, and they are usually asymmetric. Membership functions of this type are specified using 2 parameters: modal value and stretch coefficient. These parameters are very similar to the parameters of the well-known normal distribution function—mathematical expectation and standard deviation. The plot of the membership function also closely resembles the density plot of the normal distribution, as it is bell-shaped. Figure 1 shows graphs of the membership function of the terms of the input variable $\mu\_j(x)$.

The parameters of the membership function of the input variable are presented in Table 1. From Fig. 1 and Table 1, it can be seen that doubts during the identification of the level of material intensity of the project products will arise when its value is recognized as average, high, very high, or critical. After all, the ordinates of the points of intersection of the indicated pairs of graphs are not lower than 0.5 (Fig. 1). In particular, the amount of material consumption in the amount of 0.98 hryvnias/hryvnias can be considered both very high and critical at the same time. This can be verified by drawing a vertical line with the beginning on the abscissa axis at point 0.98. This line will cross at once 2 graphs corresponding to the above terms "very high" and "critical". The measure of acceptance of this material density for each

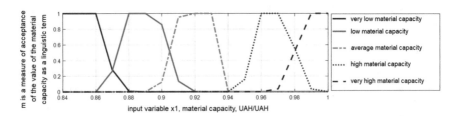

Fig. 1 Functions of appropriateness of material intensity of project products (Author's development)

of the terms can also be determined analytically, by substituting x1 = 0.98 in the corresponding formulas of lines 4 and 5 of the Table 1 [13].

Due to the features of the Anfisedit editor, the design of new fuzzy algorithms involves the same number of terms and input and output variables. The names of the linguistic terms for the output variable match the names of the terms of the input variable. Table 2 shows the results of Anfisedit-or aimed at determining the parameters of linear functions of the output variable [14].

In general, the quantitative values of the parameters of the ownership functions of the output variable can be explained as the intensity of influence on the coefficient of rationality of the design products of the factor of material intensity and the rest

Table 1 Analytical record of the membership functions of the term set "material intensity of project products" (Author's development)

H/h = j	Linguistic term	Font size and style
1	Acceptable	$$\mu_1(1) = \begin{cases} e^{-\frac{1}{2}\cdot\left(\frac{x_1-0.834}{0.0066}\right)^2}, & \text{if } x_1 < 0.834; \\ 1, & \text{if } 0.834 \leq x_1 \leq 0.860; \\ e^{-\frac{1}{2}\cdot\left(\frac{x_1-0.860}{0.0065}\right)^2}, & \text{if } x_1 > 0.860. \end{cases}$$ [0.006557 0.834 0.006457 0.8598]
2	Average	$$\mu_2(1) = \begin{cases} e^{-\frac{1}{2}\cdot\left(\frac{x_1-0.876}{0.0039}\right)^2}, & \text{if } x_1 < 0.876; \\ 1, & \text{if } 0.876 \leq x_1 \leq 0.896; \\ e^{-\frac{1}{2}\cdot\left(\frac{x_1-0.896}{0.0069}\right)^2}, & \text{if } x_1 > 0.896. \end{cases}$$ [0.003912 0.8764 0.006882 0.8962]
3	High	$$\mu_3(x_1) = \begin{cases} e^{-\frac{1}{2}\cdot\left(\frac{x_1-0.912}{0.0058}\right)^2}, & \text{if } x_1 < 0.912; \\ 1, & \text{if } 0.912 \leq x_1 \leq 0.931; \\ e^{-\frac{1}{2}\cdot\left(\frac{x_1-0.931}{0.0018}\right)^2}, & \text{if } x_1 > 0.931. \end{cases}$$ [0.005764 0.9118 0.001825 0.9311]
4	Very high	$$\mu_4(x_1) = \begin{cases} e^{-\frac{1}{2}\cdot\left(\frac{x_1-0.951}{0.0057}\right)^2}, & \text{if } x_1 < 0.951; \\ 1, & \text{if } 0.951 \leq x_1 \leq 0.973; \\ e^{-\frac{1}{2}\cdot\left(\frac{x_1-0.973}{0.0066}\right)^2}, & \text{if } x_1 > 0.973. \end{cases}$$ [0.005724 0.9511 0.006645 0.9728]
5	Critical	$$\mu_5(x_1) = \begin{cases} e^{-\frac{1}{2}\cdot\left(\frac{x_1-0.988}{0.0070}\right)^2}, & \text{if } x_1 < 0.988; \\ 1, & \text{if } 0.988 \leq x_1 \leq 1.012; \\ e^{-\frac{1}{2}\cdot\left(\frac{x_1-1.012}{0.0066}\right)^2}, & \text{if } x_1 > 1.012. \end{cases}$$ [0.006977 0.988 0.006557 1.012]

Table 2 Linear combinations of the outputs of the empirical model of the connection of material intensity (x1j) and the coefficient of the rationality of design products (yj)

H/h = j	Linguistic term	The linear combination as a function of the membership of the original variable
1	Acceptable	$\hat{y}_1 = 266.66 \cdot x_{11} - 222.61$
2	Average	$\hat{y}_2 = -701.03 \cdot x_{12} + 636.77$
3	High	$\hat{y}_3 = 385,91 \cdot x_{13} - 347.09$
4	Very high	$\hat{y}_4 = 137,07 \cdot x_{14} - 126.10$
5	Critical	$\hat{y}_5 = -45,51 \cdot x_{15} + 45.85$

of the factors that are not studied in detail in this algorithm. The impact of material intensity, similar to "traditional" linear regressions, reflects the slope coefficient—the coefficient near x1j. The rest of the factors that were not included in the models, since their connection with the quality and cost of the project products was not established a priori, still affect the price of the project works and its acceptability for the Customers. The size of this influence is quantified using a free constant.

Analysis of the formulas from Table 2 showed that the influence of material intensity and other factors on the quality and price of project products is multidirectional. After all, the slope and the free constant have different signs for different terms. Even the modules of numerical indicators of linear combinations do not change monotonically. After all, in Table 2. for the "average" term, the parameters by the module are maximum, they are larger than the rest of the terms. In the future, for the terms "high", "very high", and "critical", the absolute values of the parameters are characterized by a decrease. Instead, the "acceptable" term is characterized by parameter values that are much smaller in terms of the module for both the "average" term and the "high" term, which, in fact, breaks the monotonous decrease of the modules of the constants of linear combinations.

At the same time, the coefficient constant near the clear value of the input (x1j) has a different sign for different terms, which does not make it possible to unambiguously interpret the influence of construction material intensity on the costs of construction management and organization [15–18].

Of course, as the experience of working with Designers accumulates and its accumulates in the form of databases on the technical and economic characteristics of project products, design, and estimate documentation, the parameters of the proposed fuzzy algorithm are subject to regular refinement. However, this type of empirical model makes it possible to apply the following algorithm for the identification and quality control of project products in practice:

1. According to the estimate documentation for each structural element, calculate 2 indicators: (a) material intensity of project products ($\times 1$) as a specific weight of the cost of construction materials products, set in the sum of direct costs; (b) the coefficient of rationality of project products, that is, the ratio of costs for material and technical support of the organization and construction management

to the costs of the same purpose in the operation of construction machines and mechanisms.

2. Using the membership functions of the Table 1, to identify materiality with each of the 5 fuzzy linguistic terms. As a result, calculate the values of the functions belonging to the terms ().

3. By substituting a clear value of the material density ($\times$1) into the formulas of the Table 2 calculate the possible values of the rationality coefficient $(y\_j)^{\char`\^}$ for each of the 5 terms of the initial variable.

4. Determine the theoretical value of the indicator indicator of the rationality of the project products, as a weighted average of the linear combinations established at stage 3. At the same time, the role of weighting factors is played by the values of the membership functions () from stage 2:

$$\hat{y} = \frac{\sum_{j=1}^{5} \mu_j(x_1) \cdot \hat{y}_j}{\sum_{j=1}^{5} \mu_j(x_1)} \tag{1}$$

5. The final decision regarding the quality of the project products and the compliance of the cost of the Designer's services with the results of his work should be made taking into account the presence or absence of the need for additional resources for changes in the scope or types of construction work, or corrections of already completed ones. Therefore, the quality of the design can be considered acceptable, provided that the actual coefficient of management rationality from stage 1 does not exceed the theoretical indicator established by formula (1) at stage 5:

$$y \leq \hat{y} \tag{2}$$

In this case, the cost of the Designers' services will not be inflated; it will correspond to the quality characteristics of the design products, and the expediency of its payment should not cause doubts in the Customer.

6. Since the estimated documentation contains a significant number of structural elements, i.e. types of design products, for some of them, it is possible to violate condition (2). However, the specific weight of such design products should not exceed 25%. This threshold is justified by the provisions of the theory of economic risk management, according to which a variation of a variable value of more than 25% is higher than the critical level for a reasonably cautious consumer.

We will give an example of the application of the algorithm. At stage 1, it was established that the material consumption of a certain structural element was 0.98 hryvnias/hryvnias. Performing stage 2 allowed identifying this value as:Acceptable with zero confidence, because:

- Acceptable with zero confidence, because:

$$\mu_1(x_1) = e^{-\frac{1}{2} \cdot \left(\frac{0.98-0.860}{0.0065}\right)^2} = e^{-170.414} \approx 0, \quad \text{because } 0.98 > 0.860$$

- Average with a confidence level equal to zero, also because Displayed equations are centered and set on a separate line.

$$\mu_2(x_1) = e^{-\frac{1}{2} \cdot \left(\frac{0.98-0.896}{0.0069}\right)^2} = e^{-74.102} \approx 0, \quad \text{because } 0.98 > 0.896$$

- High with a degree of certainty, as for the previous terms, at the zero level:

$$\mu_3(x_1) = e^{-\frac{1}{2} \cdot \left(\frac{0.98-0.931}{0.0018}\right)^2} = e^{-370.525} \approx 0, \quad \text{because } 0.98 > 0.931$$

- Very high with a confidence level of 0.57, because:

$$\mu_4(x_1) = e^{-\frac{1}{2} \cdot \left(\frac{0.98-0.973}{0.0066}\right)^2} = e^{-0.562} = 0.570, \text{ because } 0.98 > 0.973$$

- Critical with a confidence level of 0.519, because:

$$\mu_5(x_1) = e^{-\frac{1}{2} \cdot \left(\frac{0.98-0.988}{0.00698}\right)^2} = e^{-0.657} = 0.519, \quad \text{because } 0.98 < 0.988$$

The zero values of the functions of the first three terms indicate that the material intensity of 0.98 hryvnias/hryvnias is too high, and other production costs will account for only 2 kopecks from every hryvnia of the cost of the project products. Instead, such a level of material intensity should be identified as very high, up to a critical level. Given the results of the calculation of the membership functions, for the last two terms, there is considerable uncertainty as to whether 98 cop. the cost of construction materials in each hryvnia of the cost of structural elements is a very high indicator of will it already meet the critical level. After all, the degree of acceptance of each of the terms does not significantly exceed 0.5. At the same time, the total value of all membership measures exceeds 1, and this is the main difference between the theory of fuzzy sets and the theory of probability, where a complete group of events has a single probability.

At the 3rd stage, the possible values of the terms of the coefficient of rationality of the design products were calculated:

acceptable$(\hat{y\_1}) = 266.66 \cdot 0.98 - 222.61 = 38.72$

average$(\hat{y\_2}) = -701.03 \cdot 0.98 + 636.77 = -50.22$

high$(\hat{y\_3}) = 385.91 \cdot 0.98 - 347.09 = 31.10$

is very high$(\hat{y\_4}) = 137.07 \cdot 0.98 - 126.10 = 8.23$

critical$(\hat{y\_5}) = -45.51 \cdot 0.98 + 45.85 = 1.25$

The 4th stage made it possible to establish for (1) the theoretical value of the coefficient of rationality of the design products:

$$\hat{y} = \frac{\sum\limits_{j=1}^{5} \mu_j(x_1) \cdot \hat{y}_j}{\sum\limits_{j=1}^{5} \mu_j(x_1)}$$

For this project product, the quality and price should be considered acceptable only when the rationality of the project product from stage 1 does not exceed $\hat{y}$, i.e. 4.9.

$$\hat{y} = \frac{\sum_{j=1}^{5} \mu_j(x_1) \cdot \hat{y}_j}{\sum_{j=1}^{5} \mu_j(x_1)}$$
$$= \frac{(38.72 \cdot 0 - 50.22 \cdot 0 + 31.1 \cdot 0 + 8.23 \cdot 0.57 + 1.25 \cdot 0.519)}{(0 + 0 + 0 + 0.57 + 0.519)} \approx 4.9$$

Along with the proposed algorithm, the Customer should not neglect the "traditional" rules of conducting business relations with Designers and Contractors. After all, the calculation of project works contains a list of tasks to which great attention is paid before the start of complex planning of any object. This makes it possible to determine the price of costs for performing the necessary tasks, as well as the profitability and expediency of the work. And, first of all, the Customer is interested in this process.

In the beginning, it may not be necessary to develop a complete set of documentation. An opportunity for a little money to get closer to understanding the scope of work—that's what pre-project work means. Among the goals of the pre-project calculation, the following stand out:—determination of payback and financial justification of investments;—Study of objects;—processing of documentation [15–18].

For many years, qualified specialists of project enterprises in Ukraine have been engaged in the preparation of documentation: factual data is being collected. Tables, diagrams, sketches, etc., are formed on the basis. The efficiency, correctness, and quality of subsequent construction processes and other complex tasks depend on this documentation. Indispensable work that allows you to determine the nuances of the subsequent erection of structures in construction is the calculation of the design cost, the result of which is the efficiency, reliability, and durability of the object under construction. The future project is made up of the above. There are many nuances, and sometimes, depending on the situation, they can change. Awareness of the specifics of construction, as well as the experience of the Customer, is the best guarantee that the collected data will be accurate and error-free. Based on them, it will be possible to make a profitable decision for the business.

Design and construction companies of Ukraine conducted a qualitative study among specialists in private construction and repair and identified the main mistakes of the Customers [19–21]. Five main mistakes made by the Customer during the repair and construction season: saving on quality, neglecting the timely conclusion of the contract, planning the work execution process, developing a design project, and monitoring the execution of rough finishing. Let's take a closer look at the consequences of each of them and ways to eliminate such omissions.

(1) Save on quality. Misunderstanding between the Contractor and the Customer during the construction or repair. No customer will save money, understanding what this can lead to. The main problems arise as a result of the customer's misunderstanding of the technological process—56 and 44% of respondents say that the main mistake is the customer's desire to save on work or materials. That is, at the initial stage, the Customer needs to choose a professional construction team that will advise on which materials or works can be saved, and where it is categorically impossible to do so. And will also be able to competently explain why certain solutions are needed. According to the authors of the study [21, 22], with which we completely agree, when choosing Contractors, special attention should be paid to: (a) reviews and recommendations, (b) portfolio (see photos from former facilities); (c) experience (do workers correctly name the stages of this or that technology, how do they orient themselves in terms of construction, how accurately and completely do they list the necessary consumables). The key issues for the Customer, in our opinion, are quality, terms, number of performers.

Cost: after inspecting the object and determining all the works, the estimate must be accurate, so the words of the master about the approximate cost of construction (repair) do not bode well. Terms: if the Customer knows that it will take approximately a month for the screed to rise, and the team promises to finish the repair within this period, there are two options: either they are deliberately misinforming you, or they are going to neglect technology.

It is better for the customer not to save when choosing a construction team, because an unprofessional approach in this field can lead to the fact that he will not only have to hire another team but also re-purchase consumables, losing time and money. The human factor is an ineradicable phenomenon, sometimes also determining, the choice should be given special attention. The number of builders depends on the composition of the works, but there is a certain regularity. The presence of three workers at the facility is considered optimal. If there are more of them, there is a high probability that labor productivity will drop: the more people there are, the higher the chance that they will be distracted by each other, chat, take smoke breaks, etc. If there are fewer of them, the work will be done more slowly. The crews are people with whom the Customer will have to interact directly in order not to make mistakes and not overpay for the work.

(2) Failure to timely conclude the contract. Since the contractor will not be legally responsible for the work performed. In this case, in the future, the Customer remains deprived of any guarantees. Even if the contract is prepared and ready to be signed, it is advisable to take your time. A detailed contract must be concluded before the start of all repair work, affirming 86% of respondents. First, at the initial stage, it will allow us to determine the exact cost of all works and terms. Secondly, it will protect the Customer or the construction team in case of non-compliance with the agreements of one of the parties.

The contract should include a detailed estimate of all necessary materials and repair services. As well as a detailed plan of work indicating the terms of delivery of all necessary materials and coordination of each stage by the Customer. It is precisely to improve the results of this stage that we have developed the above methodical support.

For high-quality performance of works, it is necessary to clearly imagine what conditions the Customer is obliged to provide under the contract. Special attention should be paid to the guarantees given by the team, namely: which works are covered by the guarantee, during which period it is valid.

(3) Customers often forget about the detailed work plan. First of all, before starting construction or repair, it is worth drawing up a preliminary plan for all works: indicate the cost of works and materials and their delivery terms—this will allow you to calculate the cost of construction or repair in advance and, if necessary, reduce some costs. Consistent work of the team will be ensured by the timely delivery of the necessary materials, at the same time will reduce the work period to a minimum, and the Customer will not have to overpay the workers for waiting. Secondly, the Customer should immediately coordinate with the Designers and Contractors the procedure for payment, which should preferably be done in stages. It is better to pay according to the fact of completed work; you can after each stage, weekly or monthly: received, checked, and paid. The majority of large project and construction companies in Ukraine that respect both themselves and the Customer work according to this principle. A full prepayment "ties" the Customer to the crew—regardless of whether they like its work or not and often becomes the very credit that spoils the relationship between the Customer and the Contractor. Thirdly, a list of all necessary materials and services should be clearly written down, indicating the cost. There are a number of ways that can help reduce the cost of construction (repair) if the collected estimate is beyond the Customer's pocket.

(4) They neglect the design of the project. Before starting construction (renovation), special attention should be paid to the coordination of the project design between the designer and the construction team. Does the construction team need a design project: 55% of specialists answered yes, 43%—not always, and 2%—answered that a design project is not needed. To the question: "does a situation often occur when a design solution conflicts with construction technologies or capabilities, 73% of respondents gave a positive answer, and 27% answered negatively.

(5) They do not pay attention to the rough decoration. The customer must carefully accept rough finishing work and pay special attention to the quality of execution of leveling floor screeds; for electrical and plumbing works; and for waterproofing and insulation works. 78% of respondents said that the maximum number of disputes and disagreements occurs at the stage of final decoration, while 28% noted that the greatest difficulties are at the stage of rough work. For rough decoration, it is better to entrust the choice of emerging materials to the masters. They have experience working with specific brands. But the materials for finishing (tiles, paint, wallpaper, floor covering) are advised by masters to

choose together with the designer or independently. It is important that the result is not just correct, but primarily liked by the Customer.

Any scientific research—from a creative idea to the final design of a scientific work—is carried out very individually—this is purposeful knowledge, the results of which appear in the form of a system of concepts, laws, and theories. To its implementation, it is still possible to determine the general methodological approaches. The hallmarks of scientific research include:

– necessarily a purposeful process, clearly formulated tasks, and the achievement of a consciously set goal;
– a process aimed at searching for something new, at discovering the unknown, at creativity, at new coverage of the issues under consideration, at putting forward original ideas;
– is characterized by systematicity: here, both the research process itself and its results are ordered brought into the system;
– inherent strict evidence, and consistent substantiation of the generalizations and conclusions made.

The main means of scientific and theoretical research:

– a set of scientific methods, comprehensively justified and consolidated into a single system;
– a set of concepts, strictly defined terms, interconnected and forming a characteristic language of science.

The results of scientific research are embodied in scientific works (articles, monographs, textbooks, dissertations, etc.), and only then, after their comprehensive evaluation, are they used in practice. The effectiveness of scientific work, to the greatest extent, depends on the choice of the most effective research methods, since it is they that allow achieving the goal.

The ability to make correct decisions in the face of incomplete and fuzzy information is the most impressive property of human intelligence. Today, one of the most important problems of science is the construction of models that reproduce human thinking and their use in computer systems. Knowledge about the specific subject area for which the system is being created when developing intelligent systems is rarely complete and completely reliable. The quantitative data obtained by fairly accurate experiments have statistical estimates of probability, reliability, significance, etc. The information obtained as a result of a survey of experts, which is filled in expert systems, may diverge since their opinions are subjective. The contradiction between fuzzy knowledge and clear methods of inference arises when processing knowledge using rigid mechanisms of formal logic. This contradiction can be resolved by using special methods for representing and processing fuzzy knowledge or, if possible, by overcoming the fuzzy knowledge. The main components of the multi-valued term fuzziness are as follows: indeterminacy of conclusions, inaccuracy, unreliability, incompleteness, and polysemy [23–25].

In 1965, Professor Lotfi Zadeh of the University of Berkeley published his seminal work Fuzzy Sets in the journal Information and Control. From this moment, the theory of fuzzy sets (fuzzy sets theory) originates. And 1975 is considered the beginning of the practical application of fuzzy set theory, when E. Mamdani built the first fuzzy industrial controller based on fuzzy linguistic rules "If—then" led to a surge of interest in fuzzy set theory among mathematicians and engineers.

The possibility of using fuzzy logic is based on the results of research by such foreign scientists as:

- 1992—V. Kosko—proved the Fuzzy Approximation Theorem. Which said that any mathematical system can be approximated by a fuzzy logic-based system;
- 1992—L. Wang—the fuzzy system is a universal approximator. It can approximate any continuous function to arbitrary accuracy if it uses a set of n(n–x) If– Then rules, Gaussian membership functions, Larsen implications, and centroid refinement, product composition;
- 1995—J. Castro—the Mamdani logic controller is a universal approximator for symmetrical triangular membership functions, composition using the minimum operation, the implication in the Mamdana form, and the centroid method of reduction to clarity.

If expert knowledge about an object or process can be formulated only in a linguistic form, then systems with fuzzy logic can be used, and it is also advisable to use them for complex processes when there is no simple mathematical model. If an adequate and easily researched mathematical model has already been found for an object or process, and also if the required result can be obtained in some other (standard) way, it is inappropriate to use systems based on fuzzy logic. The main disadvantages of systems with fuzzy logic [26, 27]: (a) the initial set of postulated fuzzy rules is formulated by a human expert and may be incomplete or inconsistent; (b) the type and parameters of membership functions that describe the input and output variables of the system are chosen subjectively and may not reflect reality.

Fuzzy logic controllers, with slightly different operations from conventional controllers, are a successful application of fuzzy set theory [28, 29]. Instead of differential equations, expert knowledge is used to describe the system. This knowledge can be expressed using linguistic variables described by fuzzy sets [28, 29].

With the help of a fuzzy knowledge base and operations on fuzzy sets, fuzzy inference is an approximation of the relationship between inputs and outputs of the system. The knowledge base, consisting of a set of rules, acts as a reflection j of the set of states Y into the set of decisions D. The optimality of the decision depends on the accuracy of the membership functions of the quantities and the knowledge base. Acceptable accuracy of solutions, in most cases, is achieved by adjusting the parameters of the functions and the weighting coefficients of the rules based on a sample of experimental data.

In Fig. 2 shows the general structure of a microcontroller using fuzzy logic and containing the following components [28, 29]: Fuzzification block; Knowledge base; Decision Block, and Defuzzification Block.

In Fig. 3, the main stages of fuzzy inference are shown.

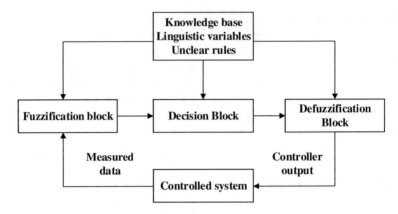

Fig. 2 General structure of a fuzzy microcontroller

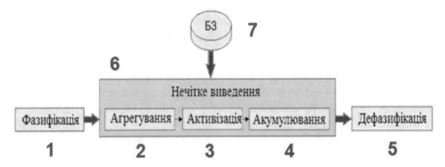

Fig. 3 Main stages of fuzzy inference: Phasification (1), Aggregation (2), Activation (3), Accumulation (4), Defazification (5), Fuzzy output (6), Knowledge base (7)

Fuzzy inference algorithms differ in the type of rules used, logical operations, and the type of defuzzification method. Fuzzy inference models were developed in the works of scientists Mamdani, Sugeno, Larsen and Tsukamoto [30–33].

Traditional logical systems are not as close in spirit to human thinking and languages used as fuzzy logic, on which fuzzy control is based. An effective means of reflecting the uncertainties and inaccuracies of the real world, this is what fuzzy logic mainly provides. The presence of mathematical means of displaying the fuzziness of the initial information makes it possible to build a model that is adequate to reality.

In fuzzy logic and fuzzy control theory, the concept of fuzzy inference occupies a central place—it includes all the basic concepts of set theory: membership functions, linguistic variables, fuzzy implication methods, etc. The development and application of fuzzy inference systems include a number of stages (Fig. 2).

From the above, we can conclude that the method under consideration is a process of cognition, according to which thinking ascends from the concrete in reality to the abstract in thinking and from it to the concrete in thinking.

According to the understanding and analysis of the essence of the design problem at any stage of the creation of the project, it is advisable to have some subsystem of means to help the Designer in his informational and constructive activities. In our opinion, such a subsystem should be qualified, at the level of an experienced member of the expert group, to "give out" advice on the dynamics of emerging design situations at the request of the Designer. "Complex" design objects can be recognized by the following distinguishing features:

1. Not all goals and principles for choosing design solutions and the conditions affecting this choice can be expressed in the form of quantitative ratios;
2. There is no formal description of the design object or it is unacceptably complex;
3. A significant part of the information necessary for the mathematical description of the object exists in the form of wishes and ideas of specialists—experts in the field of architectural and construction design.

There are two directions that are not mutually exclusive for searching for methods of mathematical modeling of complex design objects. The first is to build a model of an object suitable for implementation, an attempt to use an unconventional mathematical apparatus. Due to the specifics of objects of this class, unfortunately, this direction turns out to be unpromising. Either the model turns out to be inadequate to the design object due to natural abstraction in order to simplify, or the model turns out to be unnecessarily cumbersome and therefore unacceptable for economic reasons. The second direction is an attempt to build not a model of an object, but a model of its design as a controlled process. That is, rational and purposeful reflections of the Design Engineer in the process of selecting and comparing alternative design solutions are modeled, and not the object itself to be designed. In the general case, the mode of action is understood as ways of extracting the knowledge of an engineer and ways of searching for and making design decisions, and in the long term also methods of human thinking. A way to solve is the Designer's thoughts, considered in a simplified schematic representation that have arisen and systematized to solve some specific problem. At each phase: the acquisition and processing of incomplete information, the evaluation of complex objects, approximate reasoning, conclusions, and decision-making in unreliable situations, etc., the process of human thinking has its own characteristics. In everyday activities, the mind of a person manifests itself due to the fact that he "thinks in words", that is, every time a person expresses his thoughts in natural language.

Such a comparison of the Designer thinking with the theory and practice of automation is fruitful, which is based on identifying patterns of thinking that allow, at least in principle, their mathematical description in a formalized language accessible to an automaton of a certain "development level". This is shown by serious studies in the works [11–18, 23, 24, 29–33].

Misconceptions about fuzzy logic are common. Fuzzy logic is imprecise: At its core, fuzzy logic is no more imprecise than standard arithmetic. In fact, it is much more accurate when dealing with imprecise information. Fuzzy logic is based on probabilistic reasoning. Probability deals with the chances of occurrence of certain events, and fuzzy logic deals with these events themselves. Generally, fuzzy logic deals with ambiguity rather than uncertainty. Fuzzy logic is built on the basis of a number of heuristic assumptions. Although, due to the intuitive nature of fuzzy logic, at first glance it may seem that the rules underlying it are chosen arbitrarily or based only on common sense, in fact, these rules have been rigorously proven to be correct. In our opinion, in the aspect of the proposed work, these studies have a convincing experimental basis and are accepted as starting points.

In the designer's creative activity, situations almost constantly arise when ready-made methods of action, standard or typical solutions, instinctive reactions; and automatic forms of behavior are ineffective. In these situations, the Designer is forced to look for new design solutions to develop new forms of behavior and new sequences of actions. Such situations are called problems or problem situations. Research in the field of fuzzy modeling of objects and design processes is relevant. They make it possible to provide the system design of construction objects with a single theoretical and model base, create common logical and mathematical foundations for the formalization of intelligent procedures, and clarify, in particular, the place of fuzzy logic in system analysis.

In the modern world, elements of fuzzy logic can be found not only in design and construction but also in many industrial products—from control systems for electric trains and combat helicopters to household appliances. Without the use of fuzzy logic, modern situational centers of the leaders of Western countries are unthinkable, where all kinds of crisis situations are modeled and key political decisions are made. Active consumers of fuzzy logic are specialists in the field of political and economic analysis, as well as financiers and bankers. Their tasks require making the right decisions every day in a challenging and unpredictable market. They use fuzzy systems to create models of various economic, political, stock exchange situations. Firms specializing in fuzzy logic, having solid financial "support", were able to adapt their developments for a wide range of applications.

3   Conclusions

Pricing of project products should be carried out taking into account the "price-quality" ratio. The quality of the Designers' work affects not only the durability, strength, reliability, and operational characteristics of the building but also the duration and cost of construction. The latter is manifested along with the final indicators of the estimates by the proportions between the individual components of the costs for the construction of the object. In the research, with the help of statistical analysis and neural network modeling, the hidden regularities of changes in construction cost components were studied, and, based on them, a methodological toolkit for the identification and quality control of project products was developed. The developed toolkit is based on two indicator coefficients:

- The indicator of the material intensity of the project products, determined in terms of individual construction works or structural elements. It is suggested to calculate it as a specific weight of the cost of construction materials, products, and kits in the amount of direct costs, determined according to the data of the sections of local estimates.
- Coefficient of the rationality of project products, which is a ratio of such value indicators as costs for material and technical support of the organization and construction management, as well as costs for material and technical support for the operation of construction machines and mechanisms.

The proposed toolset for identification and quality control of project products contains an empirical model of the relationship between the material intensity of project products and the rationality of managing its implementation at the stage of project implementation. Based on this model, it becomes possible to establish the risk of additional cost overruns by the customer and the contractor during construction. The higher the risk of cost overruns, the lower the quality of the design products, and therefore their cost will be overestimated, and the Designer will be characterized by low competitiveness.

As an empirical model of communication, apply the system of artificial intelligence, in particular, the Sugeno-type fuzzy logical inference algorithm. Linear models were used in this study.

Making a decision in favor of high-quality design documentation is appropriate only if the actual ratio of the rationality of construction management, calculated based on the data of the estimate documentation, will not exceed the theoretical value for ¾ sections of the estimate. This theoretical value makes it possible to define the toolkit proposed in this work.

Along with scientific developments, in order to avoid an unjustified increase in construction costs, the Customers must follow "traditional" recommendations, in particular: do not skimp on the quality of construction materials used, enter into contracts with Designers and Contractors for all types of work, carefully develop construction plans and schedules, project design and control of their implementation.

Prospects for further research in this direction should be considered to be the study of the impact on the cost of project work of the duration of design and construction, their labor intensity, planning, and constructive and technological solutions of the project.

References

1. Mashoshina, T.: The main economic laws affecting the process of developing design and estimate documentation and the financial results of project organizations. Bull. Econ. Transp. Ind. Collect. Sci. Pract. Art. **42**, 315–318 (2013)
2. Quality system in the implementation of innovations in a construction company. Economic-management and information-analytical innovations in construction. In: II International Scientific-Practical Conference, pp. 88–89 (2020). http://library.knuba.edu.ua/node/38

3. Gritcenko, O., Tsyfra, T., Zapechna, Y.: Development of economic approaches to the formation and definition of a strategy for building enterprises. Technol. Audit Prod. Reserves **1/4**(39), 70–76 (2018). http://journals.uran.ua/tarp/article/view/124542/119361
4. Troynikova, O., Mashoshina, T.: Assessment of endogenous factors to increase the investment attractiveness of enterprises. Priazovsky Econ. Bull. **2**(13), 184–189 (2019)
5. Bielienkova, O., Novak, Y., Matsapura, O., Zapiechna, Y., Kalashnikov, D., Dubinin, D.: Improving the organization and financing of construction project by means of digitalization. Int. J. Emerg. Technol. Adv. Eng. **12**(8), 108–115 (2022)
6. Zeltser, R.Y., Bielienkova O.Y., Novak, Y., Dubinin D.V.: Digital transformation of resource logistics and organizational and structural support of construction. Sci. Innov. **15**(5), 38–51 (2019)
7. Reznik, N., Ijaz, Y., Kushik-Strelnikov, Y., Barabash, N., Stetsko, M., Bielienkova, O.: Systems Thinking to Investigate the Archetype of Globalization. Springer International Publishing (2022). https://doi.org/10.1007/978-3-031-08087-6_9
8. Guidelines for determining the cost of design, scientific design, research works and examination of project documentation for construction. Entered into force in 2021 (2021)
9. Estimated norms of Ukraine. Guidelines for determining the cost of construction. Taking into account Changes No. 1, No. 2, Effective Date 01.01.2023
10. Bolila, N.V.: Economic and management tools for ensuring the anti-crisis potential of construction enterprises. Author's abstract/dissertation. К.: КKNUBA (2020)
11. Belenkova, O.Y.: Strategy and mechanisms for ensuring the competitiveness of construction enterprises based on the model of sustainable development: monograph, p. 512. Lira-K, Kyiv (2020)
12. Sorokina, L.V., Goyko, A.F., Stetsenko, S.P., Izmailova, K.V. and other.: Econometric toolkit for managing the financial security of construction enterprises: Monograph/under science. ed. Doctor of Economics, Prof. L.V. Sorokina. KNUBA, Kyiv (2017)
13. Zheldak, T., Koryashkina, L., Us, S.: Fuzzy sets in management and decision-making systems: educational. Manual. Dnipro Polytechnic **387** (2020). http://ir.nmu.org.ua/handle/123456789/156356
14. Avramenko, S., Zheldak, TA, Koriashkina TA.: Rubryl hybrid genetic algorithm for solving global optimization problems. Radio Electron. Comput. Sci. Control **2**, 174–188 (2021). https://doi.org/10.15588/1607-3274-2021-2-18
15. Levy, L.: Using fuzzy logic to automate the operation of irrigation systems BULLETIN of the Poltava State Agrarian Academy 2 (2018). https://doi.org/10.31210/visnyk2018.02.25.
16. Bielienkova, O., Stetsenko, S., Oliferuk, S., Sapiga, P., Horbach, M., Toxanov, S.: Conceptual model for assessing the competitiveness of the enterprise based on fuzzy logic: social and resource factors. In: IEEE International Conference on Smart Information Systems and Technologies (SIST), pp. 1–5 (2021). https://doi.org/10.1109/SIST50301.2021.9465923
17. Ryzhakova, G. et al.: Modern structuring of project financing solutions in construction. In: IEEE International Conference on Smart Information Systems and Technologies (SIST) (2022)
18. Article Errors during the construction of a house. How to avoid mistakes when building a house? Eurometa https://eurometa.ua/
19. Stetenko, S., Sorokina, L., Izmailova, K., Bielienkova, O., Tytok, V., Emelianova, O.: Model of company competitiveness control using architectural intelligence tools. Int. J. Emerg. Trends Eng. Res. **9**(2), 60–65 (2021). https://doi.org/10.30534/ijeter/2021/08922021
20. Nastich, I.: Stattya Neprozora "Prozoro" Budivelniy Portal No. 1 "News" Look, analytics and hot topics. https://www.ukr.net›economics
21. Mashoshina, T.: Monitoring the pricing system in construction and its impact on the activities of design organizations. **1**, 245–248 (2014)
22. Mashoshina, T.: Transparency of the economic nature of contract prices for design work. In: XIIIth Scientific and Practical International Conference, pp. 125–126. Kharkiv (2017)
23. Zahorsky, V, Borschuk, E., Zholobchuk, I.: Ensuring sustainable development of the national economy: social and ecological aspects. Effi. Public Adm. **44** (2015). http://www.lvivacademy.com/vidavnitstvo_1/edu_44/fail/ch_2/3.pdf

24. Manjos, T, Melnyk, O., Lutsyshina, Z.: Inventory management model with fuzzy triangular demand. Bus. Inf. **11**, 174–179 (2018)
25. Uduak, U., Udoinyang, G.I., Emmanuel E.N.: Interval type-2 fuzzy logic for fire outbreak detection. Int. J. Soft Comput. Artif. Intell. Appl. (IJSCAI) **8**(3), 27–46 (2019)
26. Al-Maitah, M., Semenova, O.O., Semenov, A.O., Kulakov, P.I., Kucheruk, V.Y.: A hybrid approach to call admission control in 5G networks. Adv. Fuzzy Syst. **2018**, 1–7 (2018). https://doi.org/10.1155/2018/2535127
27. Semenova, O., Semenov, A., Voitsekhovska, O.: Neuro-fuzzy controller for handover operation in 5G heterogeneous networks. In: 2019 3rd International Conference on Advanced Information and Communications Technologies (AICT) (2019). https://doi.org/10.1109/AIACT.2019.8847898
28. Semenova, O., Semenov, A., Voznyak, O., Mostoviy, D., Dudatyev, I.: The fuzzy-controller for WiMAX networks. In: 2015 International Siberian Conference on Control and Communications (SIBCON) (2015). https://doi.org/10.1109/SIBCON.2015.7147214
29. Semenova, O., Semenov, A., Koval, K., Rudyk, A., Chuhov, V.: Access fuzzy controller for CDMA networks. In: 2013 International Siberian Conference on Control and Communications (SIBCON) (2013). https://doi.org/10.1109/SIBCON.2013.6693644
30. Sarwar, B., Bajwa, I.S., Shabana, R., Ram-zan, B., Kausar, M.: Design and Application of fuzzy logic based fire monitoring and warning systems for smart buildings. MDPI Symmetry **10**(11), 615 1–24 (2018)
31. Sowah, R., Ampadu, K., Ofoli, A., Kou-madi, K., Mills, G., Nortey, J.: Design and implementation of a fire detection and control system for au-tomobiles using fuzzy logic. In: IEEE Industry Applications Society Annual Meeting, pp. 1–8. Portland, OR, USA (2016)
32. Sowah, R., Ampadu, K.O., Ofoli, A.R., Koumadi, K., Mills, G., Nortey, J.: Fire-detection and control system in automobiles: implementing a design that uses fuzzy logic to anticipate and re-spond. IEEE Ind. Appl. Mag. **25**(2), 57–67 (2019)
33. Kushnir, A., Kopchak, B.: Development of intelligent point multi-sensor fire detector with fuzzy correction block. In: Perspective technologies and methods in MEMS design (MEM-STECH): proceedings of the XVth international conference, pp. 41–45. Polyana, Ukraine (2019)

Methodical Approach to Assessment of Real Losses Due to Damage and Destruction of Warehouse Real Estate

Lesya Sorokina⊕, **Yurii Prav**⊕, **Sergii Stetsenko**⊕, **Volodymyr Skakun**⊕, and **Nadiia Lysytsia**⊕

Abstract The article examines the problem of determining the market value of industrial real estate as a component of calculating the amount of damage caused by the military aggression of the Russian Federation. The main difficulties at the stage of assessing the market value of real estate at the time of destruction are identified. The theoretical basis for assessing the market value of real estate has been developed by substantiating an additional assessment principle related to the market environment— the principle of information asymmetry. The essence of the proposed principle and its role in the temporal instability of assessment results are revealed. Information on the estimated value of 192 objects of warehouse real estate as of 21.01.22—04.02.22 has been summarized in terms of the systematization of the main price-forming characteristics and their categories, and clustering of objects by price criteria has been carried out. The structure of the supply of warehouse real estate was studied, and the characteristics of warehouse real estate objects typical for the national market were established. The expediency of using neural network modeling to study the impact of price-forming characteristics on real estate value is substantiated. An economic interpretation of the parameters of a two-layer perceptron with three neurons in the hidden layer is proposed. On the basis of the neural network built during the research, an algorithm for a methodical approach to determining the market value of warehouse premises was developed, which makes it possible to establish to what extent the value

L. Sorokina · S. Stetsenko · V. Skakun · N. Lysytsia (✉)
Kyiv National University of Construction and Architecture, Povitroflotskyi Ave. 31, Kyiv 03037, Ukraine
e-mail: lysytsia.nv@knuba.edu.ua

L. Sorokina
e-mail: sorokina.lv@knuba.edu.ua

S. Stetsenko
e-mail: stetsenko.sp@knuba.edu.ua

Y. Prav
National Aviation University, Lubomyr Huzar Ave. 1, Kyiv 03058, Ukraine
e-mail: vsegda_prav@ukr.net

197

of the evaluated object will differ from the average market value. The information provision of assessment procedures for making corrective amendments has been improved by substantiating the quantitative values of percentage adjustments for the location of the warehouse, its category, age, number of floors and the material of the walls of the building, the availability of engineering support, as well as the technical condition of the premises.

Keywords Market value · Valuation principles · Information asymmetry · Neural network · Hyperbolic tangent · Price-forming characteristics · Corrective amendments

1 Introduction

The aggression of the Russian Federation against Ukraine caused catastrophic losses: the deaths of innocent citizens, mutilated lives, and destroyed houses, enterprises, and infrastructure facilities. Along with the vital need to protect national security, the problem of compensating the victims is becoming more acute. However, solving this problem is possible only after calculating the damage caused by the audacious invasion.

A number of scientific works [1–27] are devoted to the problems of economic assessment of losses in the national economy caused by man-made, ecological, natural, and anthropogenic factors. Among them, the article [6] deserves special attention, which examines the features of damage calculation due to hostilities [28]. However, in all the mentioned publications, insufficient attention is paid to the problem of determining the market value of the property before the damage is caused as a starting point for assessing the amount of damage.

At the time of writing the article, methods for assessing the losses caused to the objects of the national economy as a result of the military aggression of the Russian Federation have not yet been developed, but this work is actively being conducted in accordance with the resolution [29]. In the project of the methodology aimed at determining real losses due to damage and destruction of immovable property, it is assumed to calculate the number of costs necessary for the repair and replacement of damaged or destroyed immovable property as the difference between the values of the market value before the damage was caused and the residual value of replacing such an object after damage. However, this, at first glance, elementary approach is associated with a number of difficulties. First of all, the methodology of determining the value of the market value of damaged and lost objects at the time before the destruction, which was not determined for all objects on a certain date, raises the question. That is why a significant part of the problem still remains unsolved.

In this regard, the purpose of the article is to improve the theoretical basis for determining the market value of real estate, to develop an innovative methodical approach for express evaluation of warehouse real estate objects, and to improve the information support of traditional evaluation procedures carried out within the framework of a comparative approach.

2 Research Results

In accordance with NSA 1 [30], the market value is defined as the value for which it is possible to sell the object of assessment on the market of similar property on the date of assessment according to the agreement concluded between the buyer and the seller, after conducting the appropriate marketing, provided that each of the parties acted with knowledge of the matter, judiciously and without coercion.

Therefore, the main condition for determining the market value is an active market for the purchase and sale of property similar to the appraised property. However, the houses of Ukrainian citizens located in the rural areas of the peripheral regions, the demand for which was insignificant even in peacetime, suffered considerable destruction. Accordingly, the amount of their value, determined according to the normatively regulated evaluation procedures, may not correspond to the most probable sale price—the maximum that the buyer is ready to pay and, accordingly, the minimum at which the seller will agree to sell the property.

The same can be noted for the damaged property of business entities; however, in this context, the specific nature of industrial real estate should be taken into account. After all, many buildings and structures that house production facilities: workshops, warehouses, cooling towers, transformers, pumping stations, reactors, towers, bridges, and overpasses are almost never sold on the market. They have value only as part of the operating integrated property complex of the enterprise; instead, they are not able to generate cash flow, being separated from the rest of the component production complexes. Of course, it is impossible to establish a market value for specialized property, it can only be a matter of replacement value: either the replacement value or the reproduction value. Although the project the methodology envisages the use of such types of cost, a number of averaged indicators of the consolidated cost of works are provided for their determination, the clarification of which is currently ongoing.

At the same time, based on the definition of the market value [30], the prerequisite for calculating the value of damaged or destroyed real estate is the systematic updating of information on the market value of such property. In this connection, it is worth citing the provisions of Clause 22 of the Property Valuation Methodology No. 1891 [31], where it is stated that the conclusion on the value of the privatization object is valid for 18 months after the valuation date. Clause 22 of this Methodology [31] states that the term of validity of the conclusion on the value of the property drawn up for other cases of property appraisal provided for by this Methodology may be determined by the body that uses the results of such an appraisal, by the subject of appraisal activity—the subject by the economic entity conducting the assessment, taking into account the period of exposure of the object of assessment, if such a period is not determined by legislation. The term of validity of reports on expert assessment of the value of real estate is limited to 6 months. Of course, in the absence of a need to determine the value, the customers or owners of the property do not carry out its voluntary assessment. The need for such activity is primarily related to random, unsystematic factors: purchase and sale of real estate, registration of inheritance, division of property due to conflicts or the breakup of families, and provision of property as a pledge or insurance. For legal entities, periodic revaluation of assets, including real estate, is also economically unjustified, since the excess of their fair value over the residual value is reflected in the financial accounting as additional income. The consequence of recognizing such income will usually be an increase in financial results and tax liabilities. Therefore, business entities are forced to resort to expert property appraisal services only in case of urgent need, namely: the entry into the composition of business associations of new participants, or the exit of existing ones; business restructuring, as a result of which there is a merger, merger, separation or division of enterprises, finally, the determination of the collateral value of the property during lending, or the calculation of the insurance value of the property. However, the latter two cases are often based on non-market valuation bases, as the lender or insurer is interested in minimizing the risks of loan defaults and insurance payouts, respectively. Even in court decisions, the type of value is clearly defined, for the determination of which an expert assessment is carried out. According to clause 11 of NSA 1, the market value of the property is determined only in the absence of a clear definition of the type of value in the court decision [30].

With regard to the methodology for determining the market value of the real estate, it is advisable to give preference to a comparative methodical approach to evaluation, in accordance with which [30] it is necessary: first, to collect a sample of similar objects, for which the conditions of purchase and sale agreements, as well as payment, corresponding to the typical for the market of a similar property. Secondly, to take into account how the conditions on the market of similar property, which initially determined the formation of sale or offer prices, changed before the date of assessment. Only the prices of similar objects, the information about which meets the above conditions, have not changed significantly on the date of evaluation, or the changes that have occurred can be taken into account. It becomes possible to make adjustments for the differences between the objects of comparison and the object of evaluation is allowed only for the prices of similar objects, the information about which meets the above conditions.

The dynamic nature of changes in the value of the same object over relatively short periods is mainly determined by such an assessment principle as the principle of supply and demand. This principle, together with the substitution principle, forms the basis for the comparative approach.

In the aspect of real estate evaluation, this principle is transformed into five partial principles (Fig. 1) [32–37], which, in our opinion, should be supplemented with the principle of information asymmetry.

The principles related to the market environment derive from such a fundamental feature of real estate as externalism—the inextricable connection between real estate and its environment, manifested at all levels of economic interaction, from local to

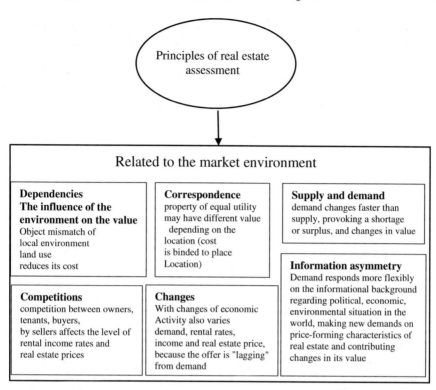

Fig. 1 Principles of real estate valuation related to the market environment. *Source* Compiled by the authors based on [32–37]

transnational. In particular, the principle of dependence is related to the assimilation of the appraised property to the surrounding land use, the principle of conformity gravitates to the location of the appraised object; the principle of supply and demand characterizes cyclical fluctuations in economic activity caused by investment requests of potential home buyers; the principle of change reflects the impact on the value of the real estate of the inertia of its consumer characteristics in the long term; the principle of competition also captures the time dimension, but at the same time it is closely related to the conditions of the most efficient use of property—to maximize the value of land, it is advisable to provide real estate for rent, differentiating rental rates for different tenants in different periods of time, even during one day. We suggested considering the principle of information asymmetry (Fig. 1), according to which permanent changes in the political, military, economic, and environmental conditions of local, regional, state, and continental economies and the world economic system are much faster reflected in the volume and structure of demand for real estate. At the same time, the offer of developers is characterized by much greater inertia, since a longer time interval is needed to change design solutions, substantiate new technologies, and select building materials, products, and kits, compared to the formation of solvent demand from buyers. Accordingly, the imbalance between the needs of consumers and the capabilities of producers will be reflected in housing shortages and price increases, but the creation and absorption of excess supply does not always significantly slow down the price growth of square meters. The named principles of real estate value formation, caused by the market environment, determine the evaluation process both in peacetime and in wartime.

Directly, the evaluation procedures according to the comparative approach are described in clause 49 of NSA 1 [30]; namely, it is established that the main elements of the comparison are the characteristics of similar properties based on its location, physical and functional features, terms of sale, etc. Adjustment of the value of similar property is carried out by adding or subtracting a monetary amount using a coefficient or percentage to the sale/offer price of the specified property or by combining them. In turn, it is advisable to justify correction coefficients or percentage adjustments with the help of mathematical tools, since, with its help, it becomes possible to correlate changes in the sale price of real estate objects with their price-forming characteristics. Today, among the methods of mathematical modeling of economic phenomena and processes, Data-mining and Big-data methods have gained increasing popularity, among which neural networks deserve special attention. In particular, in the active real estate market, when hundreds of sales transactions are carried out every day, there is a considerable amount of information about residential, commercial, and industrial objects, which is sufficient for the creation and training of neural networks. By means of the interpretation of the parameters of neural network models in combination with "traditional" methods of statistical analysis, it becomes possible to substantiate the ranges of corrective amendments, which, in fact, will make it possible to estimate the market value of the real estate before the moment of damage.

With respect to specific real estate, the value of which before damage or destruction can currently be estimated only as replacement value, NSA 1 directly allows the application of statistical methods. After all, according to clause 41 of this standard [30], along with the initial data about the object of evaluation, it is advisable to use average statistical indicators that summarize the conditions of its reproduction or replacement in modern prices.

In order to solve the problem of substantiating the assessment of the value of damaged or destroyed real estate before the damage was caused, we consider it expedient to conduct a study of the formation of the price of industrial real estate and substantiation of the main price-forming characteristics as of January–February 2022. The study is focused on such a segment of industrial real estate as warehouse real estate.

The initial data for the study was information from the unified database of reports on the evaluation of warehouse real estate, accumulated in the State Property Fund of Ukraine [38] from 01.21.22 to 02.04.2022. A total of 192 reports on the evaluation of such facilities as warehouse buildings, hangars, etc., and their parts were analyzed. In order to study the explicit and implicit relationships between value indicators and the main price-forming characteristics, the following information about the evaluation objects was taken into account:

- the price of a unit area of objects, which is defined as a dependent indicator in the modeling. To do this, we classified the sample into 3 groups.

  1. Objects with a high price.
  2. Objects with an average price.
  3. Objects with a low price.
     The rest of the given indicators are factorial, in particular:

- belonging of the settlement of the location of the object of assessment to the regional center, x_8:

  1. The facility is located in the regional center or suburban area;
  2. The location of the facility is far from the regional center

- area of the settlement, x_1:

  1. central;
  2. median;
  3. peripheral;
  4. suburbs;
  5. outside the settlement.

- classification of warehouses by classes, x_2:

  1. class "A";
  2. class "B";
  3. class "C";
  4. class "D";

- year of construction was finished;
- the area of the evaluated part of the warehouse building;
- floor space of the assessed building;
- wall material, x_5:

  1. brick, stone;
  2. monolithic;
  3. expanded clay concrete, perlite concrete, limestone blocks;
  4. others, concrete;

- engineering equipment, x_6:

  1. there is heating, electricity supply, water supply;
  2. there is electricity supply, water supply, no heating;
  3. there is water supply, there is no heating, electricity supply;
  4. without heating, electricity supply, water supply.

- physical condition of the premises, x_7.

  1. good;
  2. satisfactory;
  3. without decoration;

4. unsatisfactory.

For further neural network modeling, the serial number of the category was used, which ensures proper identification of each of the objects. Based on the date of commissioning of the building, the actual age of the building was calculated (x_3). The number of floors in the building (x_4) and the natural logarithm of the area of the evaluated part of the warehouse were included in the model (x_9). Logarithmization was carried out to ensure comparability of the values of quantitative values used in modeling, and, therefore, to increase the accuracy of the obtained dependencies.

First of all, the studied sample of indicators of the cost of a square meter of warehouse real estate was classified into 3 groups (Table 1). The first included 28 objects with high-cost indicators, which is 15% of the studied sample. On average, the price of a square meter of warehouse space for the group is UAH $12,218.3/m^2$, the standard deviation for the cluster is UAH $3,345/m^2$, the range of specific value variation is from 9,599.1 to 25,612.9 UAH/m^2. The second group, which includes 69 objects, or 36% of the sample, is characterized by an average price range, i.e. from UAH 4,565.2 to UAH $9,228.3/m^2$. At the same time, the cluster average value is UAH $6,757.2/m^2$, and the standard deviation is UAH $1,345.1/m^2$. Finally, the largest group was formed from 95 warehouses with the lowest specific value indicators. The share of this group in the total sample is 49%. This group is characterized by an average value of UAH $2,296/m^2$, and a standard deviation of UAH $1,121.2/m^2$. The limits of price variation in this part of the total sample are from UAH 4,565.2 to UAH 9,228.3/m^2. Thus, 3 non-intersecting cluster groups were obtained. The statistical significance of the cluster breakdown of warehouse real estate objects (Table 1) is also evidenced by the high value of the F-criterion, which is 407.6323 and exceeds the critical level of significance $p = 0.01$ ($F_{0.01;3;188} = 3.887$). Therefore, when determining the market value of a warehouse facility at the time before the school was opened, it should be taken into account that, depending on the location and technical characteristics of the building, warehouses of category C and below can be characterized by a high, average, or low specific cost per square meter. As of February 4, 2022, the high price is more than UAH $9,600/m^2$, the low price does not exceed UAH $4,500/m^2$, and the price between UAH 4,500 and $9,600/m^2$. should be considered average.

Table 1 The results of the cluster analysis of indicators of the cost of 1 m^2 of warehouse real estate (Calculated by the authors)

Cluster	Number of observations	Minimum value, hryvnias/m^2	Maximum value, hryvnias/m^2	The average value, hryvnias/m^2	Standard deviation, hryvnias/m^2
High prices 1 m^2	28	9,599.1	25,612.9	12,218.27	3,344.954
Average prices 1 m^2	69	4,565.2	9,228.3	6,757.183	1,345.126
Low prices for 1 m^2	95	556.6	4,363.6	2,295.974	1,121.192

The market-averaged value of the unit must be adjusted for discrepancies between the evaluated object compared to the typical value for the market. To justify the technique of making adjustments and the corrective amendments themselves, the distribution of the existing sample of objects by categories of the main price-forming characteristics was studied (Fig. 2).

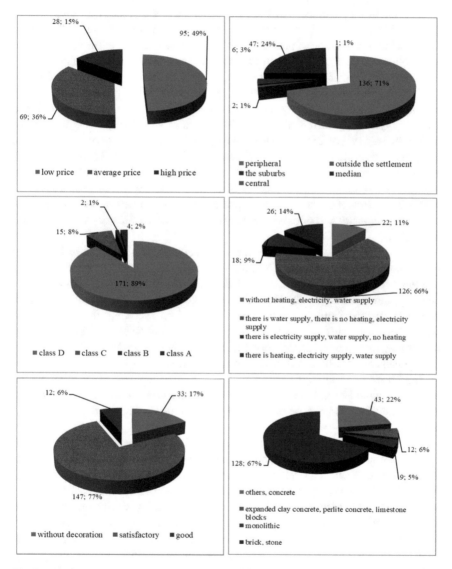

Fig. 2 Distribution of sample objects by categories of main price-forming characteristics (Compiled by the authors)

From the charts, it can be seen that a significant number of the evaluation objects belong to the low price range—49%, however, many objects are classified as having an average price, and this is 36% of the sample. the lion's share of offers—136, or 71%—was located outside the settlement, but almost a quarter of the warehouses, i.e. 24%, were located in the middle zone of the city. The vast majority of warehouses belong to the "D" category—89% of the sample, i.e. 171 objects, the "C" category is represented by only 8 objects; on the other hand, the offers of the best warehouses in terms of their consumer characteristics are single—only 3% for "B" and "A" together. Good technical condition of premises was observed in only 6% of the sample (these are 12 objects), which is almost 3 times less than the number of analogs in satisfactory condition—17% (or 33 objects). The vast majority of market properties, 146 out of 192, or 77%, are undecorated. Almost 2/3 of warehouses, i.e. 126 unheated and not equipped with water supply systems, much less—9% (or 18 objects) are equipped with all engineering communications. On the contrary, in a slightly larger number of facilities—11% (or 22 warehouses) there is no water, electricity or heating. Thus, the share of unheated warehouses in the market offer exceeds 90%. Also, 2/3 of the supply of storage facilities is located in brick or stone buildings, but the share of objects that are located in buildings of composite concrete blocks turned out to be significant—22%. The latter, in their quantity, exceed the part of the offer located in buildings constructed of expanded clay concrete and perlite concrete blocks and monolithic.

At first glance, according to the diagrams (Fig. 2), it is possible to imagine a typical representative of the offer on the warehouse real estate market as an object of a low-price category, class "D", located outside the settlement, in a brick or stone building, and in such a the object lacks decoration and engineering communications. However, the distribution of warehouses by category according to various price-forming characteristics does not clearly match. Thus, any object can simultaneously fall into the most numerous category, for example, by location factor, and into a less widespread category by engineering support or another factor.

The above considerations are in favor of the development of a mathematical model that would be able to establish hidden connections between the quality and cost parameters of warehouse buildings in order to justify their market value before damage.

Management in conditions of uncertainty makes high demands on the materiality of information and methods of its processing. In particular, attention is being paid to the use of mathematical models for forecasting changes in value and its factors. This is manifested in the sharpening of attention not only to the statistical characteristics of the model (confidence interval, root mean square error, coefficient of multiple determination) or the economic meaning of the constants of the prognostic equation. A significant number of potential users of economic-mathematical prognostic models will agree to use this powerful tool of strategic management of monitoring and controlling cost changes only under the conditions of simplification of the model formation procedure and ease of use. That is, the excessive complexity of the model with mathematical operations (degrees of variables, exponents, or logarithms) and a considerable number of independent arguments, which are necessary for qualitative approximation of the original set of observations, do not contribute to the ease of use of this model. That is why interest in such technologies as neural network modeling, data mining, and fuzzy logic tools has recently been growing. Providing approximately the same accuracy of approximation of initial information and short-term forecasts, compared to traditional linear regression equations, the latter simplifies the procedure for calculating predictive values of cost indicators, reduces time spent, and expands opportunities for choosing priorities when making important management decisions. Let's dwell in more detail on the principle of non-networks as an algorithm for the process of forecasting changes in the cost characteristics of an enterprise in unstable operating conditions.

The growing interest in neural networks is explained by their successful application in a wide variety of fields of activity to solve the problems of forecasting, classification, and management [39, 40]. Such characteristics of neural network methods as the possibility of nonlinear modeling and relative simplicity of implementation often make them indispensable for solving the most complex multidimensional problems. Neural networks are non-linear in nature and represent an extremely powerful modeling method that allows for the reproduction of extremely complex dependencies [39, 40]. For many years, linear modeling has been used as the main method in most financial studies, as it has well-established optimization procedures. Where linear approximation is unsatisfactory and linear models perform poorly, neural network models become the primary tool. In addition, neural networks overcome the "curse of dimensionality", which does not allow modeling linear relationships in the case of a large number of variables.

Neural networks learn from examples. A neural network user selects a sample and then runs a learning algorithm that automatically perceives the structure of the data. At the same time, only a heuristic set of knowledge is required from the user on how to collect and prepare data, choose the necessary network architecture, and interpret the results. However, the level of knowledge required for the successful application of neural networks is much lower, compared to the case of using traditional statistical methods.

Neural networks are based on a primitive biological model of nervous systems. The idea of neural networks appeared in the course of research in the field of artificial intelligence, namely as a result of attempts to reproduce the ability of biological nervous systems to learn and correct errors. Ratterson (1996) Neurons (the brain is made up of them) are special cells capable of transmitting electrochemical signals. A neuron has a branched input structure (dendrites), a nucleus, and a branched axon. The axons of the cell are combined with the dendrites of other cells with the help of synapses (Fig. 3). When activated, a neuron sends an electrical impulse along its axon. Through synapses, this signal reaches other neurons, which can also be activated. A neuron is activated when the total level of signals that have reached its nucleus from dendrites exceeds a certain level (activation threshold).

The intensity of the signal received by the neuron (hence the power of its activation) strongly depends on the activity of synapses. Each synapse has a length and special chemicals transmit signals along it Reflecting the essence of biological neural systems, the definition of an artificial neuron is as follows:

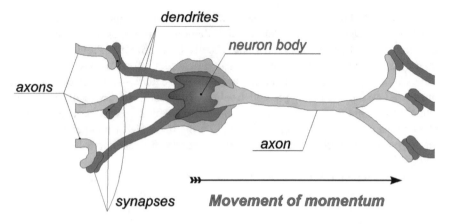

Fig. 3 Biological neuron

– It receives input signals (output data or output signals of other neurons in the
  network) through multiple input channels. Each input signal passes through a
  connection having a defined intensity (or weight); this weight corresponds to the
  synaptic activity of a biological neuron. Each neuron is associated with a certain
  threshold value. The weighted sum of the inputs is calculated, the threshold value
  is subtracted from it, and the neuron activation value is obtained as a result. It is
  also called the postsynaptic potential of the neuron—PSP);
– The activation signal is transformed using the activation function (or transfer
  function) and the result is the input signal of the neuron.

Usually, in practice, threshold functions are used quite rarely in artificial neural
networks. Therefore, the basic model of an artificial neuron can be created using
linear and non-linear functions.

The simplified functioning of the neuron is shown in Fig. 4 and involves the
following stages:

(1) the neuron receives a set (vector) of input signals from the dendrites $[x_1, \ldots x_n]$;
(2) in the body of the neuron, the total value of the input signals is evaluated.
    However, the inputs of the neuron are unequal. Each input is characterized
    by some weighting factor that determines the importance of the incoming infor-
    mation. Thus, the neuron does not simply sum the values of the input signals
    but calculates the scalar product of the vector of input signals and the vector of
    weighting coefficients.
(3) the neuron forms an output signal, the intensity of which depends on the value
    of the calculated scalar product. If it does not exceed some given threshold, then
    the output signal is not formed at all—the neuron "does not fire";
(4) the output signal enters the axon and is transmitted to the dendrites of other
    neurons.

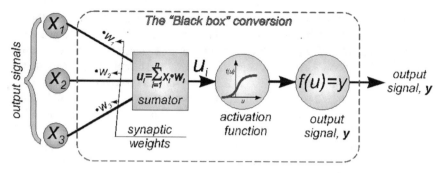

Fig. 4 Artificial neuron

The practice of building neural networks in various fields of knowledge has made it possible to formulate the following rule: if the activation function of the intermediate layer is linear, the output function should be chosen to be non-linear and vice versa, although in some cases both functions can be non-linear.

The behavior of an artificial neural network depends both on the value of weight parameters and on the excitation function of neurons. The three main types of activation (excitation) functions are most common: threshold $(if\ x\ (or\ u) \geq 0,\ f(x) = 1;\ if\ x > 0,\ f(x) = 0)$ linear and sigmoidal. the latest version of the package "Statistica 8" additionally provides an exponential function and a hyperbolic tangent. It is the function of the hyperbolic tangent that is quite successfully used to predict the initial indicators and ensures the accuracy of the linear model compiled on the initial layer. This has an important applied value since it is quite easy to give an economic interpretation to the original linear equation of the forecast of cost indicators.

One of the advantages of neural networks is their similarity to a "black box"— the application of nonlinear activation functions in the body of a neuron and the application of a linear function of transformation into an output signal remain out of sight of the user, who is presented with the already calculated result. However, the active use of neural network models in the practice of market value management largely depends on the understanding of their essence by the business personnel of operating enterprises, therefore, the "hiding" of procedures for calculating the final result can be attributed to the advantages of the method with great doubt. The activation function of the input variables chosen in this study—the hyperbolic tangent—is recommended to be used when the values of the indicators fluctuate with a significant amplitude (have high values of variation and deviations):

$$th(x) = \frac{e^x - e^{-x}}{e^x + e^{-x}}$$

Each of the input variables is taken into account in the exponent with the corresponding weight factor, the value of which is calculated during network training. The obtained values of hyperbolic tangents for each of the variables are summed up, as a result of which an enlarged argument of the output function will be obtained, if the chosen type of neural network is a three-layer perceptron, or the input variable for the next layer, if the architecture of the neural network has the form of a four-layer or more perceptron. The network architecture is a sequence of layers of neurons and their connections. Creating a network consists in determining the architecture and parameters of the neurons. To solve most problems (economics, technology, medicine), it is enough to create a three-layer perceptron, which has one intermediate layer between the input and output variables. The number of variables of the intermediate layer is determined by the volume of the input data sample: for large arrays of input information, the number of intermediate layer neurons should be an order of magnitude lower, with small samples, the optimal number of intermediate neurons is 2 or 3 times less. In our study, only 2 intermediate neurons were enough for the intermediate layer. From an economic point of view, they can be interpreted as aggregated factors of change in the result, which is similar to the statistical method of analysis of the main components. The principle in developing a high-quality prognostic model is the training of the network—the selection of free parameters in order to adapt the calculation results to the provided free parameters (sets of independent variables, $x_1, \ldots, x_n$ and their corresponding actual values of the dependent outcome variable y).

The learning process can be carried out both with a teacher who controls the learning process using examples with known solutions (supervised learning) and learning without a teacher (unsupervised learning) [39, 40]. When learning with a teacher, a signal from the external environment arrives at the input of the learning system. After processing the signal by the "student" system, the obtained result is compared with the "correct answer of the teacher"—the available value of the resulting variable. The difference between the correct solution and the network response represents an error that should be reduced by adjusting the free parameters—the model coefficients. For this, the sum of squared errors is constructed, which is a function of the free parameters of the network. The process of learning the network, which consists of finding the minimum of this function, can be repeated many times, that is, it represents a certain number of cycles—epochs, which are adjustable by the user. Schematically, the learning process of the neural network is shown in Fig. 5.

Too many epochs cause the network to "overtrain" as the system adapts its response to the input data. At each epoch, the raw data is supplied to the trained system not in a strict sequence, but in a chaotic order, which provides better adaptive capabilities of the obtained model for making forecasts in conditions of uncertainty. From the point of view of the practical use of neural networks in the management of the value indicators of enterprises, this allows spending less time on the preparation of input information, because the available observations can be submitted to the system in a non-chronological order and without taking into account the subject of the occurrence of independent variables, if the analyzed enterprises have a certain common classification feature: number of employees, level of profitability, cost and structure of assets. From the provided input information, the system randomly selects a user-defined share for training, on the basis of which independent coefficients are calculated. The rest of the sample is the "testing" part: the initial data are inserted into

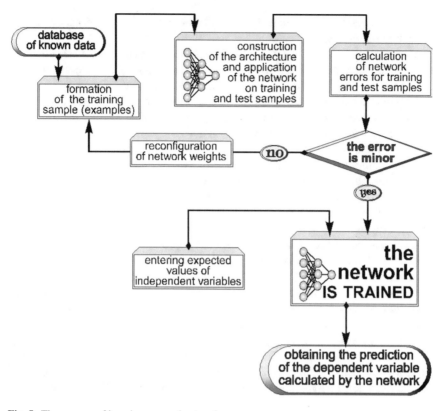

Fig. 5 The process of learning a neural network

the model created during training, and the calculated results are compared with the initial values of the result. A neural network is considered well-trained and of better quality if the error (most often the root mean square deviation) of the test subset is lower than the error in the training subset. Accordingly, the correlation coefficient of test results should be higher than the correlation of training results. The training process is considered complete if the correlation and error indicators do not improve during several cycles of network training. More attractive from the user's point of view is a neural network with a less complex structure (it will reduce the time of information processing to obtain a forecast) and higher approximation indicators of input results (correlation of the output data of the training system with the calculated ones) [39, 40]. With a relatively small array of input data, the statistical characteristics of neural networks approach the similar characteristics of linear regression equations, but this does not mean abandoning one in favor of the other. On the contrary, in the conditions of uncertainty caused by the instability of external factors, the enterprise should diversify not only the structure of assets or types of activities, but also methods of management, control, and forecasting. Therefore, when justifying strategic management decisions, it is worth taking into account various options for estimating important parameters.

Modern application packages for the development of neural networks allow you to simultaneously develop and train a considerable number of networks, and then choose the best one based on aggregated statistical characteristics.

As a neural network model, we propose a two-layer neural network with three neurons in the hidden layer, and one output (Fig. 6). Similar models are also called multilayer perceptron, and our development makes it possible to determine to what extent the value of the evaluated object differs from the average market value. After all, the output of the perceptron is the logarithm of the ratio of the value of 1 square meter of the object of assessment and the average specific value determined from the market sample. The natural logarithm of the ratio of economic indicators, in particular the value, can be characterized as the power of its growth for the evaluated object, compared to the average value. In this way, it was possible to ensure not only the sufficient reliability of the neural network approximation but also to take into account the dynamic nature of the value indicators. After all, in any state of the real estate market, regardless of whether there will be an increase in prices in the event of excess demand, or whether the value of square meters will "sag" due to economic stagnation, with the help of a two-layer perceptron, it becomes possible to determine to what extent the value of the object of evaluation will not coincide with the cost of an analog typical for the warehouse real estate market.

Neural network calculation involves a number of computational procedures. The activation function of 9 input variables is implemented on the first hidden layer ($x_1 - x_9$). This function is a hyperbolic tangent, and its argument is a linear combination of

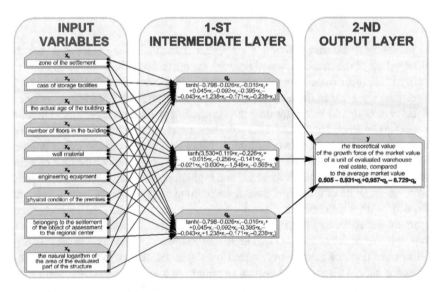

Fig. 6 Neural network model for determining the degree of excess of the value of the estimated warehouse over the average market value (Author's development)

input values containing a free constant, or the displacement of the first layer according to the terminology of neural networks:

$$q_1 = \tanh(-0,798 - 0.026 \cdot x_1 - 0.016 \cdot x_2 + 0.045 \cdot x_3 - 0.092 \cdot x_4$$
$$-0.395 \cdot x_5 - 0.043 \cdot x_6 + 1.238 \cdot x_7 - 0.171 \cdot x_8 - 0.238 \cdot x_9) \tag{1}$$

$$q_2 = \tanh(3.530 + 0.119 \cdot x_1 - 0.226 \cdot x_2 + 0.015 \cdot x_3 - 0.256 \cdot x_4$$
$$-0.141 \cdot x_5 - 0.021 \cdot x_6 + 0.600 \cdot x_7 - 1.54 \cdot x_8 - 0.565 \cdot x_9) \tag{2}$$

$$q_3 = \tanh(-3.864 + 0.293 \cdot x_1 + 9.474 \cdot x_2 - 1.836 \cdot x_3$$
$$-0.428 \cdot x_4 + 0.435 \cdot x_5 + 1.789 \cdot x_6 - 3.634 \cdot x_7 - 13.666 \cdot x_8 - 2.854 \cdot x_9) \tag{3}$$

The choice of the hyperbolic tangent as the activation function of the input layer is due not only to an attempt to improve the approximation properties of the network but also to the need to provide an economic interpretation of the model. The second layer of the neural network implements a linear combination of the 3 neurons of the first layer, while also taking into account a free constant - the displacement of the second layer:

$$y = 0.505 - 0.931 \cdot q_1 + 0.957 \cdot q_2 - 8.729 \cdot q_3 \tag{4}$$

The output of the second layer, y, is precisely the theoretical value of the growth force of the market value of a unit of evaluated warehouse real estate, compared to the average market value.

The training of the network was carried out in the MATLAB software environment, during which 1000 iterations were performed. As a result, the root mean square error of the network was 0.258. At the same time, the "traditional" linear regression for these data is characterized by an approximation error of 0.767, which makes it unsuitable for practical use.

Given the high reliability of the two-layer perceptron approximation, it is worth providing an economic interpretation of its components. Of course, it is much easier to interpret the parameters of the 2nd layer (4). Its displacement is equal to -5.569, respectively, for zero values of intermediate neurons, that is, intermediate variables $q1$, $q2$, and $q3$, it can be argued that the influence of factors of sales conditions, as well as price-forming characteristics of warehouses, which were not included in the model, leads to the fact that the cost of a specific object may be only 0.34% of the market average ($0.0034 = \exp(-5.569)$). The cost of a square meter of warehouse space may additionally decrease due to the inconsistency of the technical parameters of the object with the requirements of consumer demand. This will be observed if the values of the first (1) and third (3) neurons of the intermediate layer increase by 0.1 point—accordingly, the cost of a square meter will be adjusted by the coefficients of $0.54(0.5387) = \exp(-6.186 \cdot 0.1)$ and $0.9(0.899) = \exp(-1.061 \cdot 0.1)$. The value of the adjustment coefficients was obtained on the basis of the weighting factors of the 2nd layer of the neural network (2). However, the cost of the composition will increase with high values of the second neuron of the intermediate layer. When the value of this neuron increases by one-tenth of a unit, the cost of a square meter of the warehouse can increase by 7% with zero values of the remaining neurons $(1.071 = \exp(0.688 \cdot 0.1))$.

On the contrary, if the value of the first (q_1) and third (q_3) intermediate neurons increases by 1 tenth, the value of the object of evaluation will increase against the average market indicator by 1.86 and 1.11 times, respectively, or decrease by 7%, when by 0.1 the value of the second neuron will decrease. For the economic interpretations of the weighting factors and the displacement of the 2nd layer of the neural network, such a feature of the activation function of the input layer, that is, the hyperbolic tangent, that the values of the three intermediate neurons are in the range from -1 to $+1$, is taken into account. In turn, the values of these intermediate neurons reflect the rate of overtaking or inhibition of the value of the object of evaluation against the average market indicator.

The first and third neurons are stimulators of the increase in real estate value since the negative values of q_1 and q_3 ensure its multiplicative growth against market indicators. Of course, these neurons have negative values for warehouses with better consumer characteristics, that is, in the case of input values $x_1 \ldots x_9$, not greater than 2. At the same time, the maximum increase in value that the network can predict is 7 times the average market value.

Instead, the second neuron of the intermediate layer is a destimulator, since the value of q_2 increases as the inputs of the neural network increase, as a result of which the cost of 1 square meter of the worst warehouse real estate offers for consumers will be only 46.88% of the average market value.

Based on the neural network model (1)–(4), the following algorithm for a methodical approach to determining the market value of warehouses has been developed:

Stage 1: identification of the object of evaluation and determination of its main price-forming characteristics, each of which is numbered. The list and numbering of categories for each characteristic are given above.

Stage 2: calculation of values of neurons of the intermediate layer, q_1–q_3, according to formulas (1)–(3).

Stage 3: calculation of the output of the neural network, y, according to formula (4), that is, the "growth force" of the value of the warehouse real estate object, compared to the average market value.

Stage 4: determination of the estimated cost of one square meter (Vo_1) according to (5) and the entire warehouse (Vo) according to (6):

$$Vo_1 = \exp(y) \cdot Vsr.r \qquad (5)$$

$$Vo = Vo_1 \cdot S \qquad (6)$$

To do this, you must first multiply the exponent of the output of the neural network $(\exp(y))$, i.e. the coefficient of excess, by the average market value of the unit of area (Vsr.r) according to the formula (5). By multiplying the obtained result (Vo_1) by the area of the object of assessment (S), the market value of the object of warehouse real estate is established according to (6).

Stage 5: Conduct a verification calculation of the value of warehouse real estate in the "traditional" way, that is, by making percentage adjustments to the average market value. As of February 4, 2022, or at the time before the damage occurred, we established the number of percentage adjustments for warehouse real estate objects on the basis of the first cluster of the studied sample. The narrowing of the initial array of information is caused, firstly, by the fact that the objects with the highest value index are characterized by the maximum heterogeneity in the main price-forming characteristics. Secondly, most of the approximation errors of the neural network are due to the underestimation of the theoretical value of the cost per square meter of the 1st cluster versus the actual value. All percentage corrections were justified by comparing the cost of a square meter, averaged across the differences in the values of the main price-forming characteristics. First of all, the influence of the "area of the settlement" factor was taken into account: the location of the object in the peripheral zone causes an increase in the value of the object by 5% (4.73%), compared to the location in the middle zone. After making such an amendment, the amount of the percentage adjustment for the class of the object was substantiated: warehouses of class "C" are more expensive than their counterparts of class "D" by almost 26% (25.73%). The next adjustment involved the building age factor: warehouses located in buildings under 12 years old are 6.5% (6.47%) more expensive than others. After adjusting the cost of objects for this factor, the impact of differences in the wall material was determined: the most expensive objects built from monolithic reinforced concrete are ahead of analogs from prefabricated reinforced concrete and metal by 15%, on the other hand, analogs from brick and natural stone are inferior to monolithic ones already by 22%. In turn, stone and brick objects lag behind the cost of metal, block, and other materials warehouses by 8.5%. When these amendments were made, it turned out that the presence of heating or water supply does not significantly affect the cost of warehouse premises, the corresponding amendments amount to 2%. On the other hand, the availability of electricity supply is fundamental, due to which warehouse real estate objects can become more expensive by 5%. Finally, the correction for physical condition was justified: warehouses in good condition are more expensive than those in satisfactory condition by 5%, and the lack of decoration reduces the cost by another 14%. Thus, the difference in the cost of a square meter of an unfurnished warehouse compared to real estate in good condition reaches 19%.

Finally, to determine the amount of damage caused by the aggression of the russian federation against Ukraine, it is worth taking the larger of the market value indicators determined in stages 4 or 5.

It is quite possible to implement the proposed algorithm "manually", but the success in the process of digitalization of all spheres of national economy management and the active use of cloud technologies make it possible to create a software application capable of instantly performing all the proposed calculations with minimal costs.

The brutal, inhumane aggression of the Russian Federation caused large-scale destruction of housing, industrial buildings, and infrastructure facilities. Restoration of damaged and lost objects is associated with considerable costs, the amount of which is almost impossible to estimate at present. Despite the unprecedented steps taken by the Government of Ukraine, primarily the simplification of procedures for the import of building materials, the problem of rising construction costs will unfortunately not be avoided. With the goal of the fastest possible post-war reconstruction of Ukraine, there will be an increase, first of all, in the demand for construction materials, products, and kits in most international markets. At the same time, world prices will certainly increase. In turn, the cost of construction and the need for increased investment will increase. Such a problem will be felt most acutely by all participants in housing construction, since, in post-war conditions, such projects will be deprived of a commercial component. However, for commercial real estate, in particular warehouses, this problem does not lose its acuteness.

Losses from damage and destruction of objects, as discussed earlier, are determined by the amount of costs for their elimination. However, as with any depreciation of assets, the expediency of capital investments aimed at eliminating physical and functional wear and tear is determined primarily by the amount of growth in the value of the restored property. Under the conditions of a constant threat of new destruction, a catastrophic shortage of electricity, broken logistics connections, and a considerable risk of devaluation of the national currency, it is natural for prices to rise in the business sector. In other words, we should expect an increase in the prices of construction materials, products, and kits both during the development of design and estimate documentation for the elimination of destruction and damage, and during the execution of works. Of course, such conditions will negatively affect the investment opportunities of real estate owners, construction customers, and external creditors. Therefore, already at the pre-investment stage, it is necessary to substantiate the financial needs of each construction project as accurately as possible, and considerable attention should be paid to determining the amount of additional investment capital required due to the increase in the cost of construction resources. To provide informational and methodological support for this stage of the development of the damage elimination project, we also consider it appropriate to develop a neural network model.

The fundamental basis for the proposed methodical approach was the theory of induced investments, the interpretation of its postulates for the post-war recovery of Ukraine is presented in Fig. 7.

The methodical approach is developed in the form of a two-layer neural network (Fig. 8), the inputs of which are market data on the prices of construction materials, which are a priority for eliminating the destruction of partially habitable buildings. First of all, these are decorative materials that are the first to lose their properties

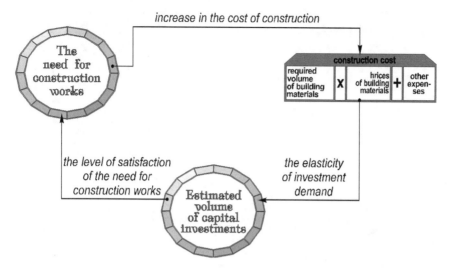

Fig. 7 Relationship between the cost of construction materials, products, kits, and the need for investment capital (author's development)

due to destruction caused by combat operations. The output of the network is the forecast value of the growth rate of capital investment in residential construction in the six-month perspective. To ensure the correctness of neural network calculations, as well as the proportionality of the results both at the input and at the output, they are determined in standard deviations from the average value for Ukraine.

The training and test samples were formed based on the data of the construction materials market of all regions of Ukraine for the 1st quarter of 2021 and statistical information on the volume of capital investments in housing construction based on the results of the three quarters of 2021. The training of the network was carried out by performing 395 iterations, as a result of which the root mean square error of the network was determined at the level of 0.237. At the same time, the "traditional" linear regression for these data is characterized by an approximation error of 0.491. Despite the certain complexity of the calculations (Fig. 8), the outputs of each of the layers of the neural network can be given an economic interpretation: after linear transformation of the input data, the argument of the hyperbolic tangent is determined, which is the activation function and the output of the 1st layer. Given its mathematical notation, in particular, the use of an exponent as the basis for a continuous interest rate, in Fig. 8, the concepts of "reserve of growth power" and "accelerator" of the growth power of capital investment in housing construction are proposed.

The second layer of the neural network (Fig. 8) involves a linear transformation of the output of the 1st layer, which provides a standardized value of the predictive index of capital investment in residential construction as a result.

To determine the additional need for capital investments after six months, the result of the neural network (Fig. 8) must be multiplied by the standard deviation of this index in the current period and increased by its current value, averaged across

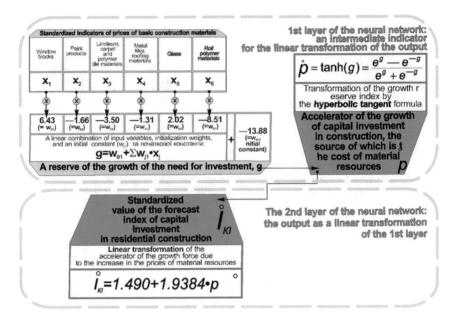

Fig. 8 A neural network for predicting the increase in the need for capital investments in the medium term (6 months) (author's development)

Ukraine as a whole. The obtained indicator should be reduced by one. The result will be the percentage value of the additional need for the investment resource.

3 Conclusions

In order to objectively assess the damage caused to real estate objects as a result of military aggression, it would be necessary to ensure a systematic revaluation of the value of all buildings and structures in the country. Practically, it is impossible to implement this condition, and therefore, the problem of clarifying the value indicators of even those real estate objects whose expert evaluation has already been carried out more than once remains relevant. The short-term validity of existing evaluation reports is explained by the operation of evaluation principles: dependence, conformity, supply and demand, competition, and change. Given the current economic, social, political, environmental, and informational realities in Ukraine and the world, we consider it expedient to justify another principle of real estate valuation related to the market environment—the principle of information asymmetry. Its essence is that the demand responds more flexibly to the informational background regarding the political, economic, and environmental situation in the world, making new demands on the price-forming characteristics of real estate and causing changes in its value. In order to determine the market value of the real estate at the time preceding the damage, it is advisable to use a comparative approach and to substantiate the impact of the main price-forming characteristics on the value of the real estate, it makes sense to use statistical methods, as well as technologies of intelligent data analysis. The study of reports on the evaluation of warehouse premises made it possible to divide this type of real estate into three cluster groups: objects with a high value: as of 02.04.2022—more than 9,600 UAH/m^2, objects with a low value that did not exceed 4,500 UAH/m^2. At the same time, the price in the range from 4,500 to 9,600 hryvnias/m^2 should be considered average.

The research proposed a model in the form of a two-layer perceptron with three neurons in the hidden layer. The output of the neural network is the value of the growth force of the value of a unit of the area of the evaluated object of warehouse real estate against the average market indicator. Such a model takes into account the dynamic nature of the price of warehouse real estate and is characterized by a high level of approximation reliability. however, in order to improve such a traditional procedure of the comparative approach as the introduction of percentage corrections, the following indicators were substantiated:

- the location of the object in the peripheral zone causes an increase in the value of the object by 5%, compared to the location in the middle zone;
- "C" class warehouses are 26% more expensive than "D" class counterparts;
- warehouses located in buildings under 12 years old are 6.5% more expensive than others;
- the most expensive objects built from monolithic reinforced concrete are ahead of analogs from prefabricated reinforced concrete and metal by 15%, on the other hand, analogs from brick and natural stone are inferior to monolithic ones already by 22%. In turn, stone and brick objects lag behind the cost of metal, block, and other materials warehouses by 8.5%;
- the presence of heating or water supply does not significantly affect the cost of warehouse premises, the corresponding corrections amount to 2%. Instead, it is fundamental to have an electricity supply, due to which objects of warehouse real estate can become more expensive by 5%;
- warehouses in good condition are more expensive than those in satisfactory condition by 5%, and the lack of decoration reduces the cost by another 14%.

Prospects for further research are the construction of similar neural network models and the justification of corrective amendments for other types of real estate: administrative, commercial, residential, and land plots.

References

1. Voytko, O.: Main statistical indicators of economic damage assessment from natural emergencies [Electronic resource]. Effi. Econ. **11** (2023). http://nbuv.gov.ua/UJRN/efek_2015_11_128
2. Kolmakova, V.: Peculiarities of ecosystem assessment of damage from salvo and other accidental pollution (on the example of water resources) [Electronic resource]. Econ. Nat. Use Environ. Prot. **15**, 98–105 (2017). http://nbuv.gov.ua/UJRN/epod_2017_2017_15
3. Kraynyk, O.: The dominant role of assessing ecological and economic damage of the national economy in the conditions of regulating the destructive effects of economic activity [Electronic resource]. Econ.: Realities Time **3**, 206–212 (2015). http://nbuv.gov.ua/UJRN/econrch_2015_3_32
4. Kuzmin, O., Bublyk, M., Rybytska, O.: Model of economic assessment of man-made losses in the national economy by factors of influence, perception and evaluation [Electronic resource]. Econ. Innov. (58), 138–143 (2014). http://nbuv.gov.ua/UJRN/ecinn_2014_58_15
5. Tretyak, A.M., Tretyak, V.M., Kapinos, N.O.: Assessment of damage caused to the nature reserve fund located on the territories of the illegally annexed Autonomous Republic of Crimea and temporarily occupied parts of the Donetsk and Luhansk regions [Electronic resource]. Land Manag. Cadastre Land Monit. (3), 65–71 (2016). http://nbuv.gov.ua/UJRN/Zemleustriy_2016_3_11
6. Kharitonov, K.: Regarding the formation of a system of indicators for assessing losses of the national economy in the period of armed conflict (war) [Electronic resource]. Soc. Dev. Secur. **10**(6), 119–126 (2020). http://nbuv.gov.ua/UJRN/socdevsec_2020_10_6_13
7. Nepomnyashchyy, O., Radchenko, O., Marusheva, O., Prav, Y., Shandryk, V.: The public policy implementation mechanisms in the context of digitalization: economic affairs (New Delhi) this link is disabled **64**(4), 787–794 (2022)
8. Bielienkova, O., Stetsenko, S. Oliferuk, S., Sapiga, P., Horbach, M., Toxanov, S.: Conceptual model for assessing the competitiveness of the enterprise based on fuzzy logic: social and resource factors. In: IEEE International Conference on Smart Information Systems and Technologies (SIST), pp. 1–5 (2021). https://doi.org/10.1109/SIST50301.2021.9465923
9. Fedun, I., Stetsenko, S., Tsyfra, T., Valchuk, B., Valentyna, A.: Innovative software tools for effective management of financial and economic activities of the organization. Lect. Notes Netw. Syst. **485**, 17–38 (2023)
10. Bielienkova, O., Novak, Y., Matsapura, O., Zapiechna, Y., Kalashnikov, D., Dubinin, D.: Improving the organization and financing of construction project by means of digitalization. Int. J. Emerg. Technol. Adv. Eng. **12**(8), 108–115 (2022)
11. Vorobec, S., Voytsekhovska, V., Zahoretska, O., Kozyk, V.: The context of the circular economy model implementation, based on indicators of the european union in/for Ukraine by means of fuzzy methods. In: Kryvinska, N., Greguš, M. (eds.) Developments in Information and Knowledge Management for Business Applications. Studies in Systems, Decision and Control, vol. 421. Springer, Cham (2022). https://doi.org/10.1007/978-3-030-97008-6_4
12. Ryzhakova, G.V. et al.: Modern structuring of project financing solutions in construction. In: 2022 IEEE International Conference on Smart Information Systems and Technologies (SIST) (2022)
13. Reznik, N., et al.: Systems Thinking to Investigate the Archetype of Globalization. Springer International Publishing (2022). https://doi.org/10.1007/978-3-031-08087-6_9
14. Chupryna, I., et al.: Designing a toolset for the formalized evaluation and selection of reengineering projects to be implemented at an enterprise. Eastern-Euro. J. Enterp. Technol. (2022). https://doi.org/10.15587/1729-4061.2022.251235
15. Akselrod, R., Shpakov, A., Ryzhakova, G., Honcharenko, T., Chupryna, I., Shpakova, H.: Integration of data flows of the construction project life cycle to create a digital enterprise based on building information modeling. Int. J. Emerg. Technol. Adv. Eng. **12**(01), 40–50 (2022). https://doi.org/10.46338/ijetae0122_02

16. Li, Y., Biloshchytska, S.: The problem of choosing a diversification strategy for a building enterprise in risk condition. Sci. Bull. Uzhhorod Univ. **2**(35), 119–126 (2019)
17. Romali, N.S., Yusop, Z.: Flood damage and risk assessment for urban area in Malaysia. Hydrol. Res.. Res. **52**(1), 142–159 (2021)
18. Brémond, P., Grelot, F.: Review article: economic evaluation of flood damage toagriculture– review, and analysis of existing methods. Nat. Hazards Earth Syst. Sci. **13**(10), 2493–2512 (2013)
19. Mohd Mushar, S.H., Kasmin, F., Syed Ahmad, S.S., Kasmuri, E.: Flood damage assessment : a preliminary study. Environ. Res. Eng. Manag. **75**(3), 55–70 (2019)
20. Sorokina, L.V.: Improving the procedure of forecasting changes in financial condition in construction works by means of two-stage model of fuzzy inference. Actual Probl. Econ. **120**(6), 285–293 (2011)
21. Dottori, F., Figueiredo, R., Martina, M.L.V., Molinari, D., Scorzini, A.R.: INSYDE: a synthetic, probabilistic flood damage modelbased on explicit cost analysis. Nat. Hazards Earth Syst. Sci. **16**(12), 2577–2591 (2016)
22. Tytok, V., Bolila, N., Ryzhakov, D., Pokolenko, V., Fedun, I.: CALS–technology as a basis of creating modules for assessment of construction products quality, regulation of organizational, technological and business processes of stakeholders of construction industry under the conditions of cyclical and seasonal variations. Int. J. Adv. Trends Comput. Sci. Eng. **10**(1), 271–276 (2021). https://www.warse.org/IJATCSE/static/pdf/file/ijatcse381012021.pdf
23. Stetsenko, S., Bolila, N., Moholivets, A., Gavrilyuk, V.: Statistical and analytical aspect of the construction industry development. Shliakhy pidvyshchennia efektyvnosti budivnytstva v umovakh formuvannia rynkovykh vidnosyn **46**, 188–196 (2020)
24. Kishchenko, T.E., Gusarova, L.V., Bolila, N.: Development methodology for implementation of construction investment projects. [Electronic resource]. Effi. Econ. **6**, 57–61 (2018). http://www.economy.nayka.com.ua/?op=1&z=6407
25. Sieg, T., Vogel, K., Merz, B., Kreibich, H.: Tree-based flood damage modeling of companies: damage processes and model performance. Water Resour. Res.Resour. Res. **53**, 6050–6068 (2017)
26. Koppe, E.: Compensation for war damage resulting from breaches of jus ad bellum (2021). https://doi.org/10.1007/978-94-6265-439-6_24
27. Wenning, R.J., Tomasi, T.D.: Using US natural resource damage assessment to understand the environmental consequences of the war in Ukraine. Integr. Environ. Assess. Manag.. Environ. Assess. Manag. (2023). https://doi.org/10.1002/ieam.4716
28. Veklych, O.: Methodological basics for assessing economic damage from loss of ecosystem services in consequence of military aggression of the russia. Environ. Econ. Sustain. Dev. 48–55 (2022). https://doi.org/10.37100/2616-7689.2022.12(31).5
29. On the approval of the Procedure for determining damage and losses caused to Ukraine as a result of the armed aggression of the Russian Federation.—Decree No. 326 dated 20.03.2022 of the Cabinet of Ministers of Ukraine: .[electronic resource]: access mode https://zakon.rada.gov.ua/laws/show/326-2022-%D0%BF#Text
30. General principles of valuation of property and property rights: national standard No. 1: approved. by the resolution of the Cabinet of Ukraine dated September 10, 2003 No. 1440 [Electronic resource]. Access mode: zakon.rada.gov.ua/go/1440-2003-p.
31. Methodology of property valuation dated 10.12.2003 No. 1891 On approval of the Methodology of property valuation, approved by Resolution of the Cabinet of Ministers of Ukraine dated December 10, 2003 No. 1891. Access mode: http://zakon2.rada.gov.ua/laws/show/1891-2003-p
32. Drapikovskyi, O., Ivanova, I., Krumelis, Y.: Real Estate Valuation: Teaching, p. 424. Kyiv: [Company SIK GROUP Ukraine] (2021)
33. Li, Y., Biloshchytska, S.: Diversification of activity as a component of adaptive strategic management of construction enterprise. Manag. Dev. Complex Syst. **37**, 173–177 (2019). https://doi.org/10.6084/m9.figshare.9783233(2019)

34. Siniak, N., Zróbek, S., Nikolaiev, V., Shavrov, S.: Building information modeling for housing renovation-example for Ukraine. Real Estate Manag. Valuation **27**(2), 97–107 (2019)
35. Tugay, O.A., Zeltser, R.Y., Kolot, M.A., Panasiuk, I.O.: Organization of supervision over: construction works using uavs and special software. Sci. Innov. **15**(4), 23–32 (2020)
36. Honcharenko, T., Chupryna, Y., Ivakhnenko, I., Zinchenco, M., Tsyfra T.: Reengineering of the construction companies based on BIM-technology. Int. J. Emerg. Trends Eng. Res. **8**(8), 4166–4172 (2020). https://doi.org/10.30534/ijeter/2020/22882020
37. Zeltser, R.Y., Bielienkova O.Y., Novak, Y., Dubinin, D.V.: Digital transformation of resource logistics and organizational and structural support of construction. Sci. Innov. **15**(5). 38–51 (2019)
38. Data on assessment objects from the Unified database of assessment reports of the FSMU (in machine-readable format): [electronic resource]: access mode: https://www.spfu.gov.ua/ru/con tent/spf-estimate-basereport-dani-z-edinoi-bazi.html
39. Semenova, O., Semenov, A., Voitsekhovska, O.: Neuro-fuzzy controller for handover operation in 5G heterogeneous networks. In: 2019 3rd International Conference on Advanced Information and Communications Technologies (AICT) (2019). https://doi.org/10.1109/AIACT.2019.884 7898
40. Al-Maitah, M., Semenova, O., Semenov, A., Kulakov, P., Kucheruk, V.: A hybrid approach to call admission control in 5G networks. Adv. Fuzzy Syst. **2018**, 1–7 (2018). https://doi.org/10. 1155/2018/2535127

Development of Information Processes as a Prerequisite for the Sustainable Development of Agricultural Enterprises

Svitlana Zaika⬤, Jacek Skudlarski⬤, Oleksandra Mandych⬤, Oleksandr Hridin⬤, and Olena Zaika⬤

Abstract There are results of the study of the prerequisites and features of the processes of informatization of agricultural enterprises in the article. We have studied the essence, and main components of the digital economy and the prospects of the digital transformation of agriculture. The priority factors that determine the specificity of the digital transformation of agriculture and determine the directions of informatization of the agricultural production system have been elaborated. We have established the informatization on the agricultural sector. It is characterized by fragmentation, which causes differentiation in the level of use of information technologies by both individual agrarian enterprises and branches of agriculture, and information technologies that are successfully used in agriculture to ensure digitalization are given. We have analyzed the main elements of digitalization, and an organizational and economic mechanism for the development of agricultural enterprises was proposed based on it. This ensures a synergistic effect of the informatization of commodity producers on the growth of the efficiency of the agricultural sector of the country's economy. In addition, it helps to overcome the digital gap between agriculture and high-tech sectors of the economy and contributes to the revival of rural areas and sustainable development of agricultural enterprises.

Keywords Informatization · Information technologies · Digital transformation · Digitalization · Agricultural enterprises

S. Zaika (✉) · O. Mandych · O. Hridin · O. Zaika
State Biotechnological University, Kharkiv, Ukraine
e-mail: zaika.svitlana@biotechuniv.edu.ua

J. Skudlarski
Warsaw University of Life Sciences, Warsaw, Poland
e-mail: jacek_skudlarski@sggw.pl

© The Author(s), under exclusive license to Springer Nature Switzerland AG 2024
A. Semenov et al. (eds.), *Data-Centric Business and Applications*, Lecture Notes on Data Engineering and Communications Technologies 194,
https://doi.org/10.1007/978-3-031-53984-8_10

1 Introduction

There is considerable interest of producers in the latest information technologies, the intensive development of which at the beginning of the XXI century at the current stage of the development of agricultural production. It significantly accelerated the processes of globalization and led to significant changes in various branches of agriculture.

Innovations in improving productivity and digital technologies for optimizing resource consumption have become the primary trend in the development of agribusiness in many countries of the world [1, 2]. Today, global economic growth is taking place within the framework of the transition to the Industrial Revolution, which has received the name "Industry 4.0" [3]. This applies to almost all sectors of the national economy, including the agricultural sector. In order to remain competitive in foreign and domestic markets, agribusiness must perceive these changes.

In the agricultural sector of Ukraine, there is already a positive experience in the implementation of digital technologies, but so far, they are used mainly by large agricultural holdings, and in small agricultural enterprises, informatization processes are practically absent. At the same time, the agricultural enterprises of Ukraine have formed prerequisites for further qualitative transformations and the application of innovations and modernization, because the state, even in this difficult time, maintains leading positions in the world in terms of export of many types of agricultural products. Therefore, the agricultural sector is an important component of the national economy, the strategic importance of which lies not only in the formation of the country's food security but also in developing the agricultural market and increasing export potential.

Thus, the purpose of our research is to study the prerequisites and features of the informatization of agricultural enterprises as a determining factor of their sustainable development.

It was solved following tasks to realize the goal of the research; that is reflecting its logic:

– the essence of the digital economy and the conditions of digital transformation are considered;
– the peculiarities of informatization of agricultural enterprises were investigated;
– the organizational and economic mechanism of the development of agricultural enterprises based on digitization was worked out.

Legislative and normative acts and numerous works of foreign and domestic scientists on the problems of informatization of agricultural enterprises and digitalization of the agricultural sector of the economy as a priority direction of its development made up the theoretical and methodological basis of the research.

The studies were based on a systematic approach to the subject area being studied, as well as on the use of monographic, dialectical, abstract-logical research methods, personal observations, etc.

2 Digital Economy: Essence, Components, and Prerequisites for Development

Recently, digital technologies have been rapidly integrated into the agricultural sector of the national economy. Innovative high-tech developments aimed at creating a "smart" farmer are being comprehensively implemented, from farm management applications to milking robots in animal husbandry, and from unmanned machinery with minimal or no human intervention to drones to detect diseases in the soil for the crop industry. Scientists describe smart farming as agricultural production sites where "smart" technologies and "big data" are used as software-driven systems consisting of associations, processes, and data flows based on organized methods of collection, analysis, and transmission of information.

At the same time, the prospects for the use of digital technologies in agriculture are ambiguous. On the one hand, they help minimize risks, contribute to the creation of highly productive workplaces, and increase the efficiency of production. On the other hand, digital technologies are labor-saving, so they can cause the release of a part of workers and thus cause problems such as reduced employment or lack of sustainability [4].

The level of economic, technological, and informational development of agriculture is lower compared to other branches of the economy. This level and the specifics of the agar production system itself require research into the prospects of the digital transformation of agriculture, substantiation of the conditions for the initiation of mass processes of its digitalization, and an assessment of the readiness of commodity producers in the agrarian sector to radically modernize their material and technical base and the system of inter-subject relations.

The term "digital economy" is relatively new. In 1994, it was first formulated by the Canadian business guru Don Tapscott [5] as an economic activity that, unlike the traditional economy, is defined by network consciousness and dependence on virtual technologies. He reflected the defining features of the new society: orientation to knowledge, digital form of representation of objects, virtualization of production, innovative nature of technological processes, integration, and convergence of systems of all levels, elimination of intermediaries, and simplification of the "producer–consumer" chain, dynamism, and globalization of communications.

D. Tapscott singled out five basic elements based on which the digital society functions and develops:

I. An effective person is an individual who owns a modern computer connected to the global Internet.
II. A high-performance team is a working group of employees that uses digital technologies to perform their tasks.
III. An integrated enterprise is a company in which all business processes organized in a digital information environment.
IV. An extended enterprise is a company with an extensive network of branches connected by a distributed computer network.

V. Business activity in an inter-network environment—a global digital community [6].

In 1999, Neil Lane characterized the digital economy as "the convergence of computer and communication technologies on the Internet and the flow of information and technologies that stimulate the development of electronic commerce and large-scale changes in the organizational structure" [7].

In turn, Eric Brynjolfson and Brian Kahin noted that the digital economy is an incomplete transformation of all sectors of the economy due to the digitization of information using computer technologies [8].

Thomas Mesenburg singled out three components of the digital economy:

(1) e-business infrastructure, which is part of the entire economic infrastructure used to carry out electronic transactions and electronic trade;
(2) electronic business, which is any operation carried out by the enterprise using computer networks;
(3) electronic trade, that is, the volume of goods and services that are sold using computer networks [9].

The author focused on measuring the phenomenon of e-business and e-commerce.

So, although the concept of "digital economy" has been the subject of active discussion by many researchers for almost three decades. Neither a unified interpretation of its content nor a generally recognized theoretical base for studying the essence and problems of implementing digital technologies in various branches of the economy is formed to this day.

Today, there are several approaches to defining the "digital economy" category:

- a set of markets organized taking into account the wide use of information and communication technologies;
- the branch of public production, which ensures the creation of an elemental base of electronic devices and the development of a complex of technical and software means of informatization of society;
- the method of organization and formalization of the system of social relations with the help of means of informatization of various spheres of life;
- a system of information technologies for ensuring economic activity and managing the processes of socio-economic development;
- a tool for generating transformational effects arising from the use of digital technologies and digitalization of the economy;
- a new concept of social development, which provides for a fundamentally different model of digitization of the processes of coordination of production and consumption;
- a specially created virtual environment that allows ensuring the improvement of the efficiency of reproductive processes and the quality of life of the population;
- a method of organizing economic activity based on electronic commerce and electronic money circulation technologies;
- a set of economic sectors and market segments in which added value is created based on the use of digital technologies [10–18].

The official definition of the digital economy is enshrined in the "Concept of the Development of the Digital Economy and Society for 2018–2020", where it is described as an activity in which the main means (factors) of production are digital (electronic, virtual) data, both numerical and text [19].

Therefore, the digital economy is not a new sector of the economy but the transformation of all industries, taking into account the processes of informatization, which involves the integration of information resources and information systems into a single information management system that integrates all technological solutions.

3 Factors and Priority Areas of Digital Transformation of Agriculture

Along with the term "digital economy", the term "digital transformation" entered scientific circulation. Digital transformation, in the broad sense, is characterized as the process of transition of socio-economic systems to a qualitatively new level of use of digital technologies. These technologies have the ultimate goal of the transformation of digitalization objects within the framework of the strategy of transition to a new model of development and implementation of priority directions for the formation of the digital economy.

Products created based on the use of digital technologies, processes related to the implementation of digital technologies; people whose activities are related to the use of digital technologies, and systems within which people interact are traditionally distinguished as objects of digital transformation, processes, and products. That is, digital technologies are the core of the digital economy.

Digital technologies, in this context, include technologies focused on working with discrete information signals. It ensures consistently high quality of basic information procedures (collection, transmission, processing, storage, search of information) and minimization of time and resource costs for their implementation, improving the quality of communication channels and the speed of information exchange. At the same time, digital technologies are inherently infrastructural concerning basic production technologies. And if basic production technologies are at a low technological level, then their digitalization will not provide the expected effect.

The importance of the processes of transition to a new technological system objectively determines the need to fulfill certain provisions. These provisions determine the success of the initiation of a new model of social development:

– the scale of the tasks of digital transformation requires the active participation of the state as a macro-regulator of digitization processes, as an entity that controls a significant part of the country's economy and an entity that provides a significant amount of digital services;

– the strategy of digital transformation and the scale of digitization processes must correspond to the level of economic development and financial capabilities of the state, regions, and producers;

– the beginning of a large-scale digital transformation is impossible without achieving the required level of information infrastructure development, which ensures the formation of a single information space and the possibility of radical modernization of the system of interaction of economic entities;

– integration into a single information space increases the requirements for ensuring the level of information security of subjects of the digital economy and necessitates the formation of effective mechanisms for countering cyber threats;

– the effectiveness of digital transformation processes is largely determined by the quality of the innovation system of society and its ability to generate and use in practice solutions that allow continuous improvement of the technical and technological basis of the social reproduction system;

– the implementation of the digital transformation strategy requires coordination and synchronization of the digital development of individual territories and industries, as well as preventing the emergence of a technological gap between them, capable of sharply reducing the efficiency of the functioning of the entire macroeconomic system;

– the mass introduction of digital technologies in all spheres of activity requires the modernization of the domestic electronic industry, which forms the basic basis of the digital economy and minimizes the level of technical and technological dependence on highly developed countries;

– digital transformation objectively affects both the growth of labor productivity in the real sector of the national economy and the change in the structure of employment of the population, requiring the development of a strategy to minimize the consequences of a mass reduction of workers in traditional sectors of public production;

– a mass transition to the use of digital technologies is impossible without the modernization of the professional education system, which is associated with a significant change in the content of the competencies of employees at various levels and the need for their constant self-education and self-development [20].

In addition, to understand the possibility of certain negative factors caused by digital transformations, it is necessary to define some preventive tasks at different levels of execution:

- **at the state level:**

    – development of legal support for digital transformation processes;
    – development of the concept of ensuring the employment of the population in the conditions of the expected reduction of jobs in traditional sectors of the economy;
    – development of information infrastructure, provision of economic and physical access to information resources and digital goods of all subjects of the economy;

- ensuring information security of subjects of the digital economy;
- modernization of professional and technical education to form the necessary competencies in specialists for comfortable integration into the processes of digital transformation.

- **at the level of product manufacturers:**

  - formation of the potential for modernization of the technical and technological base of production and introduction of innovative production technologies;
  - development of new digital platforms;
  - introduction of new technologies and forms of communications within digital ecosystems;
  - use of new forms of labor organization and increase in the social responsibility of business.

- **at the specialist level:**

  - formation of competencies that allow not only to be a qualified user of digital platforms but also to participate in their development and support;
  - understanding of the functionality of digital services;
  - formation of skills for working with digital technologies that ensure the satisfaction of own information needs;
  - formation of the ability for self-development and self-realization, readiness to use new forms of socialization and communications;
  - formation of skills to ensure one's information security and protection of private life in the information space.

The formation of effective mechanisms for countering potential threats is a mandatory and necessary condition for the success of digital transformation and obtaining a significant effect sufficient to satisfy the interests of all subjects of the digital economy.

Each branch and field of activity has specific essential features that determine the specifics of their development and the features of the processes of technical and technological modernization, including digital transformation. Digitization as a stage of the evolutionary development of socio-economic systems is objectively related to the formation of a set of conditions necessary and sufficient for the initiation of digital transformation processes, which ensure their mass nature. A large-scale transition of enterprises to digital economy technologies can be ensured only on the condition of achieving such a level of technical and technological development that guarantees the possibility of their full integration into a single information space.

Agriculture belongs to the branches of the national economy, the level of development of which significantly lags behind high-tech spheres of activity such as communications and telecommunications, the banking sector, the chemical industry, etc. Therefore, in these conditions, the implementation of digitalization of agricultural production is complicated by several factors.

Determining the basic factors that determine the specifics of the digital transformation of agriculture allows us to systematize them into three main groups. These

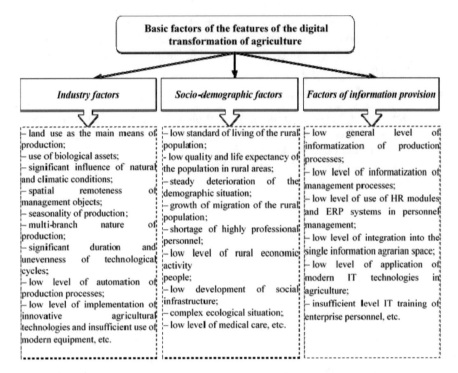

Fig. 1 Basic factors of the features of the digital transformation of agriculture

groups take into account the sectoral features of agricultural production, sociodemo-graphic features, as well as the features of information support in the industry (see Fig. 1).

The first group of factors reflects industry specifics and conditions that require the development of unique technologies for the digitalization of production processes and management of agricultural production.

In particular, the need to interact with land and biological objects (plants and animals) significantly complicate the algorithms for processing arrays of information about the state of management objects. It is associated with the constant adjustment of numerous parameters of the enterprise's functioning and the variety of options for management decisions determined by individual, variable characteristics of land plots and biological objects.

The dependence of agriculture on natural and climatic conditions requires the formation of a system of information support for risk management and the presence of a powerful apparatus of predictive calculations that allow for to reduction in the level of risks. Minimize their consequences and ensure the sustainable development of the enterprise.

The spatial dispersal of management objects places increased demands on the level of development of telecommunications and information infrastructure, which ensures the exchange of operational information and guarantees the minimum level of its distortion. This factor leads to significant costs for the formation of a unified information space of the enterprise, which is exacerbated by the mobile nature of a large part of the fixed assets involved in the process of agricultural production.

The duration of production cycles of agricultural production and the unevenness of their flow significantly affect the organization of reproductive processes in agricultural production and their management. The need to ensure the interaction of industries with different lengths of production cycles requires the presence of effective strategic, tactical, and operational management tools.

The low level of automation of production processes objectively determines the complexity of their informatization. In conditions of insufficient level of development of intelligent control systems, complex digitalization of production processes is possible only in industries where production processes can be maximally automated. In agriculture, especially in animal husbandry, the rate of automation of a significant part of work operations and elements of technological processes has significantly increased in recent years. However, the possibilities of using complex automation technologies are still significantly limited for many enterprises.

The multi-sectoral nature of most agricultural producers requires the use of numerous technologies and a fairly wide range of technical means of their implementation. That is associated with an increase in costs not only for the formation of technical and technological support but also for ensuring an effective management system. At the same time, a significant number of enterprises in the agricultural sector have a relatively small scale of production and existing auxiliary and service industries. Without digitalization of which it is impossible to ensure the transition of agricultural producers to a digital development model.

The second group of factors describes the limitations of the quality of rural labor resources involved in digitalization processes. The experience of large integrated agro-industrial formations regarding the technical and technological modernization of agricultural production formulated a number of serious problems of staffing in almost all spheres of activity within a significant part of rural areas in various regions of the country. This is primarily due to:

- low level of general professional training of a significant part of the working-age rural population;
- a shortage of labor resources of the necessary qualifications due to the existing surplus of the working population, which does not have the financial means, and sometimes the desire, to improve their qualifications and professional level;
- conservatism of thinking, unpreparedness for radical changes, and low level of economic activity;
- deterioration of the demographic situation and extremely low rates of growth in the level and quality of life in rural areas, which increase the outflow of young people from the countryside.

The third group of factors includes factors that reflect the features of information provision. A digital transformation is a specific form of informatization. These form an informational and infrastructural basis that allows the process of mass implementation and use of digital technologies to begin and forms the prerequisites for the success of digitalization. The lower level of technological development of agriculture compared to other industries and spheres of activity objectively determined the lower level of informatization of agricultural production. This is manifested in the low level of informatization of both production processes and the management system of agricultural production. The low quality of the information infrastructure of agricultural producers causes problems that limit the possibilities of their integration into a single information space. Despite the use of information and telecommunication technologies by large agro-industrial holdings, comprehensive informatization of the agricultural production system has not yet been developed, which also limits the possibilities of developing a universal digital platform for agricultural producers. In addition, the low level of IT training of managers, specialists, and employees acts as an additional limitation.

The priority directions of informatization of the agricultural production system, which ensure the formation of conditions for the digitalization of agriculture, are:

- the implementation of digital technologies for informatization of production processes related to equipping agricultural machinery with microprocessors and sensors, which ensure the improvement of its management, robotization, and automation;
- implementation of platform information solutions focused on the complex solution of a set of production-technological and organizational-economic tasks facing agricultural producers;
- the formation of a multifunctional system that provides information support for the management of agricultural production by supporting the optimal structure of information resources and their quality and modernizing the set of tools and methods that allow the implementation of the set of tasks of agricultural production management;
- modernization of the information infrastructure by the information needs of users, increasing the reliability and quality of communication channels and increasing the speed of information exchange, creating conditions for the availability of basic information and communication technologies, and expanding the range of received electronic services;
- rationalization of processes of information interaction of agricultural business entities with their counterparties (suppliers of resources and services, consumers of products) and expansion of electronic trade opportunities.

It should be noted that digital technologies do not guarantee sustainable competitive advantages. The effectiveness of their implementation is, to a large extent, determined by the level of technical, technological, and informational development of specific productions and processes. Digital technologies are not a source of additional economic effect, they are a tool to increase it. In cases where industries or spheres of activity are characterized by a low level of economic development. Their digital transformation requires a primary solution to the problem of overcoming their technical and technological backwardness and the formation of technological assets that can become a driver of digital transformation and a key factor in its initiation.

That is why the initial condition for initiating the processes of digital transformation of agriculture is the level of development of the information infrastructure. It allows agricultural producers to fully integrate into a single information space, get free access to information resources necessary for effective management of their development processes,

So, one of the effective forms of technical and technological modernization, which forms the basic conditions for the transition to digitalization, is informatization.

4 Informatization and Information Technologies in Agriculture

In a broad sense, informatization of the processes of social development is a set of processes related to the organization of optimal conditions that ensure the satisfaction of the complex information needs of various users. Concerning economic systems, informatization has two aspects:

(1) informatization of management of system development processes;
(2) informatization of production and technological processes.

The basis of informatization is formed by information technologies that ensure the implementation of such functions as collection, transmission, systematization, storage, and primary and analytical processing of information. Its interpretation and generation of new information, assessment of its relevance, information support for processes of justifying managerial influences, etc.

In terms of the level of informatization and the use of information technologies, agriculture still lags behind the rest of the branches of the national economy, which is due to some factors:

- the presence of small producers who have limited financial capabilities, use outdated technologies, and are characterized by low innovative activity;
- a relatively low level of concentration of production, which limits the effectiveness of the use of a significant part of IT solutions for the informatization of production and technological processes, which, in turn, shifts the informatization processes to the sector of large-scale production;
- insufficient level of IT training of agricultural workers, ignorance and misunderstanding of the possibilities of information technologies as a tool for increasing the efficiency of production processes and managing the reproduction process;
- the fragmentation of the innovation system of the agricultural sector of the economy, due to the low demand for innovative developments from agricultural producers, the decrease in the innovation potential of agricultural science, the destruction of the links between science and production, and the deformation of the innovation infrastructure;

- the absence of effective software complexes that provide the possibility of a systematic solution to the set of management and technical tasks of agricultural production and the formation of the information base of the management system;
- the low quality of the information structure and the limited possibilities of using modern telecommunication technologies, which ensure the formation of a single information space for product manufacturers and their integration into a single information space of a higher level;
- low efficiency of the existing network of information and consultation centers, which have a weak influence on the processes of informatization of agricultural production [21, 22].

So, today the informatization of the agrarian sector is characterized by fragmentation, which causes differentiation by the level of use of information technologies of both individual agrarian enterprises and branches of agriculture.

Fundamental changes in the field of informatization of agriculture began with the entry into the industry of large capital. This capital is ready to invest significant funds in the technical and technological modernization of production and to introduce innovative technologies. That ensures increased efficiency of agricultural production and the formation of sustainable competitive advantages, including those related to the use of information technologies.

The consolidation of enterprises led to an increase in the level of concentration of production, the complication of the organizational structure of agro-industrial formations and the management system, and an increase in the complexity and scale of functional tasks. The consolidation of enterprises contributed to an increase in the demand for effective IT solutions for agriculture. Sensing an increase in demand, large IT companies intensified the processes of adapting their traditional software products to the specifics of agriculture. It began to promote new tools for solving traditional tasks of agricultural production management and informatization of individual production processes. But, today, due to the lack of a single strategy for the informatization of agriculture. That would reflect the set of typical tasks prioritized for the industry and the functionality of individual information technologies. There are still no complex platform solutions that would meet the information needs of agricultural producers of various categories.

Today, the informatization of the agricultural production system can be considered in the context of the development of precision agriculture and precision animal husbandry technologies.

UN experts predict that by 2050 there will be 9.8 billion people on the planet. Providing food for such a large population on the planet requires an increase in their production by at least 70%. And this means that agricultural producers all over the world, regardless of their place of residence and activity, must radically change their production processes and increase their efficiency to the highest possible level.

Already, experts note that thanks to the technologies of precision agriculture, based on the technologies of the Internet of Things. It is the process of combining data from various devices, meter readings, sensors, and the like into a common information system. It is possible to achieve a significant increase in the production of agricultural products [23].

In the "Concept of the Development of the Digital Economy and Society of Ukraine for 2018–2020", it is noted that for the development of agriculture. It is important to implement digital agriculture. It is a fundamentally new management strategy based on the application of digital technologies, and a new stage of the development of the agricultural sector, related to using geographic information systems, global positioning, onboard computers, and smart equipment. As well as management and executive processes capable of differentiating methods of processing, and applying fertilizers, chemical meliorates, and plant protection agents [19].

In a broad sense, precision agriculture technologies involve the implementation of crop productivity management processes by taking into account the level of variability of their growing environment due to the complex use of a Global Positioning System, Yield Monitor Technologies, Variable Rate Technology, Remote Sensing of the Earth and Geographic Information System.

The main technological trends in the development of precision agriculture shortly are:

– development of parallel driving systems;
– mass digitization of fields;
– the possibility of differentiated soil cultivation and the use of seed sowing norms, application of mineral fertilizers, plant protection agents, and growth stimulants taking into account the quality of soils in local areas;
– expanding the capabilities of the Internet of Things;
– massive use of big data processing technologies and improvement of operational management quality;
– transition to the use of unmanned tractors and self-propelled agricultural machines;
– implementation of artificial intelligence technologies [24, 25].

Precision agriculture provides many opportunities for the use of artificial intelligence in the optimization of agricultural production, which allows commodity producers to transform raw agricultural data into practical conclusions that contribute to improving the quality and quantity of crops. This allows agricultural producers to choose crop varieties optimal for their region and to use farm automation to minimize resource costs.

In turn, the emergence of Agriculture 4.0 contributed to the emergence of some tools for automating livestock farms and data management solutions that help increase the resources and productivity of the agricultural industry [26]. Enterprises that have implemented innovative technologies have shown a clear transition from the use of traditional, time-consuming processes to advanced, effective technological solutions.

In addition, the digitization of agricultural production comes down to the fact that it provides producers with accurate data in real time. A combination of intelligent farm data and satellite imagery provides actionable performance information based on a wide range of growing conditions, enabling growers to better plan farm operations and manage upgrade resources more effectively.

At the same time, the data collected at various points of the supply chain allow us to understand the needs of the market and adequately manage the production of crops. Both private enterprises and public institutions can use this knowledge to reduce risks, improve crop management, and minimize crop losses and food waste [27].

Thanks to digitalization, government institutions have the opportunity to create a centralized database with information about farmers. Information on the yield of crops and the production of products enables credit and insurance companies to calculate possible risks, develop effective policies, and ensure their rapid implementation. Real-time monitoring of agricultural production in combination with satellite data allows producers to receive information on productivity, which helps to overcome climate and other environmental problems.

Examples of information technologies that are successfully used in agriculture to ensure digitization are:

AEPO is a system that can detect the main places of pest accumulations and spot-treat them with insecticides: a digital survey from the air in the ultraviolet range is carried out with the help of drones.

CropCare is a constantly updated database of various pest control data. The nomenclature of crops and GPS data of the fields are entered into it, after which the optimal drugs are selected.

AgroGuard is a system of guard posts equipped with infrared sensors. In the event of a violation of the boundaries of the plot or any other event, the agro-entrepreneur promptly receives a message on the phone and makes a prompt decision.

DrT-Tech is a program that allows you to systematize all the data collected from the sensors and the fields into one structure. To view information, a corresponding program installed on a smartphone is used.

HerdGrow is a program used for livestock business, the essence of which is to automatically select a ration for cattle based on data from their passports.

Fractal is a designer of intelligent processes, with the help of which you can maximally automate work processes in the agricultural industry: accounting of working hours, regulation of certain mechanisms, etc. The program connects all devices to a single local network with an uninterrupted power supply.

AgromaxEffect is a program that simulates the future harvest based on a specific crop and site characteristics. Agricultural entrepreneurs, insurance companies, and banks for risk assessment use the program.

The trading bot is a way to find the best offers for agricultural products of interest to agricultural companies. It makes it possible to optimize the work of the procurement department, etc. [28].

AgroMonitoring is a specialized agricultural service that combines bulletin boards for buying and selling agricultural machinery with analysis of logistics and distances from elevators and settlements to port terminals.

EMA-i is an early warning application developed by FAO (Food and Agriculture Organization) that allows farm veterinarians to transmit information about animal diseases in real-time. It is integrated into the Global Animal Disease Information System (EMPRES i), which ensures reliable storage and use of data by different countries. EMA-i is compatible with national reporting systems for animal diseases, provides real-time information transmission, and contributes to increasing the efficiency of the early warning and response system for animal diseases, which has a positive impact on food safety.

MyCrop is a platform for the joint work of small farmers, aimed at expanding their opportunities by providing information, expertise, and resources to increase the volume of production and the efficiency of its production. MyCrop helps farmers make optimal decisions and implement them: the platform allows you to map land, plan crop selection, create work plans for individual farms, and automate work taking into account weather conditions, soil quality, data on diseases, pests, and yield in almost real-time.

ET-Agricultural Brain is a system that allows you to determine the state of health of each animal on a pig farm based on recognition by appearance, temperature, and voice. In addition, after tracking how the sow sleeps, and what she eats, the artificial intelligence will tell if she is pregnant. Artificial intelligence also makes it possible to identify sick boars and reduce their number to a minimum. Many sensors are installed on the farm, collecting information, and taking into account which optimal conditions are created for raising livestock while reducing the possibility of human error. According to the developers' calculations, the use of artificial intelligence technologies in pig farming will allow farmers to reduce labor costs by 30–50% and reduce the need for feed. Thanks to the optimization of animal breeding conditions, shorten the fattening period by five to eight days.

IoT & AI is a digital farming technology to collects and analyzes agricultural and environmental data. That allows even inexperienced farmers to monitor critical information such as humidity, air temperature, and soil quality using remote sensors to improve yield and predict crop yields [24, 29].

Cropio is a digital platform, an integrated complex software solution that provides satellite monitoring of the condition of crops, keeping track of indicators, and tracking equipment and machinery to maximize the effectiveness of decision-making. Cropio allows remote monitoring of agricultural land, including auto-documentation, forecasting, and planning of agricultural operations.

PROD is a mobile application for working with the database of prices for vegetables and fruits in Ukraine. The user receives only current prices for fruit and vegetable products, at the same time there is no need to visit many non-target sites. Everything is carefully collected in one place. In a few clicks, you can find out how much a certain type of product costs and determine the best price for your purchase. In the application, it is possible to post your ads for the purchase/sale of vegetables and fruits, quickly contact the author of the ad directly (without intermediaries) to make a deal, get detailed information about the price and features of the product, etc.

AgroUA is an agricultural information and communication platform that has a wide range of applications—from news and current prices for agricultural products to posting commercial and informational ads. AgroUA allows you to be at the epicenter of the country's agrarian life and keep abreast of new events and technologies. You

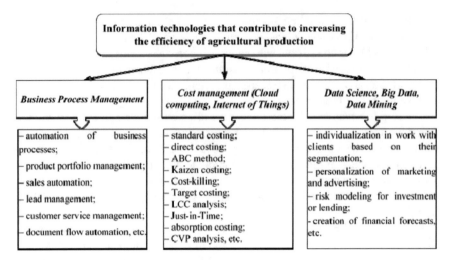

Fig. 2 Information technologies that contribute to increasing the efficiency of agricultural production

can announce yourself to all market participants, publish in catalogs, advertise yourself and your products, and announce events and innovations. It is also possible to announce tenders, place buy/sell ads, communicate in the forum, follow offers and vacancies on the market, introduce yourself to employers, share experience, follow tenders of agricultural holdings, read news, visit all agricultural groups in social networks, enter to all popular agricultural sites, user guides for some agricultural activities and many other possibilities [30].

So, modern information systems and technologies, which can be conventionally grouped into three groups (see Fig. 2), are able not only to fully satisfy the requirements of agricultural commodity producers but also act as an important prerequisite for their development and contribute to increasing the efficiency of the agrarian business.

A market research report predicts that the digital transformation market will reach a peak of $3,294 billion by 2025, with an average annual growth rate of 22.7% compared to 2019 [4].

At this stage of informatization development, agricultural enterprises that do not use information technologies in their activities cannot fully develop. Because many buyers do not have the opportunity to find a product on their own, compare prices with competitors and evaluate its characteristics, product manufacturers lose a significant part of potential consumers.

In Ukraine, there are a few electronic stores that work with agricultural products—one of the most popular is the online site prom.ua. Also, the catalog of Ukrainian manufacturers "Made in Ukraine" (https://www.madeinua.org), includes electronic stores of farms.

Free software, web services, and mobile settings are promising means of information technology that can be used in agricultural production. Among applications for business in the agricultural sector, the following software is available: office packages OpenOffice or LibreOffice, programs for project management OpenProj or GanttProject, statistical analysis packages PSPP or OpenStat2, and others.

A web service is a software system with standardized interfaces identified by a web address. Useful for agricultural enterprises:

1. Search services based on specialized portals, enterprise catalogs, rating publications, and survey results. The Google Alerts service is capable of monitoring thematic news and sending selections to e-mail. Google Trends can be used to monitor and analyze user inquiries about a brand or product.
2. Electronic mailing services are intended for the formation of a database of addresses and the implementation of informational mailings. Yes, the Streak service allows you to turn your Gmail inbox into a CRM contact tracking subsystem.
3. Online office based on Google documents. Limited opportunities for preparing documents, calculations, presentations, and projects are compensated by free use and the possibility of group work with documents.
4. SMM marketing in social networks. In particular, PromoRepublic provides a toolkit for automating the promotion of products in social networks, which is promising for targeting sales to the end consumer.
5. Order processing and sales. The Bitrix service will be appropriate for farms and small private enterprises as a CRM system. The Agro Yard online platform (https://agroyard.com.ua/) was created for conducting tenders, purchases, and cashless settlement of farms, there is also a mobile version of the platform.

The main advantages of using Internet applications include:

- saving on software and technical equipment;
- access to the service through a browser, including from mobile devices;
- positive perception of web applications by clients.

The risks of using web applications are:

- there is no guarantee of long-term service operation;
- dependence on imported developments and mainly English-language interfaces [21].

5 Digitization of the Agricultural Sector of the National Economy

The main components of the digital environment of the agrarian sector of the economy are the system of infrastructural support, legal protection of intellectual property; development of wireless communication infrastructure, staffing of information and

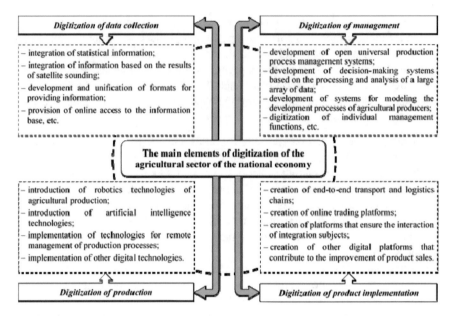

Fig. 3 The main elements of digitization of the agricultural sector of the economy

communication technologies, transition to an innovative and investment model of development, use of service-oriented approaches to studying consumer needs; formation of a system of anticipatory modernization of the technical and technological base and digital technologies. At the same time, the defining elements of the digitalization of the agricultural sector of the economy. That can ensure the balanced development of its digital ecosystem, are the digitalization of data collection, digitalization of management, digitalization of production, and sales of products (Fig. 3).

In recent years, the Ukrainian economy has made significant progress in many directions of digitalization. The decisive legal act that affects these processes is the "Strategy for the Development of the Sphere of Innovative Activity for the Period Until 2030". That states that the main driver of economic growth soon should be the agrarian sector, which has a high potential for modernization, introducing the latest technologies and increasing the level of processing of its own products, although its prospects are limited [31].

The development and implementation of an effective organizational and economic mechanism for the development of agricultural enterprises based on digitalization. They are necessary to determine the direction and size of the main financial flows that must be formed for the successful large-scale digitization of the agricultural industry (see Fig. 4).

Based on the possibility of using different organizational models and approaches to financing, the mechanism provides for the formation of an investment budget and a budget of current expenditures from several sources—individual funds of agrarian enterprises, funds of collective institutions, state support, and loan funds. Assessment

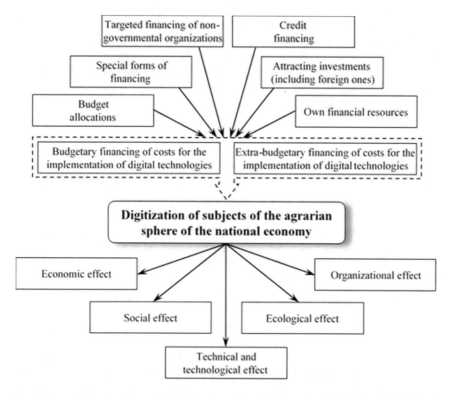

Fig. 4 Organizational and economic mechanism of development of agricultural enterprises based on digitization

of the potential of these sources will allow us to determine the need for the total amount of funding for the implementation of digital technologies.

The next block of the mechanism is obtaining some effects from the implementation of digitalization of agricultural producers, including economic, organizational, technical, technological, ecological, and social.

The economic effect of the implementation of digital technologies includes a decrease in cost and at the same time an increase in revenue (mainly due to an increase in productivity). This effect is a source of repayment of investment costs and satisfaction of the economic interests of commodity producers. This effect is distributed among the participants of digitization projects in different proportions determined by their organizational and legal relations. So, for example, when expanding the use of geo-information systems, using services for the provision of unmanned equipment provided by suppliers of digital solutions based on a contract with product manufacturers. There will be current costs that are covered by reducing the cost price and increasing revenue, and when buying own equipment, and software—initial investments will be covered for several years with a certain payback period.

The organizational effect of digitization consists, first of all, of the possibility of optimizing administrative and management costs, speeding up business processes, and strengthening the relationships and interaction of interested parties.

The technological effect consists of the development of information and communication infrastructure, while the technological efficiency of the implementation of digital technologies will be measured by comparing the values of cost reduction and the costs of digitalization itself.

The ecological effect, in turn, is characterized by a reduction in environmental damage due to a reduction in the impact on soils, plants, air, and water; optimization of resource use and adaptation to climate change; preservation and restoration of natural soil fertility, which can be estimated based on data on the change in the content of nutrients and the cost of fertilizers necessary to bring their level to the norm.

The social effect consists of facilitating the solution of socially significant problems and includes the impact of digital technologies on the incomes of industry workers and the budgetary effect due to changes in the tax base and changes in food security due to increased productivity.

The above effects of the implementation of digitalization of agricultural commodity producers have a synergistic effect on the efficiency of the agricultural sector of the country's economy as a whole and lead to the reduction of the digital gap between it and the high-tech sectors of the economy.

The agricultural sector of the economy is an ideal area for the application of information technologies. Given this, the use of the latest information technologies in agriculture will allow increasing the production of products and will have a powerful positive impact on the effective sustainable development and functioning of agricultural enterprises.

6   Conclusion

As a result of the conducted research, it was established that the digital economy is a certain stage of the development of society. This is associated with the growth of the role of information as a strategic resource and the introduction of information and communication technologies in various sectors of the economy to increase their efficiency. The possibilities of the latest information technologies allow not only to search, process, and store information, but also to analyze it to develop options for making further decisions, which facilitates the work of specialists in conditions of excess information. At the same time, the main task of informatization of agricultural production is to reduce the cost of production and increase its quality and competitiveness based on the effective use of resources.

Digitization is the direction of transformation of the agricultural sector of the economy, which requires a deep rethinking in terms of finding opportunities to apply individual component technologies depending on the direction of the company's work.

The introduction of modern digital technologies is an objective requirement for the successful management of the agrarian business because these information technologies enable detailed monitoring of land resources. That contributes to the formation of sound management decisions with the prevention of negative consequences of economic activity; increasing the efficiency of land use and the quality of produced agricultural products, reducing the cost of its production due to the rational use of all types of resources, etc.

In general, information technologies ensure compliance with the requirements of technological, ecological, and economic security of the agricultural industry. At the same time, digital agricultural technologies are a system of new means of production interacting with each other, including software, information and control systems, and networks. Those under the condition of the formation of appropriate organizational and economic relations contribute to increasing the economic, organizational, technical, technological, ecological, and social effects of agricultural enterprises, increasing the efficiency of agricultural production and sustainable development of the industry as a whole.

Therefore, due to the synergistic effect of informatization of agricultural enterprises in the agricultural sector, the level of economic development will increase. That allows for preserving labor resources in rural areas, ensuring the overcoming of the digital divide, and contribute to the socio-economic revival of rural areas.

References

1. Engelhardt-Nowitzki, C., Kryvinska, N., Strauss, C.: Strategic demands on information services in uncertain businesses: a layer-based framework from a value network perspective. In: 2011 International Conference on Emerging Intelligent Data and Web Technologies (2011). https://doi.org/10.1109/EIDWT.2011.28
2. Fauska, P., Kryvinska, N., Strauss, C.: The role of e-commerce in B2B markets of goods and services. IJSEM **5**, 41 (2013). https://doi.org/10.1504/IJSEM.2013.051872
3. Polaschek, M., Zeppelzauer, W., Kryvinska, N., Strauss, C.: Enterprise 2.0 Integrated communication and collaboration platform: a conceptual viewpoint. In: 2012 26th International Conference on Advanced Information Networking and Applications Workshops (2012). https://doi.org/10.1109/WAINA.2012.73
4. Pishchenko, O.: Strategies of the digital agricultural sector in conditions of environmental and economic security. Bull. Khmelnytskyi Natl. Univ. **5**(1), 303–310 (2022). https://doi.org/10.31891/2307-5740-2022-310-5(1)-50
5. Tapscott, D.: The Digital Economy: Promise and Peril in the Age of Networked Intelligence. McGraw-Hill, NY (1994)
6. Strutynska, I.V.: Definitions of the term "Digital transformation." Black Sea Econ. Stud. **48–2**, 91–96 (2019). https://doi.org/10.32843/bses.48-47
7. Lane, N.: Advancing the digital economy into the 21st century. Inf. Syst. Front. **1**(3), 317–320 (1999). https://doi.org/10.1023/A:1010010630396
8. Brynjolfsson, E., Kahin, B.: Introduction. Understanding the Digital Economy–Data, Tools, and Research, pp. 1–10. MIT Press, MA, Cambridge (2000)
9. Mesenbourg, T.L.: Measuring the Digital Economy. U.S. Bureau of the Census (2001)

10. Kling, R., Lamb, R.: IT and organizational change in digital economies. understanding the digital economy. In: Brynjolfsson, E., Kahin, B. (eds.), pp. 295–324. MIT Press, MA, Cambridge (2000)
11. Bahl, M.: The Work Ahead: The Future of Businesses and Jobs in Asia Pacific's Digital Economy. Cognizant (2016)
12. The Digital Economy. British Computer Society, London (2014)
13. Dahlman, C., Mealy, S., Wermelinger, M.: Harnessing the Digital Economy for Developing Countries: Working Paper, vol. 334. OECD, Paris (2016)
14. Rouse, M.: Digital Economy. Techtarget, Newton (2017)
15. Digital Economy. Oxford University Press, Oxford (2017) Oxford Dictionary
16. Tymoshenko, N.Y., Melekh, N.V.: Global causes and current trends in the development of digital innovations in Ukraine and the world. Pryazovsky Economic Bulletin. Econ. Manag. Natl. Econ. 6(17), 84–89 (2019). https://doi.org/10.32840/2522-4263/2019-6-16
17. Zaika, S., Kuskova, S., Zaika, O.: Trends in the development of services in the conditions of digitalization of the economy: digital economy as a factor of economic growth of the state. In: Haltsova, O.L. (ed.). Kherson, pp. 258–281 (2021)
18. Vyshnevskyi, V.P., Harkushenko, O.M., Kniaziev, S.I., Lypnytskyi, D.V., Chekina, V.D.: Digitization of the economy of Ukraine: transformational potential, p. 188. Kyiv (2020)
19. Resolution of the Cabinet of Ministers of Ukraine "On the approval of the Concept for the development of the digital economy and society of Ukraine for 2018–2020 and the approval of the plan of measures for its implementation" dated January 17, 2018 no 67
20. Sidenko, V.R., Skrypnychenko, M.I., Ponomarenko, V.S., Chuhunov, I.Y., et al.: Institutional transformation of the financial and economic system of Ukraine in the conditions of globalization. In: Sidenko, V.R. (ed.) Kyiv, p. 648 (2017)
21. Zelinska, O.V., Hovorukha, V.R.: Increasing the efficiency of information systems in agriculture. Effi. Econ. (11) (2019). https://doi.org/10.32702/2307-2105-2019.11.47
22. Pavliuk, T., Volontyr, L.: Use of modern information technologies in agriculture. Formation Market Econ. Ukraine 38, 122–127 (2017)
23. Svynous, I.V., Havryk, O.Y., Tkachenko, K.V., Mykytiuk, D.M., Semysal, A.V.: Organizational and economic principles of the use of digital technologies in the activities of agricultural enterprises. Agroworld 16, 9–14 (2020). https://doi.org/10.32702/2306-6792.2020.16.9
24. Sherstiuk, L.M., Nezdoimynoha, O.Y.: Digital agriculture: foreign experience and peculiarities of implementation and use in Ukraine. In: Kalashnyk, O.V., Makhmudova, K.Z., Yasnolob, I.O. (eds.) Economic, Organizational and Legal Mechanism of Support and Development of Entrepreneurship, pp 310–318. Poltava (2019)
25. Horobets, N.M.: Digital technologies in the system of strategic management of agricultural enterprises. Agroworld 1, 36–43 (2022). https://doi.org/10.32702/2306-6792.2022.1.36
26. Skudlarski, J.: O rolnictwie 4.0 w Kamieniu Śląskim. Agro Profil. Magazyn Rolniczy 3, 88–91 (2019)
27. New technologies and digitization transform agriculture and open up new opportunities for policy improvement. OECD
28. Horobets, N.M., Khomiakova, D.O., Starykovska, D.O.: Prospects for the use of digital technologies in the activities of agricultural enterprises. Effective Econ. (1) (2021). https://doi.org/10.32702/2307-2105-2021.1.90
29. Digital farming makes agriculture sustainable. The Government of Japan
30. Mobile agricultural store in a mobile application Agro UA. Agroexpert
31. Law of Ukraine "On the Strategy of Sustainable Development of Ukraine until 2030"

Development and Increasing the Value Added Scenarios for the Woodworking Industry of Ukraine in the Context of the Circular Economy

Iryna O. Hubarieva⬛, Olha Yu. Poliakova⬛, Viktoriia O. Shlykova⬛, Dmytro M. Kostenko⬛, and Stanislav Buka

Abstract The woodworking industry becomes the basis for starting a circular economy due to the universality of the use of wood and the ability to be completely absorbed by nature without polluting it when used rationally. To form woodworking industry development scenarios for Ukraine, the contribution of individual types of products to the formation of the gross value added of the industry in 35 countries was investigated. The series of linear models of the dependence of the gross value added on the natural indicators of the output of products was built for 2000–2018. The relative contribution of wood-based panels in the gross value added was shown to be almost twice as high as the contribution of sawn wood, but the volume of veneer sheet production is not statistically significant. The export orientation and import dependence by individual types of products were used to confirm the specialization in the export and import of products of the woodworking industry among the countries, considered in the study, and indicate of degree of circular economy in the industry. Three scenarios were built for the development woodworking industry in Ukraine about substitution of imports, development for domestic consumption,

I. O. Hubarieva · O. Yu. Poliakova (✉) · V. O. Shlykova · D. M. Kostenko
Research Center for Industrial Problems of Development of the National Academy of Sciences of Ukraine, Kharkiv, Ukraine
e-mail: polya_o@ukr.net

I. O. Hubarieva
e-mail: gubarievairyna@gmail.com

V. O. Shlykova
e-mail: v.shlykova@ukr.net

D. M. Kostenko
e-mail: kostenko.d.n@ukr.net

O. Yu. Poliakova · V. O. Shlykova
Simon Kuznets Kharkiv National University of Economics, Kharkiv, Ukraine

S. Buka
Baltic International Academy, Riga, Latvia
e-mail: stanislavs.buka@bsa.edu.lv

and expansion of export of different products. It was proved that for the purpose of transitioning to a circular economy and increasing the gross value added, it is advisable to implement a combination of scenarios. The development of the production of wood-based panels makes it possible to approach the standards of the circular economy and causes a significant cumulative effect for the entire economy in the medium-term perspective.

Keywords Circular economy · Woodworking industry · Value-added · International trade · Scenario for Ukraine

1　Circular Economy and Woodworking Industry

The circular economy, which can be considered as the next level of the green economy, is a fairly new direction, but it has already gained some popularity among scientists and government officials. The main idea of the circular economy, in contrast to the linear one, is the full, repeated, and cyclical use of all resources [1]. Wood is one of the most versatile materials that can be used in many ways. On the other hand, wood is a material that is completely absorbed by nature without polluting it when used rationally. Therefore, the woodworking industry can become the basis for starting a circular economy.

Currently, wood and the woodworking industry are already considered a resource for implementing the principles of the circular economy in many areas: construction, packaging production, energy and energy saving, etc. [1–5].

The woodworking industry of Ukraine has a fairly wide range of products with sufficient resource potential [6], and therefore, its prospects as one of the leaders on the way to a circular economy have already attracted attention [7]. The development of scenarios for the development of the woodworking industry in Ukraine is aimed at determining the appropriate directions for the production and export of goods from the point of view of the widest possible implementation of circular economy standards and ensuring the competitiveness of the industry on the world market.

2　Formation of Value Added in the Woodworking Industry

The main tasks of industrial development in recent years are the production of products with high value-added and the approximation of the industrial structure of the countries of the world to its progressive state, which has a positive effect on sustainable economic development and the quality of life of their population [8, 9]. Therefore, before formulating the conditions of the scenarios, trends in the dependence of the gross value added of the investigated type of activity on the quantitative indicators of the qualitative composition of products, namely the number of produced sawn wood, veneer sheetsy, and wood-based panels, were identified.

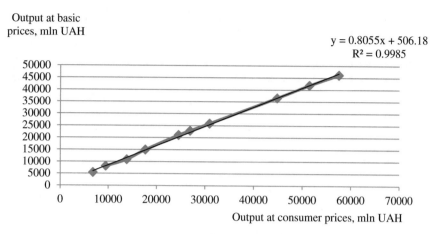

Fig. 1 Relationship model of output at basic prices and output at consumer prices for the type of economic activity "Manufacture of wood, paper, printing and reproduction" in Ukraine

For the calculations and analysis of the parameters of the dependence model of the gross value added, the data of the "Input—Output" Tables (IOT) for 2000, 2002–2018 were used [10, 11]. Data for 2000, 2002–2005 are presented in Ukrainian statistics only in consumer prices, and there are no data on gross value added. Instead of the latter, the GDP indicator by types of economic activity is used. Data on gross value added have appeared in industry statistics and input–output tables since 2006, and since 2013, data on GDP by activity have been missing from input–output tables. Therefore, intermediate models were used for data alignment. A simple linear model (Fig. 1) was used to reconstruct data on output in basic prices in 2008. A high coefficient of determination makes it possible to use this model.

In the same way, a linear model was used to restore data on gross value added in 2000, 2002–2005 (Fig. 2).

The original data, together with the restored ones, are shown in Table 1.

To analyze global trends and build a general model, we used data from 35 countries of the world, which were used to assess the competitiveness of the woodworking industry [12], which allowed us to consider this industry as one of the leaders in the implementation of the circular economy. We used data on selected countries of the world from the official website of the OECD as part of the Trade in Value Added (TiVA) database [13].

For comparison with other countries, all data were converted into US dollars at the annual average official exchange rate [14].

In the statistics of Ukraine, data on output, GDP and gross value added for the type of activity "Manufacture of wood and of products of wood and cork, except furniture; manufacture of articles of straw and plaiting materials" (code of CTEA-2010 C16) are not submitted separately, but only within the framework of the more general type "Manufacture of wood, paper, printing and reproduction" (code of CTEA-2010 C16-C18). Therefore, to extract from the data on output and gross value added, the

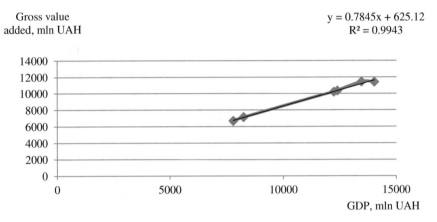

Fig. 2 Relationship model of gross value added and gross domestic product for the type of economic activity "Manufacture of wood, paper, printing and reproduction" in Ukraine

weighting coefficient of type C16 from the weighting structure by types of activities and main industrial groups was used to calculate the index of industrial products [15]. For the data of 2000, 2002–2015, the coefficient of 2010 was used—0.3, for the data of 2016–2020—0.27.

For the countries of the world, the ratio between the combined type of economic activity C16-C18 and a separate type of C16, according to data on gross value added and output, is slightly different from the domestic one. On average, the specific weight of processed wood is 0,3 (30%), with the exception of the Baltic countries (Estonia, Latvia, Lithuania), where this specific weight is 0,6–0,8 on average for the period, and Japan, where, on the contrary, the specific weight is the lowest of the considered countries—0,1. Therefore, it can be considered that the use of the calculated coefficient is appropriate.

Table 2 shows the estimated output data on gross value added and output for the woodworking industry of Ukraine.

In order to form scenarios for the development of the woodworking industry, the first step was to study the contribution of individual types of products to the formation of the gross value added of the industry. The analysis and development of models were carried out on the basis of FAO data on the volumes of production, export, and import of products in terms of three main types [16].

As for the possible dependence of the gross value added of the woodworking industry on the production of the main types of products in Ukraine, it is impossible to draw an unambiguous conclusion. Table 3 shows the correlation coefficients between industry indicators for Ukraine for 2000–2020. As can be seen from the data in the table, there is no significant correlation between the value-added and natural indicators of the industry, either in output or in exports. At the same time, the correlation between natural indicators of production and export is very significant, which does not allow for building a model that would cover all types of products, their production, and exports.

Table 1 Data of "Manufacture of wood, paper, printing and reproduction" in Ukraine

Years	Output, mln UAH		GDP, mln UAH	Gross value added, mln UAH	Output at basic prices, mln USD	Gross value added, mln USD
	At consumer prices	At basic prices				
2000	6823	5426	1818	2051*	997	377
2001	n.a	n.a	n.a			
2002	9482	8101	2704	2746*	1521	516
2003	13,824	10,860	3214	3147*	2036	590
2004	17,707	14,923	4364	4049*	2805	761
2005	24,539	21,081	6544	5759*	4114	1124
2006	26,923	22,886	7800	6684	4532	1324
2007	30,933	25,890	8249	7143	5127	1414
2008	43,313	35,394.8*	12,387	10,344	6720	1964
2009	44,800	36,545	12,231	10,197	4691	1309
2010	51,454	41,919	13,451	11,441	5282	1442
2011	57,482	46,176	14,029	11,406	5795	1432
2012		42,394	10,551	10,197	5305	1276
2013		44,890		9602	5616	1201
2014		52,457		12,034	4413	1012
2015		72,017		16,071	3297	736
2016		92,173		20,286	3607	794
2017		109,054		23,718	4100	892
2018		131,718		28,552	4842	1050
2019		133,626		29,659	5170	1148
2020		134,267		28,213	4981	1047

n.a.—no data available. So in the next tables
*modeled data

That is why a model of the dependence of gross value added on production was built for the entire sample of countries considered in the study. Preliminary analysis has shown that the scale of output and gross value added in the USA and China is not comparable to other countries. From 2000 to 2008, the USA was ahead of China in terms of gross value added, gradually losing its lead. China has been the leader since 2009. At the same time, the USA was ahead of the nearest leader (except China) by 2.6–4.6 times, and China increased the gap by 5.9 times. But these countries were not excluded from consideration, given their significant volumes of production, as well as the fact that, with the exception of the export of wood-based panels from China in recent years, these countries are not the leaders in the world market.

In the first stage, correlation matrices were calculated between natural indicators of output of various types of products and the amount of gross value added. The

Table 2 Estimation of output and gross value added for "Manufacture of wood and of products of wood and cork, except furniture; manufacture of articles of straw and plaiting materials" in Ukraine, mln USD

Years	Output in basic prices	Gross value added
2000	299.22	113.12
2001	n.a	n.a
2002	456.26	154.68
2003	610.95	177.01
2004	841.65	228.34
2005	1234.08	337.13
2006	1359.56	397.07
2007	1538.02	424.34
2008	2015.96	589.16
2009	1407.16	392.64
2010	1584.71	432.52
2011	1738.65	429.47
2012	1591.56	382.82
2013	1684.85	360.39
2014	1323.93	303.72
2015	989.03	220.71
2016	983.83	216.53
2017	1118.26	243.21
2018	1320.68	286.28
2019	1410.05	312.97
2020	1358.37	285.43

results of the calculations are shown in Table 4. Correlation analysis showed that there was a close positive correlation between the output of sawn wood and wood-based panels and gross value added throughout the period under review. As for veneer sheet output, its correlation at the beginning of the period was weak. Since 2009, the relationship with gross value added has become significant. Therefore, initially, all three types of products were included in the model.

To identify the influence of the output of various types of products of the wood-working industry, a series of models of the dependence of the gross value added on the natural indicators of the output of products was built:

$$VA = a_0 + a_1 * Sw + a_2 * VS + a_3 * WP,   \qquad (1)$$

where $a_0, \ldots a_3$ is the regression coefficients.

Sw, VS, WP is the production volumes of sawnwoods, veneer sheets, and wood-based panels in natural terms, respectively.

Analysis of the results of building regression models showed that the models have a high degree of adequacy. However, the regression coefficients for veneer sheet

Table 3 Correlation coefficients between value and natural indicators of the woodworking industry in Ukraine for 2000–2020

Indicators	VA	Sw	VS	WP	Sw_E	VS_E	WP_E	Output
Value added, (VA) mln USD	1.00	−0.19	−0.08	0.34	−0.11	0.15	0.03	0.93
Output, 1000 m³ sawn wood (Sw)	−0.19	1.00	0.81	0.67	0.90	0.73	0.26	−0.01
Veneer sheets (VS)	−0.08	0.81	1.00	0.81	0.90	0.93	0.45	0.21
Wood-based panels (WP)	0.34	0.67	0.81	1.00	0.83	0.83	0.59	0.61
Export, 1000 m³ Sawn wood (Sw_E)	−0.11	0.90	0.90	0.83	1.00	0.82	0.53	0.15
Veneer sheets (VS_E)	0.15	0.73	0.93	0.83	0.82	1.00	0.43	0.39
Wood-based panels (WP_E)	0.03	0.26	0.45	0.59	0.53	0.43	1.00	0.29
Output, mln USD	0.93	−0.01	0.21	0.61	0.15	0.39	0.29	1.00

Table 4 Correlations coefficients between gross value added and natural indicators of production output for the set of countries by years

Years	Sawnwood	Veneer sheets	Wood-based panels
2000	0.8888	0.3515	0.9248
2001	0.8626	0.3929	0.8998
2002	0.8653	0.4184	0.8562
2003	0.8604	0.2550	0.7537
2004	0.8841	0.2740	0.7196
2005	0.8926	0.2867	0.7136
2006	0.8857	0.3705	0.7379
2007	0.8582	0.5293	0.7855
2008	0.8091	0.6481	0.8386
2009	0.7857	0.7459	0.8868
2010	0.7991	0.7712	0.8935
2011	0.7871	0.7563	0.9197
2012	0.8385	0.7608	0.9253
2013	0.8756	0.8144	0.9136
2014	0.8706	0.8244	0.9327
2015	0.8732	0.8451	0.9282
2016	0.8856	0.7897	0.8885
2017	0.9081	0.9157	0.8859
2018	0.9108	0.8835	0.8855

output are statistically insignificant. Moreover, they (with the exception of 2017) are negative. The generalization of the results of model parameter estimation is shown in Table 5.

Analysis of Table 5 data allows us to note that the relative contribution of wood-based panels in the gross value added was almost twice as high as the contribution of sawn wood. Only in 2017–2018 did the situation change in favor of sawn wood, which may be related to the cyclical development of other industries that are consumers of woodworking products.

The dynamics of the adjusted coefficient of determination (Adjusted R^2) for the built models allows us to conclude that, over time, the concentration of gross added value occurs in only three selected types of products.

It should be noted that due to the high correlation between the original data series, multicollinearity is present almost everywhere in the models. However, since the task was to assess the possible contribution of different types of products to the formation of gross added value, multicollinearity was not completely eliminated. In the second step, the volume of production of veneer sheets was excluded, and the parameters of the model of the species were estimated:

Table 5 Coefficients and quality indicator of the regression of gross value added on the output of sawn wood, veneer sheets, and wood-based panels by years

Years	a_0	Sawnwood, a_1	Veneer sheets, a_2	Wood-based panels, a_3	Adj. R-squared
2000	125.574	0.105	−1.097	0.410	0.870
2001	122.395	0.109	−2.130	0.419	0.841
2002	268.961	0.143	−3.490	0.331	0.839
2003	612.670	0.103	−5.691	0.490	0.858
2004	581.684	0.157	−6.033	0.438	0.870
2005	563.895	0.198	−4.861	0.349	0.884
2006	704.756	0.193	−4.189	0.300	0.876
2007	1035.441	0.180	−3.809	0.301	0.864
2008	1091.975	0.186	−2.841	0.306	0.872
2009	969.713	0.172	−2.895	0.290	0.884
2010	843.448	0.199	−1.314	0.248	0.905
2011	979.727	0.185	−0.708	0.258	0.913
2012	669.650	0.208	−1.110	0.260	0.932
2013	626.866	0.260	−3.722	0.278	0.946
2014	759.110	0.225	−3.366	0.309	0.951
2015	793.526	0.221	−10.215	0.449	0.952
2016	999.529	0.269	−12.361	0.405	0.940
2017	120.533	0.253	0.990	0.167	0.894
2018	28.931	0.308	−0.758	0.193	0.895

$$VA = a_0 + a_1 * Sw + a_3 * WP. \tag{2}$$

The results of estimating the parameters of the reduced model by year are shown in Table 6.

Analysis of the data in Table 5 confirms the previous conclusions about the ratio of the relative contribution of the production of wood-based panels and sawnwoods. But with the exclusion of veneer sheets from consideration, the redistribution in favor of sawnwoods took place in 2016. At the same time, in the dynamics of the regression coefficients, a certain anti-synchronous cyclicality can be observed: cycle 1—2000–2008, cycle 2—2009–2014 (2015), cycle 3—from 2015 (2016). Therefore, it should be expected that during the next 3–4 years, wood-based panels will again overtake sawnwoods in terms of relative gross value added per 1000 cubic meters of products.

The specified results of the analysis allow us to identify one of the promising scenarios for the development of the Ukrainian woodworking industry the increase in the production of wood-based panels. It will also give an impetus to the production of other products of the timber industry.

Table 6 Coefficients and quality indicator of the regression of gross value added on the output of sawn wood and wood-based panel by years

Years	a_0	Sawnwood, a_1	Wood-based panels, a_3	Adj. R-squared
2000	16.908	0.104	0.401	0.873
2001	−101.609	0.112	0.384	0.840
2002	−67.384	0.146	0.272	0.829
2003	10.935	0.202	0.167	0.788
2004	30.379	0.239	0.136	0.810
2005	32.442	0.249	0.132	0.829
2006	297.285	0.228	0.131	0.830
2007	728.670	0.210	0.163	0.845
2008	868.930	0.202	0.209	0.863
2009	737.696	0.186	0.205	0.876
2010	731.513	0.203	0.213	0.906
2011	907.444	0.185	0.242	0.915
2012	567.672	0.207	0.236	0.932
2013	336.761	0.256	0.209	0.940
2014	454.134	0.228	0.243	0.947
2015	149.618	0.221	0.249	0.929
2016	127.108	0.254	0.186	0.896
2017	119.966	0.272	0.176	0.897
2018	29.011	0.292	0.188	0.898

3 Export and Import of the Woodworking Industry of Ukraine

Export orientation, if not excessive, usually promotes the development and adoption of new technologies in industry. Figure 3 shows the dynamics of Ukraine's export and import of industry products as a whole for 2009–2020 [17]. As can be seen from the figure, during 2017–2020, the exports of the industry exceeded the import, therefore, its further increase is promising, especially since there was no steady trend to increase exports in the last 3–4 years, and Russia's military aggression against Ukraine in 2022 caused a drop in total exports.

To determine the impact of the export of certain types of woodworking products, a model of the dependence of export volumes in value terms on the export of certain types in natural terms was built. The results are shown in Table 7. Due to significant multicollinearity in the model, only one coefficient is statistically significant—the regression coefficient for the volume of sawnwoods exports. The second part of Table 7 shows the results of the evaluation of the one-factor model of dependence on the volume of sawn wood exports, which can be used for preliminary forecasting of volumes.

The predominant role of sawnwoods in the formation of the value of exports is connected with the significant advantage of exports in natural terms, as can be seen in Fig. 4. Export volumes of sawnwoods are 6–7 times greater than exports of wood-based panels and 20 times greater than veneer sheets. That is why the dynamics of the export of sawn woods in natural terms is consistent with the dynamics of export value. The other two types of production do not have clearly expressed dynamics. Therefore, their influence on the formation of the value of exports was not significant. A comparison with previous results allows us to note that the export of Ukraine needs

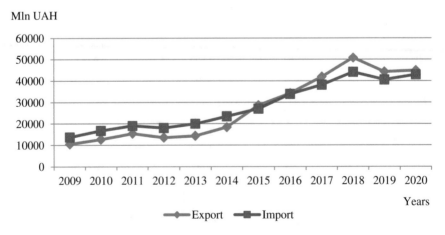

Fig. 3 The dynamics of export and import of products and services of the industry "Manufacture of wood, paper, printing and reproduction" in Ukraine

Table 7 The results of the regression analysis of the dependence of the value volume of the industry's exports on the export of certain types of wood products for Ukraine

Regression Summary for Dependent Variable: Export (UKR_Wood) R = 0.99702549
R^2 = 0.99405983 Adjusted R^2 = 0.99183226 F(3,8) = 446.25 p < 0.00000
Std.Error of estimate: 1357.2

	Beta	Std.Err.—of Beta	B	Std.Err.—of B	t(8)	p-level
Intercept			−10,968.4	2082.206	−5.26768	0.000757
Sw_E	0.94737	0.045389	15.3	0.735	20.87224	0.000000
VS_E	0.05559	0.044931	14.4	11.620	1.23726	0.251075
WP_E	0.03274	0.027810	3.3	2.819	1.17745	0.272856

Regression Summary for Dependent Variable: Export (UKR_Wood) R = 0.99607111
R^2 = 0.99215765 Adjusted R^2 = 0.99137341 F(1,10) = 1265.1 p < 0.00000
Std.Error of estimate: 1394.8

	Beta	Std.Err.—of Beta	B	Std.Err.—of B	t(10)	p-level
Intercept			−8869.67	1097.526	−8.08151	0.000011
Sw_E	0.99607	0.028004	16.13	0.453	35.56864	0.000000

to change its structure in the direction of increasing the specific weight of wood-based panels.

In the same way, the analysis of the import of Ukrainian products from the wood-working industry was carried out. Figure 5 shows the dynamics of exports by product type. A comparison of the data in Figs. 3 and 5 shows that only the dynamics of imports of veneer sheets in natural terms are consistent with the dynamics of imports in value terms. The surge in sawn wood imports in 2017–2018 distorts the dynamics.

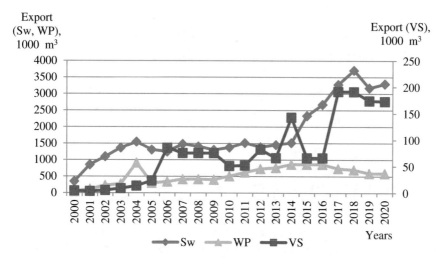

Fig. 4 The dynamics of Ukraine's export of woodworking products by types: Sw—export of sawnwoods, WP—export of wood-based panels, VS—export of veneer sheets

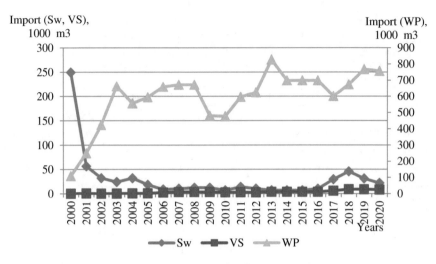

Fig. 5 The dynamics of Ukraine's import of woodworking products by types

The import of wood-based panels also does not have stable dynamics. Therefore, it can be predicted that the model of the formation of imports will also reveal the significance of only one indicator—the import of veneer sheets. This is confirmed by the results of building multifactorial and single-factorial models presented in Table 8.

The analysis of import dynamics and the built model allows us to put forward as one of the possible scenarios for import substitution in the consumption of wood-based panels.

4 Export and Import of the Woodworking Industry as an Indicator of Circular Economy

The peculiarities and degree of development of the circular economy in the wood-working industry of the countries of the world are reflected in the volume and structure of export, import, and production of various types of products. In order to identify these features and to form more detailed scenarios for the development of the wood-working industry of Ukraine, the structure and influence of foreign trade of the countries of the world were studied.

Among the countries considered in the study, there is a certain specialization in the export and import of products of the woodworking industry. This is confirmed by the results of the assessment of export orientation and import dependence by individual types of products. Export orientation determines the specific weight of exports in the production of a certain type of product and was calculated according to the formula:

$$EO_i = E_i / V_i,$$

Table 8 The results of the regression analysis of the dependence of the value volume of the industry's imports on the import of certain types of wood products for Ukraine

Regression Summary for Dependent Variable: Import (UKR_Wood) R = 0.91073142
R^2 = 0.82943173 Adjusted R^2 = 0.76546862 $F_{(3,8)}$ = 12.967 p < 0.00194
Std.Error of estimate: 5400.3

	Beta	Std.Err.—of Beta	B	Std.Err.—of B	t(8)	p-level
Intercept			−1747.32	11,399.65	−0.153278	0.881975
Sw_I	0.176063	0.370182	157.33	330.79	0.475612	0.647067
VS_I	0.670138	0.412268	3096.40	1904.90	1.625493	0.142711
WP_I	0.172920	0.204872	17.87	21.17	0.844041	0.423166

Regression Summary for Dependent Variable: Import (UKR_Wood) R = 0.90233779
R^2 = 0.81421348 Adjusted R^2 = 0.79563483 $F_{(1,10)}$ = 43.825 p < 0.00006
Std.Error of estimate: 5041.0

	Beta	Std.Err.—of Beta	B	Std.Err.—of B	t(10)	p-level
Intercept			7319.176	3463.248	2.113385	0.060696
VS_I	0.902338	0.136304	4169.287	629.796	6.620062	0.000059

Table 9 Median values of export orientation and import dependency

Products	Export orientation	Import dependency
Sawnwood	0.3814	0.2923
Veneer sheets	0.5006	0.6013
Wood-based panels	0.4239	0.4799

where EO_i is the export orientation by ith type of products, $i \in$ {Sawnwoods, Veneersheets, Wood-basedpanels};

E_i is the export of the ith type of products, 1000 m^3;

V_i is the output of the ith type of product, 1000 m^3.

Import dependence determines the specific weight of imports in the total consumption of products of a certain type and was calculated as follows:

$$I O_i = \text{Im}_i / C_i,$$

where IO_i—import dependence by ith type of products, $i \in$ {Sawnwoods, Veneer sheets, Wood-based panels};

Im_i—import of the ith type of products, 1000 m^3;

C_i—total consumption of the ith type of production in the economy, 1000 m^3.

The criterion for the classification of countries was the excess of the median value calculated for all countries for all years of observation given in Table 9.

According to the results of calculations, four groups of countries can be distinguished for each type of product:

- net (predominant) exporters (PE)—countries whose production is export-oriented during all or most of the time period (with a preponderance of recent years);
- net (predominant) importers (PI)—countries that mainly consume imported products of this type during all or most years;
- active players on the international market ("traffickers") (Tr)—countries that actively export imported products. Excessive, more than 1 (even several times) values of export orientation and import dependence are observed for them;
- without pronounced specificity (UnS)—the rest of the countries.

The generalization of the results is shown in Table 10.

The analysis of the export structure of the countries of the world makes it possible to note a certain specialization of exports. Thus, none of the countries specializes in the export of veneer sheets, and according to the results of 2020, seventeen countries mainly export sawnwoods.

Analysis of Table 10 allows us to draw several conclusions regarding the prospects for the development of the export of the woodworking industry of Ukraine. A significant part of the small European countries, despite the presence of their own wood resources, are not actually producers. They are focused on the import of products and

Table 10 Characteristics of the countries of the world according to the intensity of international trade in certain types of products of the woodworking industry

Country code	Sawnwood			Veneer sheets			Wood-based panels		
	Export orientation	Import dependence	Type	Export orientation	Import dependence	Type	Export orientation	Import dependence	Type
AUS	0	0	UnS	13	6	UnS	0	0	UnS
AUT	21	15	Tr	21	21	Tr	21	16	Tr
BEL	21	21	Tr	9	13	UnS	21	21	Tr
BGR	14	1	PE	19	17	Tr	18	7	Tr
BRA	0	0	UnS	0	0	UnS	2	0	UnS
GBR	0	21	PI	n.a	n.a	n.a	0	21	PI
GRC	0	21	PI	n.a	n.a	n.a	5	10	UnS
DNK	20	21	Tr	4	7	UnS	9	21	PI
EST	21	19	Tr	11	2	UnS	21	21	Tr
IND	0	0	UnS	0	0	UnS	0	0	UnS
IRL	17	21	Tr	n.a	18	n.a	21	21	Tr
ESP	0	20	PI	7	13	UnS	13	2	UnS
ITA	2	21	PI	0	4	UnS	0	6	UnS
CAN	21	0	PE	18	16	Tr	21	1	PE
CHN	0	14	PI	0	1	UnS	0	0	UnS
KOR	0	11	UnS	0	1	UnS	0	7	UnS
LVA	21	11	Tr	9	10	Tr	21	16	Tr
LTU	21	19	Tr	14	19	Tr	5	18	PI
MEX	0	21	PI	0	0	UnS	0	21	PI
NLD	21	21	Tr	3	21	PI	21	21	Tr

(continued)

Table 10 (continued)

Country code	Sawnwood			Veneer sheets			Wood-based panels		
	Export orientation	Import dependence	Type	Export orientation	Import dependence	Type	Export orientation	Import dependence	Type
DEU	3	5	UnS	9	9	UnS	17	2	PE
POL	0	0	UnS	0	6	UnS	0	0	UnS
PRT	6	5	UnS	20	19	Tr	20	9	PE
RUS	20	0	PE	5	0	UnS	0	0	UnS
ROU	16	0	PE	14	7	PE	21	8	PE
SVK	16	5	PE	14	17	Tr	21	21	Tr
SVN	21	21	Tr	15	10	Tr	17	18	Tr
USA	0	6	UnS	13	11	Tr	0	0	UnS
TUR	0	0	UnS	1	3	UnS	0	0	UnS
HUN	17	21	Tr	17	12	Tr	21	20	Tr
FIN	21	0	PE	21	14	Tr	21	12	Tr
FRA	0	0	UnS	15	20	Tr	17	3	PE
CZE	17	15	Tr	21	20	Tr	21	18	Tr
SWE	21	21	Tr	20	12	Tr	0	20	PI
JPN	0	1	UnS	0	17	PI	0	15	PI
UKR	20	0	PE	15	0	PE	3	2	UnS

their longer export (Baltic countries, Austria, the Netherlands, Hungary, the Czech Republic, Slovenia, Ireland, and even Sweden). Therefore, the development of trade with these countries should not be considered as a priority, since the terms of trade will not be favorable.

Secondly, the least specialized are the countries in the foreign trade of veneer sheets. This indicates the possible inexpediency of the development of separate exports of this type since such specialization is not observed in world practice.

Thirdly, there are fewer net exporters of wood-based panels than there are net exporters or traffickers of sawnwoods, so it can be assumed that this niche in world trade is less monopolized, and therefore, focusing efforts on expanding Ukraine's presence in it should give an advantage.

It can be assumed that the woodworking industry of countries where the export of sawnwoods prevails is far from the standards of the circular economy. In those countries specializing in the export of wood-based panels, circular economy standards are implemented faster. Thus, it is expedient to develop the export of both sawnwoods and wood-based panels with the advantage of the latter, which will contribute to the development of circular technologies in the industry.

The importance of export-oriented and import-substituting scenarios is confirmed by the analysis of the structure of Ukrainian exports. The group of goods "Wood sawn or chipped lengthwise..." under the UCGFEA (Ukrainian Classification of Goods for Foreign Economic Activity—compiled on the basis of the Harmonized System of Description and Coding of Goods and the Combined Nomenclature of the European Union) code 4407, i.e. those classified as sawnwoods, has the largest specific weight in terms of export value of woodworking production. However, the cost of one cubic meter of this product is small (0.141 thousand USD), so a significant specific weight is provided by the largest physical volume of export.

The second by the specific weight of the value is the group under the code 4408 "Sheets for veneering...". The products of this group are the most expensive per 1 m^3 ($1,19 thousand). Therefore, even a small increase in the export of this type of product can significantly increase export revenues.

The third group of goods in terms of contribution to the export value consists of goods of group 4418 "Builders' joinery and carpentry of wood..." with a fairly wide nomenclature, which can refer to both sawnwoods and wood-based panels. However, the unit cost of such goods is small, although they are mostly goods for final consumption or for construction purposes.

As for the import substitution scenario, its expediency in various directions is confirmed by the fact that the cost of a unit of import exceeds the cost of a unit of import in almost all groups, most notably in group 4408, i.e. wood-based panels. The largest import is the specific weight of the group of goods 4411 "Fibreboard of wood..." and 4410 "Particle board...", the most expensive is the import of goods under group 4408, and the second is the import of goods under group 4413 "Densified wood...". Thus, the development of the production of more complex and technological products in the woodworking industry will reduce the import dependence of domestic production and final consumption in many positions.

The external parameters of the scenarios, that do not relate to the development of the woodworking industry itself, correspond to the basic version of the forecast of the National Bank of Ukraine regarding the economic development of Ukraine for 2023–2024 [18]. The forecast is developed in a basic version and an alternative one, which is more pessimistic.

Figure 6 shows a brief description of scenarios for the development of the woodworking industry of Ukraine for the period up to 2030. Short-term scenarios provide for implementation during 2023–2026, or full implementation within 3 years from the end of hostilities throughout the country. However, it is assumed that some of these scenarios can be implemented during martial law, but much more slowly, due to the risks mentioned in [18]. Medium-term scenarios envisage implementation by 2030, or within 6–8 years after the end of hostilities. Regional features of the woodworking industry of Ukraine contribute to the implementation of all scenarios, despite the martial law.

In all scenarios regarding GDP and output in 2022, according to the estimates of the National Bank of Ukraine, a drop of 30–35% was assumed for all types of economic activity. Starting from 2023, a recovery rate of 5–5.5% was set for all types of economic activity except for the woodworking industry.

Scenario 1 is aimed at rapid import substitution on the domestic market of wood-based panels, at least in volumes that were imported from Russia and Belarus. The importance of the rapid implementation of this scenario is determined by the increased demand for such products in the regions of active hostilities and adjacent to them in connection with the need to carry out repair work, arrange temporary accommodation, etc. The predominant import of wood-based panels was carried out from Belarus, so according to the data of 2020, in natural terms, 55.5% of the import of particle boards (code 4410), 57.5% of fibreboards of wood (code 4411) and 58.5% of plywoods (code 4412). Imports from Russia accounted for 13.8 and 7.9% of the first two positions, respectively [17].

The main results of scenario 1 implementation are shown in Table 11. Since Scenario 1 is short-term and only focuses on import substitution, the results for 2025 are lower than those for 2020. This is because the length of the recovery period compared to the drop in 2022 is short to compensate for the drop. The most positive result of Scenario 1 is a decrease in import dependence for sawnwoods and wood-based panels by 0.9% points and 17.5% points.

The cumulative effect of increasing the output of wood-based panels for the purpose of import substitution was manifested in the fact that output by the type of economic activity "Wood production…" (code C16-18) will increase in 2025 compared to 2020 by 30.4%, and the gross value added of the industry—by 35%. In general, due to the development of the production of wood-based panels for the purpose of import substitution, the gross value added by the economy will increase by 10.1%.

Compared to a simple forecast of economic recovery at an average annual rate of 5%, Scenario 1 provides a 13.1% higher increase in gross value added by the end of 2025.

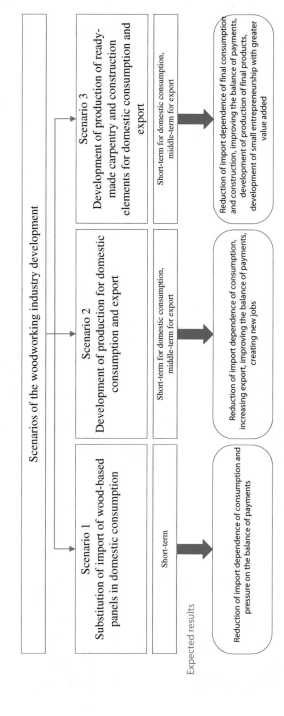

Fig. 6 Scenarios of the development of the woodworking industry of Ukraine for the period until 2030

Table 11 The main results of Scenario 1 implementation

Indicators	Years		
	2020	2023	2025
Output, 1000 m^3			
Sw	3 748.0	2 754.8	3 037.1
VS	230.1	169.1	186.4
WP	2 468.2	1 979.7	2 496.7
Export, 1000 m^3			
Sw	3 297.4	2 250.5	2 481.2
VS	173.0	118.1	130.2
WP	588.6	401.7	442.9
Import, 1000 m^3			
Sw	21.9	19.5	21.5
VS	8.2	7.3	8.1
WP	756.8	591.3	260.2
Import dependency, %			
Sw	4.6	3.7	3.7
VS	12.6	12.6	12.6
WP	28.7	27.3	11.2
Output of manufacturing C16-18, mln UAH	134 267	136 405	175 099
Increase to 2020, %		1.59	30.41
VA of manufacturing C16-18, mln UAH	29 659	29 804	40 053
Increase to 2020, %		0.49	35.05
Gross value added, mln UAH	3 626 725	3 548 036	3 992 044
Increase/decrease to 2020, %		−2.17	10.07
Estimated GVA, mln UAH		3 202 030	3 530 238
Increase to estimated GVA, %		10.8	13.1

Scenario 2 envisages not only import substitution in the domestic market, but also an increase in total production volumes and an increase in exports of all major types of products. As shown above, increasing the production of sawn woods and wood-based panels can provide an increase in value-added on the world market. At the same time, the veneer export market niche is small, but the least monopolized. Scenario 2 involves implementation in two stages. In the first (short-term) stage, the restoration of production and export of three groups of products to the maximum retrospective levels during 2023–2025 is expected. In the second stage, in 2026–2030, it is planned to increase the production and export of all types of products with average annual growth rates for the period 2000–2020, but not less than 5–6% per year. In the second stage, it is taken into account that the growth rate of output should be ahead of the growth rate of exports. The parameters of Scenario 2 are shown in

Table 12. Imports during 2023–2025 remain at the level of 2020, with imports from Russia and Belarus compensated by imports from other countries. In the second stage, imports are reduced at a rate equal to the difference between the growth rates of output and exports.

Scenario 2 is implemented in two stages; therefore, Table 13 shows the intermediate results of 2025 and the final results of 2030. This scenario most widely covers all areas of development of the woodworking industry. Output and exports for all considered positions are growing in 2025 compared to 2020 to the values that were the maximum in retrospect. Imports during 2023–2025 also grow at a small rate to ensure domestic final and intermediate consumption.

The most positive result of Scenario 2 is a significant reduction by 2030 of import dependence for all commodity items. Thus, import dependence for sawnwoods will decrease in 2030 compared to 2020 by 0,8% points, for veneer imports—twice, for wood-based panels by 8.3% points.

The cumulative effect of the implementation of Scenario 2 is manifested in an increase in the output of the type of economic activity "Wood production..." (C16-18) by 59.9% by 2030, the gross value added of this industry almost doubled (by 81.9%), and the gross of value added of the entire economy by 40.5%. Compared to a simple forecast of an increase in the gross value added of the economy in 2025, slightly less than in Scenario 1—13.05%, and in 2030—by 13.14%.

Thus, Scenario 2 at the first stage is slightly worse than Scenario 1 only according to the criterion of import dependence, but in the second stage, the recovery of the woodworking industry is replaced by development, which has a significant effect on the economy as a whole.

Scenario 3 envisages the development of the production of goods most ready for final consumption and used in construction. The perspective of this scenario from the point of view of improving the balance of payments and the development of production with a large value added is confirmed by the following. In terms of the export value of goods per 1 kg, the most expensive is the group of goods 4418 ("Builders' joinery and carpentry of wood...") excluding the export of unique goods ("Wooden frames for paintings, photographs, mirrors or similar objects", code 4414).

Table 12 Average growth rates of output, export, and import of products by type in Scenario 2

Average rates of increase (+)/decrease (−)	Groups of products		
	Sawn woods	Veneer sheets	Wood-based panels
1 stage			
Output	3.0%	6.0%	9.3%
Export	4.0%	3.4%	13.6%
2 stage			
Output	5.5%	7.6%	7.4%
Export	5.0%	6.5%	6.0%
Import	−0.5%	−1.1%	−1.4%

Table 13 The main results of Scenario 2 implementation

Indicators	Years			
	2020	2023	2025	2030
Output, 1000 m³				
Sw	3 748.0	3 860.4	4 095.5	5 352.7
VS	230.1	243.9	274.0	395.2
WP	2 468.2	2 542.3	3 037.1	4 340.0
Export, 1000 m³				
Sw	3 297.4	3 429.3	3 709.2	4 734.0
VS	173.0	178.9	191.3	262.1
WP	588.6	668.7	862.9	1 154.8
Import, 1000 m³				
Sw	21.9	23.0	25.3	24.7
VS	8.2	8.6	9.5	9.0
WP	756.8	794.7	876.1	816.5
Import dependency, %				
Sw	4.6	5.1	6.1	3.8
VS	12.6	11.7	10.3	6.3
WP	28.7	29.8	28.7	20.4
Export orientation, %				
Sw	88.0	88.8	90.6	88.4
VS	75.2	73.4	69.8	66.3
WP	23.8	26.3	28.4	26.6
Output of manufacturing C16-18, mln UAH	134 267	141 749	170 136	214 717
Increase to 2020, %		5.57	26.71	59.92
VA of manufacturing C16-18, mln UAH	29 659	30 971	38 918	53 962
Increase to 2020, %		4.42	31.22	81.94
Gross value added, mln UAH	3 626 725	3 171 033	3 990 909	5 097 815
Increase/decrease to 2020, %		−12.56	10.04	40.56
Estimated GVA, mln UAH		3 202 030	3 530 238	4 505 578
Increase to estimated GVA, %		−0.97	13.05	13.14

But the last group is small in terms of volume and cannot form the basis of export development, unlike builders' joinery and carpentry of wood.

Scenario 3 has a local character. The difficulty of evaluating the results under this scenario is explained by the lack of stable dynamics in the production and export of goods of the selected group (windows, doors, door frames, etc.). This group of goods makes up 20–48% of the export of builders' joinery and carpentry of wood. Other multi-layer products have the largest specific weight in this group, but there is no production data for them in official statistics. In Scenario 3, it is assumed that

Table 14 The main results of Scenario 3 implementation

Indicators	Years			
	2023	2025	2028	2030
Export of positions 441810, 441820, thousands USD	38 480	44 055	53 970	61 790
Forecast by trend	40 647	45 602	53 034	57 988
Increase/decrease to forecast, %	−5.3	−3.4	1.8	6.6

the export of goods by selected groups will grow at an average annual rate of 7%, starting from 2023. The results of the implementation of Scenario 3 are shown in Table 14.

According to the results, the volume of total exports under items 441810 ("Windows, French-windows and their frames") and 441820 ("Doors and their frames and thresholds") in 2030 exceeds the historical maximum of 2013 by 2.9%. At the same time, compared to the forecast based on the trend of the last five years, the increase in exports will be 6.6%.

Scenario 3 alone cannot significantly affect the development of the entire woodworking industry, but it can complement Scenarios 1 or 2 in the medium term.

5 Conclusion

Summarizing the results of the woodworking industry development scenarios allows us to conclude that from an economic point of view and for the purpose of transitioning to a circular economy, it is advisable to implement a combination of scenarios. In the short-term perspective, it is important to achieve import substitution for the most vulnerable group of products—wood-based panels, to increase the production of final products with complete processing of raw materials, and also to ensure increased construction needs. The development of the production of wood-based panels makes it possible to approach the standards of the circular economy not only in the forest industry complex but also in construction, energy and energy saving of Ukraine. In the medium-term perspective, the further increase in production of all types of products makes it possible to obtain a significant cumulative effect for the entire economy in increasing gross value added by more than 40% within eight years.

References

1. Circular economy in construction. What, why and how. Sustainability Committee, City of Aarhus. https://endelafloesningen.aarhus.dk/media/35966/circular-economy-in-construction.pdf

2. Study on circular economy principles for buildings' design. Final Report (2021). Publications Office of the European Union, Luxembourg (2021). ISBN 978-92-9460-645-7. https://doi.org/10.2826/3602

3. de Carvalho Araújo, C.K., Salvador, R., Piekarski, C.M.: Circular economy practices on wood panels: a bibliographic analysis. Sustainability11(4), 1057 (2019). https://doi.org/10.3390/su11041057

4. Pichelin, F.: Why wood is the most important material for the circular economy. Institute for Materials and Wood Technology, Bern University of Applied Sciences. https://www.innovationnewsnetwork.com/wood-circular-economy-materials/458/

5. Kromoser, B., Reichenbach, S., Hellmayr, R., Myna, R., Wimmer, R.: Circular economy in wood construction—additive manufacturing of fully recyclable walls made from renewables: proof of concept and preliminary data. Constr. Build. Mater. **344** (2022). https://doi.org/10.1016/j.conbuildmat.2022.128219

6. Hubarieva, I.O., Kriachko, Y.M.: Assessing the raw material potential of woodworking industry of Ukraine and countries of the world. Bus. Inf. **9**(C), 89–95 (2021). https://doi.org/10.32983/2222-4459-2021-9-89-95 (in Ukrainian)

7. Blahun, I.S., Mendela, Y.M.: Circular model of the development of the forestry complex of Ukraine. Sci. Perspect. **11**(29), 114–127 (2022). https://doi.org/10.52058/2708-7530-2022-11(29)-114-127 (in Ukrainian)

8. Khaustova, V.Y., Reshetnyak, O.I., Poliakova, O.Y., Shlykova, V.O.: Assessment of the Ukrainian industries' participation in global value added chains. Probl. Econ. **3**(45), 73–85 (2020). https://doi.org/10.32983/2222-0712-2020-3-73-85 (in Ukrainian)

9. Gryshova, I., Kyzym, M., Khaustova, V., Korneev, V., Kramarev, H.: Assessment of the industrial structure and its influence on sustainable economic development and quality of life of the population of different world countries. Sustainability (Switzerland) **12**(5), 1–25, 2072 (2020). https://doi.org/10.3390/su12052072

10. Input-output table (at consumer prices). State Statistics Service of Ukraine. https://ukrstat.gov.ua/operativ/operativ2006/vvp/vitr_vip/vitr_e/arh_vitr_e.html

11. Input output table for Ukraine (at basic prices). State Statistics Service of Ukraine. https://ukrstat.gov.ua/operativ/operativ2021/vvp/kvartal_new/tvv_oc/arh_tvv_oc_e.html

12. Hubarieva, I., Hubariev, O, Zinchenko, V., Pronoza, P.: Ensuring the Competitiveness of the ukrainian woodworking industry in the post-pandemic period. In: International Scientific and Practical Conference Sustainable Development in the Post-Pandemic Period (SDPPP-2021). SHS Web of Conferences, vol. 126, p. 02001. Tallinn, Estonia (2021). https://doi.org/10.1051/shsconf/202112602001

13. Trade in value added (TiVA) (ed.) Principal Indicators. OECD (2021). https://stats.oecd.org/index.aspx?queryid=106160

14. Official hryvnia exchange rate against foreign currencies (period average). National bank of Ukraine. https://bank.gov.ua/files/Exchange_r_e.xls

15. Weight structure by type of activity and main industrial groupings for the calculation of industrial production index. State Statistics Service of Ukraine. https://ukrstat.gov.ua/operativ/operativ2020/pr/vsvd/vsvd.xlsx

16. Forestry Production and Trade. http://www.fao.org/faostat/en/#data/FO

17. Monthly volumes of foreign trade in goods by countries of the world. State Statistics Service of Ukraine. https://ukrstat.gov.ua/operativ/operativ2008/zd/o_eit/arh_o_eit_e.htm

18. Inflation Report 2022. National bank of Ukraine. https://bank.gov.ua/admin_uploads/article/IR_2022-Q3_en.pdf?v=4

Methodological and Technological Solutions to Improve the Security of Ukraine's Accounting System During the Hostilities

Daria Trachova, Olena Demchuk, and Viacheslav Trachov

Abstract Ukrainian companies have been threatened by cyberattacks since the country regained its independence in 1991, and with the start of a large-scale invasion in February 2022, this has taken on even greater proportions. In such circumstances, every company must assess the vulnerability of their critical services to cybersecurity incidents and technology failures. These threats may arise from attacks on systems and infrastructure, or they may be the consequences of military action. However, these problems will remain after the hostilities end, so it is necessary to develop measures to reduce their impact, both at the enterprise level and at the macro level (at the level of the region, country, etc.) right now. To achieve this goal, this article carried out a comprehensive analysis of threats in the information space of accounting associated with the start of full-scale hostilities on the territory of Ukraine. The most important threats were identified, and the possibility of using triple-entry for their leveling in Ukrainian accounting, both at the enterprise level and at the country level, was studied. We also analyzed the existing practical solutions for the triple-entry method realization. Based on this analysis, we proposed to use blockchain technology. Blockchain has proven to be a promising, new technology that has many related technologies but requires a lot of research and practical experiments. We also proposed the use of blockchains of two levels—enterprise level and macro level. Also, a scheme for the interaction of two-level blockchains was proposed and the choice of several characteristics of these blockchains was justified. A hierarchical structure of internal blockchain users of an enterprise is proposed, considering the possibility of interaction with an external blockchain at the macro level. In the final chapter, the focus was on the successful integration of the triple-entry method and blockchain solutions into Ukraine's accounting system. The chapter aimed to provide a comprehensive overview of the steps necessary for the successful seamless implementation of the triple-entry method and blockchain in Ukraine's accounting system.

D. Trachova (✉) · O. Demchuk · V. Trachov
Dmytro Motornyi, Tavria State Agrotechnological University, Bohdana Khmelnytskoho Ave, 18, Melitopol 72312, Ukraine
e-mail: daria.trachova@tsatu.edu.ua

O. Demchuk
e-mail: olena.demchuk@tsatu.edu.ua

© The Author(s), under exclusive license to Springer Nature Switzerland AG 2024
A. Semenov et al. (eds.), *Data-Centric Business and Applications*, Lecture Notes on Data Engineering and Communications Technologies 194,
https://doi.org/10.1007/978-3-031-53984-8_12

Keywords Accounting information security · Wartime · Blockchain · Triple-entry method

1 Problems of Accounting Security in Ukraine After the Start of Full-Scale Hostilities

The start of hostilities in Ukraine in 2014 has had a profound impact on the country's economy, with ripple effects felt throughout all sectors, including accounting. In the face of such conflict and instability, the security of financial information has become a major concern, with accounting firms and individual practitioners struggling to protect sensitive data from theft, corruption, and fraud.

One of the biggest problems faced by accountants in Ukraine is the prevalence of cybercrime. With the rise of technology, cybercriminals have become more sophisticated, using increasingly advanced techniques to steal financial information from both individuals and organizations. As a result, accounting firms and practitioners in Ukraine must take proactive measures to protect themselves and their clients against cyber threats, including investing in cutting-edge security software and regularly training their employees on safe online practices.

The large-scale hostilities that began in late February 2022 add even more serious factors to the security breaches. Prior to the outbreak of hostilities, ordinary enterprises in Ukraine, when considering credential security risks, focused on internal threats, such as unauthorized access to the system and distribution of confidential information, credential corruption, etc. [1, 2]. Now, for companies in the occupied territories and in the zone of active hostilities, technical risks and problems of managing the level of access to information have come to the fore. Enterprises were not ready to lose connection with the server due to the inability to reach it physically or due to the lack of the Internet. After all, it turned out that in the occupied territories, freedom of movement and communication were the first to disappear. The scale of these problems can be seen in Fig. 1 which shows the significant increase in the number of enterprises that have been forced to relocate due to the conflict.

Simultaneously with the local problems of the company, there was a need to control access and use of data from the state system for administering the credentials of payers of taxes and fees. Thus, at the same time, technological and organizational threats arose at the micro and macro levels of the economy, which affected absolutely every company in the country.

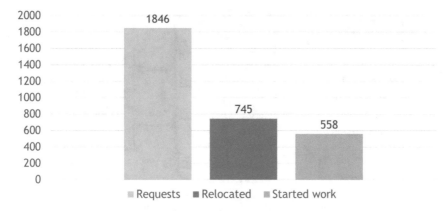

Fig. 1 The number of relocated enterprises [3]

2 Literature Review on the Problems of Accounting in High-Risk Conditions and Approaches for Their Solution

2.1 Problems of Accounting Security in Ukraine After the Outbreak of Full-Scale Hostilities

In modern literature, issues related to the specific problems of the accounting system in the conditions of full-scale military operations are practically not presented. Some authors approached security issues in accounting from the perspective of cybersecurity [2]. Others dealt with the peculiarities of this issue in the context of a hybrid war [1]. We decided to study this issue from the standpoint of accounting methodology.

At the enterprise level, security and control procedures begin with a security policy, which is a comprehensive plan that helps protect an enterprise from both internal and external threats. This plan must comply with ISO 27002 (at the time of this article writing, the current version of the standard is ISO 27002:2022)—international standards that establish best practices for information security. This standard includes 4 following main sections, which contain a total of 93 types of information security measures:

(1) organizational;
(2) human;
(3) physical;
(4) technological.

The practice of the current hostilities in Ukraine (especially their beginning) has shown that the most vulnerable in most (especially small and medium) enterprises were the measures of the physical and technological levels. Moreover, even if the company applied all the measures specified in the ISO 27002:2022 standard, it was

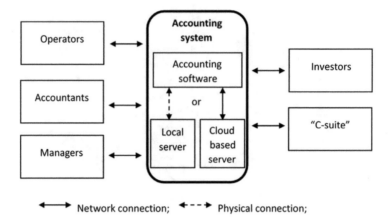

Fig. 2 Scheme of information flows when using accounting software based on the centralized principle of data storage

helpless in the face of a complete shutdown of the power supply, the internet and telephone communications. This vulnerability is primarily due to the fact that accounting and management software currently used in Ukraine store data in databases hosted on a server. Figure 2 shows that regardless of whether a local server or a cloud one is used, when the network connection is disconnected or access to the local server is limited, access to the information will be lost.

2.2 Triple-Entry Method in Accounting

2.2.1 General Aspects of the Theory of Triple-Entry

The problem of limiting physical access to information in the event of a violation of communication channels or restriction of access to the server is laid down at the methodological level of the principles of recording accounting information. Currently, the generally accepted system for entering information is the double-entry system [4], which lies in displaying one operation on two levels—debit and credit. This means that two entries must be made to record one operation. This system replaced the single-entry system when the fact of a purchase or sale must be written in the ledger only once [5]. With obvious simplicity, this mechanism led to errors and the possibility of fraud, so with a large number of transactions, it was very difficult to track them.

Of course, the double-entry system solves a number of problems related to the accidental deletion of information about the operation or its deliberate distortion. But firstly, it requires huge time and resources of auditors to search for these errors, and secondly, it does not provide reliable data storage.

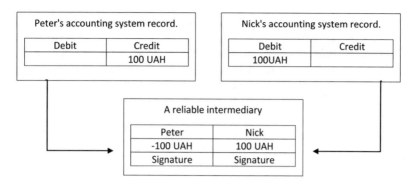

Fig. 3 Scheme of the Triple entry method by Griggs [7]

Therefore, in the context of the reliability of data storage in the accounting environment, one can increasingly hear about the triple-entry system. This term was first used in an article by Ijiri [6], where the author proposed that, in addition to debit and credit entries, a third level of entries, called credit, be included in the new set of accounts to reflect changes in income. The idea behind this "triple-entry accounting" system is to provide the organization with more timely financial information, which will enable it to make better strategic decisions.

However, in 2005, a different interpretation of the triple-entry method appeared. Financial cryptographer Ian Grigg, in his article [7], put forward a new concept, according to which each of the two parties conducting the operation reflects only their numbers in their internal documents, and the full information about the operation is placed in the public register of a certain intermediary (Fig. 3). This approach, according to the author, allows avoiding transaction fraud and reducing redundancy in internal records. This version of the interpretation of the term triple-entry has become very popular and is now used all over the world, and not the version of Ijiri [6].

In an article by Tyra [8], the author proposed using the Bitcoin infrastructure as an intermediary in the triple-entry concept proposed in Grigg [7]. The author also suggested that such an approach would be convenient for both participants in the transaction and external users—auditors, investors, etc. Since then, blockchain-related triple-entry has become a common occurrence in accounting. The industry has already witnessed the huge potential of blockchain triple-entry bookkeeping. At the moment, there are more than 10 blockchain projects related to triple-entry in accounting: Abendum, Balanc3, Pacio, and Request Network.

In 2016, Deloitte (one of the largest consulting and auditing companies) published an article [9] stating that the implementation of blockchain-enabled triple-entry accounting would be a game changer in accounting. However, they noticed that the intermediary that checks each transaction must be truly independent and reliable.

Dai J. and Vasarhelyi M. in their paper [10] outline a triple-entry system that integrates an existing enterprise resource planning (ERP) system with blockchain. This means that businesses get traditional double-entry bookkeeping, but a blockchain ledger is connected to it. As a result, it is possible to continuously record and check the ongoing transactions.

Cai [11] considers that the proposal of the previous authors does not fully correspond to the ideas set out in Grigg [7] and proposes a simplified triple-entry accounting system based on the original concept of Grigg [7]. In the system proposed by Cai [11], one digital entry in the accounting system of each organization is suffi-cient if it corresponds to an entry in the third ledger—the blockchain ledger. That is, all enterprises interacting with each other must abandon their internal accounting systems and switch to a separate platform, which must be a reliable and indepen-dent intermediary. However, the problem is that so far, few enterprises will agree to abandon their existing systems and rely on an external system.

2.3 Blockchain and Its Use in Accounting

2.3.1 Basic Elements of Blockchain Technology

Blockchain technology was conceived and demonstrated in Nakamoto [12]. The author used the blockchain to create a decentralized, public, and cryptographically secure digital currency system. The system, called Bitcoin, provides peer-to-peer digital currency trading. This eliminates the need for financial intermediaries while keeping transactions secure.

The Bitcoin blockchain can be thought of as a new type of database that records digital currency transactions in blocks. The blocks are arranged in a linear chronolog-ical order in the public network [13]. The key characteristics of the Bitcoin blockchain are:

(1) decentralization,
(2) strong authentication,
(3) burglary resistance.

In Nakamoto [12], decentralization means that each node (in this case, a user of the system) has access to the entire list of transactions (operations performed by the enterprise). Blockchain can also verify the identity of each user based on a public key encryption system [14]. These systems generate two keys: private and public (some codes are obtained using cryptographic functions). Here, the private key remains in the system, and the public key is given to the user. It is worth noting here that, although it is public, its public distribution is not recommended. Because each operation made by the user is first encoded using this public key, and only then sent over a public (that is why it is called public) communication channel. The only way to decode this message is to have the private key. In addition, in order to ensure the immutability of data in Bitcoin blocks, a system was invented that allows you to create a block only after performing calculations that require large resources. As a result, unwanted interference with blockchain records becomes almost impossible.

Thus, we see that the blockchain architecture is a decentralized public database. To date, the most common types of blockchain are public and private [15]. In a public blockchain, any network user has the right to view and create blocks with transactions in the chain. However, this is undesirable in accounting systems, since different user groups should have the right to work only with certain types of documents. The difference between a private blockchain is that only holders of certain cryptographic keys can use the information in the blockchain. Moreover, users (blockchain nodes) are divided into two or more categories according to the level of access to information. On the one hand, this allows us to protect the confidentiality of data, and on the other hand, it makes users unequal.

There is also a third type of blockchain—permission [16]. It is similar to a private blockchain, but in permission, permission to publish a transaction (record an accounting operation) is given only to certain nodes (users of the system), which are pre-selected by the central authority. This approach significantly simplifies the verification process, avoids unwanted manipulations, and makes the system more scalable [16]. The disadvantages of this approach include the fact that such a model does not prevent authorized users from colluding to create fraudulent transactions.

To extend the functionality of blockchains from simple cryptocurrency transactions to use it in e-commerce, second-generation blockchains have implemented the ability to use so-called "smart contracts" [13]. Smart contracts are computer programs running on blockchains that autonomously validate, secure, and perform operations on data [17]. Basically, smart contracts are used to encode the rules for agreeing on conditions between the parties involved in the operation.

The concept of a smart contract was first proposed by Szabo [18], who noted that the execution and monitoring of contracts mainly depend on a trusted central authority. New blockchain-based smart contracts decentralize enforcement power to every node in the blockchain network. In addition, since the trading history is distributed among all users of the system, it will be almost impossible to distort the terms of the transaction. This will help to significantly reduce counterparty risk [17]. Figure 4 shows an example where a blockchain-based smart contract is used to monitor and execute a buy/sell transaction. When the buyer and supplier agree on a deal, they note that the money can only be transferred when the goods arrive at the warehouse. This condition is encoded into a smart contract, which is then embedded into the blockchain. Once all conditions are met, the transaction information will be added to the blockchain.

But in addition to trade agreements, smart contracts can also encode other, more complex and branched terms. As the complexity and automation of smart contracts increase, their application can be greatly expanded. Applications are already being developed for self-issuance of bonds or crowdfunding with the promise of future dividends [19]. Investors can also participate in decision-making through decentralized voting to approve future strategies [20].

When designing the architecture of a blockchain-based accounting system that includes multiple enterprises, the issue of blockchain scalability must also be taken into account. The essence of this problem lies in the fact that with an increase in the number of enterprises, the security and confidentiality of information stored in one

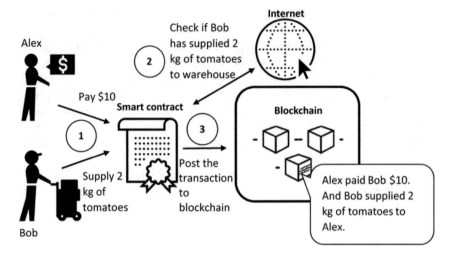

Fig. 4 Scheme of how a smart contract works on the example of a simple purchase and sale transaction

common blockchain may decrease, as well as the speed of its processing. The most common and reliable solution to this problem at the moment is to use sharding.

Sharding technology was first proposed and used in the field of databases by Corbett et al. [21]. By dividing all participating nodes in the network into multiple shards, each shard is only responsible for maintaining its own respective data. Thus, the network's processing capabilities can be scaled. As the number of nodes in the network increases, more processing capability is realized by adding more segments. Blockchain sharding was first proposed by Luu et al. [22], and this approach allowed to increase in the number of transactions processed per second. Since then, there have been many studies on blockchain sharding, such as Kokoris-Kogias et al. [23], Zamani et al. [24] and Wang and Wang [25].

In general, sharded blockchains have the following three characteristics. First, the participating nodes are divided into different shards, where the nodes in each shard only need internal communication most of the time. Clients and nodes in each shard can obtain information about the current state of the blockchain by contacting the nodes within the shard that are responsible for maintaining the blockchain. Second, each segment is only responsible for processing the corresponding transactions. Thus, transactions are processed by different segments in parallel. As the number of nodes in the network increases, more hops can be added to realize scalability. The third is storage sharing. That is, nodes of different shards need to store only the data of the corresponding shard [26]. It follows that the sharding technology satisfies all the requirements for scaling the accounting system for several enterprises.

2.3.2 The Use of Blockchain in Accounting

Although the literature in many other areas suggests blockchain applications, there is not much research on the use of this technology in accounting and auditing practice.

Yermack [27] explored the use of blockchain for real-time transaction accounting in his article. He suggested that by voluntarily disclosing transaction details via the blockchain, stakeholders would be able to have instant access to accurate financial information. Using this data, stakeholders will be able to create the reports they need much faster.

Fanning and Centers [28] also suggested that blockchain technology could be beneficial to the audit profession, as it makes it relatively easy to compare relevant accounting entries present in the records of each trading party. There is still no explicit illustration of how to achieve this goal, but such an approach would reduce the effort of auditors involved in verifying financial transactions.

Kiviat [17] illustrated the idea of blockchain-enabled "triple-entry accounting" using Bitcoin transactions as an example. This article describes the mechanism for placing records of transactions with bitcoins in the blockchain to prevent falsification of transactions. Unfortunately, this mechanism is specifically designed for the Bitcoin system and cannot be directly applied to conventional accounting systems.

2.3.3 Comparison of Blockchain with ERP Systems

At the moment, the main tool that simplifies accounting and management accounting in enterprises is the so-called Enterprise Resource Planning (ERP) system. An ERP system is software that allows a company to manage financials, supply chains, operations, trade, reporting, production, and personnel. Such systems make it possible to process almost any enterprise information in various ways [29]. ERPs are usually built on relational database management systems to automatically process various business transactions [29]. Additionally, ERP systems can provide data that can be used for analysis and management decision-making in addition to automating processes [30].

Dai and Vasarhelyi [31] concluded that the blockchain is a new type of database and can be used in combination with an existing ERP system. Unlike conventional ERP, which typically has a centralized architecture, blockchain distributes transaction verification, storage, and organization authority to a group of computers. This mechanism can significantly reduce the risk of a single-point failure [16] but, at the same time, make it difficult to manage the system.

In Swan [32], the author noted that blockchain, coupled with smart contracts, allows accountants to develop and deploy various automated controls. This minimizes human intervention and thereby reduces the likelihood of error or intentional harm.

3   The Use of the Triple Entry Method and Blockchain in Accounting in Ukraine During and After the War

As stated in the previous section, many accounting researchers and practitioners consider the triple-entry method as a paradigm for improving the security of corporate accounting systems. However, this system requires an independent and reliable component to verify and record each individual transaction. Records stored by this component may also be at risk of loss or unauthorized modification due to cyber attacks [33].

Blockchain technology can improve this mechanism and mitigate the above problems. It can act as an intermediary, distributing and automating the storage and verification process, providing a secure foundation that prevents falsification and incorrect accounts. Due to the nature of the blockchain, once a record is verified and added to the chain, it is almost impossible to change or delete it. In addition, blockchain can provide fast transaction verification by using smart contracts that are based on accounting standards or predefined business rules.

Ideally, at the level of a country, region or world (macro level), we should obtain a global blockchain in which the subject of the company's operations, according to his rights, interacts with its company's blockchain only, using smart contracts. Enterprises, concluding transactions with each other, interact within the framework of a macro-level blockchain [34].

Blockchain is essentially a distributed database with a high degree of replication. This means that every computer that is part of a blockchain network (or a specific computer with the necessary permissions) has a complete copy of all blockchain data. As a result, investors, directors, and other users with extended rights will always have access to a complete set of company information on the blockchain. So if at least one copy of this blockchain exists, it can always be restored for all users who need it. This feature greatly increases information security when some nodes are destroyed or access to them is restricted.

We have already mentioned that the blockchain only allows users to add data—it is not possible to modify or delete it. This feature seriously reduces the possibility of cyber-attacks and fraudulent activities, both from inside and outside the system. However, in real life, errors in the recording of accounting operations are quite common. Therefore, it is necessary to provide for the identification of the original operation so that a corrective one can be created later. In this context, operations can be understood not only as different payments but also as various transactions that ensure the flow of accounting data within the company [35].

Smart contracts in this system can serve as automatic controls that control accounting processes based on predefined rules [36]. They can work both for adding data to the blockchain and for selecting data and creating reports. Such analytics can be used to detect anomalies and obtain other useful information [36]. In this system, directors, investors, and optionally, other users can actively collaborate to verify transactions [37]. By combining all these components, a transparent and real-time verifiable accounting ecosystem can be achieved [37]. Figure 5 shows a diagram of

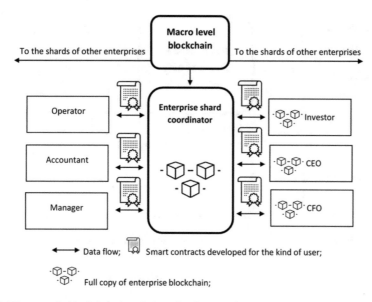

Fig. 5 Diagram of a blockchain-based macro level accounting system

a macro-level accounting system based on a permissioned blockchain that supports sharding.

Unfortunately, the implementation of such a model is extremely difficult at the present time, given the hostilities and the associated economic problems at both the corporate and state levels [38]. Therefore, we believe that at the first stage, it will be rational to use the blockchain as an additional technology to the ERP systems currently used in enterprises [39].

One of the possible options for a simplified triple-entry accounting information system is shown in Fig. 6 [40]. This system will record information about both transactions between business parties and data flows within an organization. In the system, each transaction will create an entry stored in the ledger of the blockchain, in addition to the entries that were included in the traditional double-entry system.

When a company purchases goods from its supplier, its ERP simultaneously creates a record reflecting the event on the debit and credit in the internal database and creates an entry in the blockchain ledger [39]. At the same time, all counterparties of the company must have an account in the blockchain. That is, they must be registered as users in it. In order for users to have different permissions, their accounts in the blockchain should be organized in a three-level **hierarchical** structure:

(1) user accounts:

    (a) company employees,

    (b) counterparties (if they do not have a macro-level blockchain account);

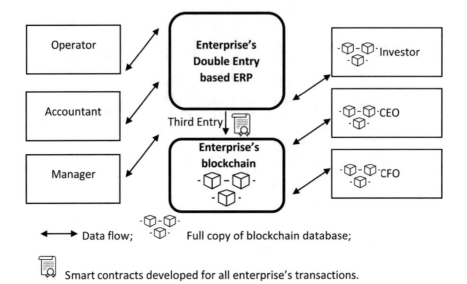

Fig. 6 Enterprise triple entry system based on ERP and blockchain

(2) accounts of the company's balance sheet items:

(a) assets,
(b) liabilities,
(c) equity;

(3) company account (for future interaction with other enterprises at the macro level).

The main advantage of a hierarchical account structure is that items on child levels can have their own special permissions while inheriting basic permissions from the parent level.

Once a request for an operation appears in the ERP system, it sends it to the blockchain as input for the smart contract corresponding to the operation. Smart contracts allow automatic fulfillment of conditions based on incoming parameters and future actions. For example, various business rules can be encoded in it, such as providing discounts for early payments or granting a bonus if the purchase amount exceeds a certain value. In addition, the smart contract must also describe the verification process for the operation performed. For example, if an auditor has doubts about a transaction, it may be deferred for confirmation by the accountants, and the CFO may decide to cancel it entirely. Such role restrictions can only be implemented in so-called **permissioned** blockchains.

Further, approved transactions will be grouped into blocks and added to the main chain, and then only authorized users will be able to view them. To protect corporate data privacy, transactions can be encrypted before being uploaded to the blockchain ledger, and only users who have the decryption key will be able to view the contents of these transactions.

Quite definitely, smart contracts can be carriers of the logic of all accounting operations and ensure their implementation. However, their creation should be carried out taking into account the requirements of all parties faced with the operation throughout its life cycle. At the same time, depending on the type of operation, the cost of creating a smart contract for it can vary greatly due to the different hourly wages of its participants. So, for example, when selling a product, the following stages and participants can be distinguished: data entry (operator)->verification (supervisor, accountant)->analysis (director). And for the planned monthly fuel purchase: request for the current fuel price, the dollar exchange rate, and the required amount of fuel (smart contract)->verification (financial director, director)->audit (auditor or owner). Therefore, at first, the cost of creating such smart contracts will be quite high, but as the libraries of such smart contracts develop, their price will gradually decrease.

Smart contracts can certainly carry the logic of all accounting processes and ensure their implementation. However, their creation must take into account the needs of all the parties who face the operation throughout its life cycle. At the same time, depending on the type of operation, the cost of creating a smart contract can vary significantly due to the different hourly wages of its participants. For example, when selling a product, we can distinguish the following stages and stakeholders: data collection (operator)->verification (supervisor, accountant)->analysis (general manager). And for the forecasted monthly fuel consumption: request for the current fuel price, the dollar exchange rate, and the required amount of fuel (smart contract)->verification (CFO, director)->audit (auditor, owner or investor). Therefore, the cost of creating such smart contracts will initially be quite high, but as the libraries of such smart contracts develop, their price will gradually decrease.

4　Implementation of the Triple-Entry Method and Blockchain into Ukraine's Accounting System

The successful implementation of the triple-entry method and blockchain in Ukraine's accounting system requires careful planning and consideration of several key factors. One of the most important aspects of this implementation is the seamless integration of the new system with existing systems, processes, and practices. Ensuring a smooth integration will help minimize disruption to daily operations and minimize any potential resistance from stakeholders.

However, it is vital to consider the potential challenges and constraints that may arise during the implementation process. Therefore, we need to focus on exploring data migration and systems integration considerations and implementing best practices to ensure a seamless integration process. Additionally, we must delve into the

importance of training and education in the integration process and address organizational and cultural barriers to adoption, recognizing potential challenges and
constraints that may arise during implementation.

4.1 Identifying Potential Challenges and Limitations

The implementation of the triple-entry method and blockchain in Ukraine's
accounting system may face a number of challenges and limitations. First of all,
these are technical restrictions that might necessitate a sizable investment in technical infrastructure and compatibility with existing systems. This entails making sure
the network infrastructure, software, and hardware are in place to support the new
technology.

Cybersecurity is a crucial factor to consider when implementing the triple-entry
method and blockchain as well. The system must be designed to protect against
unauthorized access to sensitive data, data breaches, and other security threats. Data
management and storage must be carefully managed, including the secure and efficient transfer of data between stakeholders and ensuring that data is accurate, up-to-
date, and properly secured. Training and capacity building for technical personnel is
also crucial for the successful implementation of the system. This includes system
administrators, IT support staff, and other technical personnel who will be responsible
for the successful operation of the system.

One of the biggest challenges in implementing any new technology is overcoming
resistance from stakeholders. This is especially true in the case of accounting systems,
where stakeholders may have vested interests in the status quo or may be wary of
change. In order to evaluate the potential for resistance, it is important to understand
the motivations and concerns of key stakeholders, including employees, clients, and
suppliers. This can be done through stakeholder engagement and consultation, as
well as through research and analysis of best practices and case studies from other
organizations.

Resistance from employees may arise from a lack of understanding of the new
technology and its benefits, as well as concerns about job security and changes to
existing processes and procedures. To mitigate this resistance, it is important to
provide training and education on the new system, as well as to communicate its
benefits clearly and effectively.

Clients and suppliers may also be resistant to change, particularly if they are
already familiar with existing accounting systems and processes. To mitigate this
resistance, it is important to engage with these stakeholders early in the implementation process to address any concerns they may have and to provide evidence of the
benefits of the new system.

Organizational and cultural barriers are also common challenges in the adoption
of new technology and can have a significant impact on the success of an implementation. In the case of accounting systems, these barriers can include a lack of awareness

of the benefits of the new system, a resistance to change, and an entrenched culture of doing things the way they have always been done.

In order to overcome these barriers, it is important to engage with stakeholders early in the implementation process to build awareness and understanding of the benefits of the new system. This can be done through training and education, as well as through effective communication and stakeholder engagement.

Another key factor in overcoming organizational and cultural barriers is to ensure that the new system is designed to meet the needs of the organization and its stakeholders. This may involve conducting a thorough needs analysis and involving stakeholders in the design process to ensure that the new system is aligned with the organization's goals and objectives. It is also important to address any concerns about the impact of the new system on existing processes and procedures, as well as any concerns about job security and other employment-related issues. This can be done through clear and effective communication, as well as through training and education programs that help employees understand the benefits of the new system.

Also, it is important to consider regulatory and legal considerations, as the use of blockchain technology and the triple entry method may be subject to various laws and regulations. The new system must comply with existing laws and regulations governing accounting, data privacy, and information security, as well as any new regulations that may be introduced in response to the technology.

In particular, the system must comply with relevant data privacy laws and regulations, such as the General Data Protection Regulation (GDPR) in Europe, which requires organizations to protect personal data and ensure that it is only processed for specific and legitimate purposes. Legal considerations must also be taken into account, including the requirement for contracts and agreements between stakeholders, such as clients and suppliers, to be recorded in the blockchain, and the impact of the technology on existing legal processes and procedures. It is also essential to consider the potential liability and responsibility of stakeholders in the event of a data breach or other security incident and to ensure that proper insurance is in place to protect against potential losses.

4.2 Stakeholder Training and Education

As already noted before, training and education are a critical aspects of the implementation of the triple-entry method and blockchain in Ukraine's accounting system. Effective training and education can help stakeholders understand the benefits of the new system, overcome any resistance to change, and ensure a successful implementation.

In order to be effective, training and education programs should be tailored to meet the needs of different stakeholder groups, including employees, managers, and other stakeholders. This may involve conducting a needs analysis to identify the specific knowledge and skills that stakeholders need to effectively use the new system.

Training and education programs can take many forms, including formal classroom-style training, online learning, and hands-on workshops. It is important to choose the most appropriate form of training for each stakeholder group, taking into account factors such as their learning style, availability, and existing knowledge of the technology.

In addition to training and education, it is also important to provide ongoing support and resources to help stakeholders use the new system effectively. This may include access to online resources, support forums, and regular check-ins with a dedicated support team.

4.3 Integration into Existing Systems

The integration process refers to the steps involved in implementing the triple-entry method and blockchain technology into Ukraine's existing accounting system. This process can be complex and involve many different stakeholders, including employees, managers, technology specialists, and others.

The first step in the integration process is to conduct a comprehensive analysis of the existing accounting system to identify areas that need to be upgraded or changed to support the new technology. This may involve updating hardware and software systems, as well as changing processes and procedures. A comprehensive plan should be developed to ensure that all stakeholders are aware of the changes and their role in the integration process. This includes a timeline for implementation and clear goals for the integration process.

Next, it is important to engage stakeholders and build support for the integration of the new system. This may involve communicating the benefits of the new system, addressing any concerns or objections, and working with stakeholders to identify the best ways to integrate the new technology.

Once the analysis is complete and stakeholders are on board, the next step is to implement the new system. This may involve installing new software, upgrading existing systems, and training employees on how to use the new technology.

Throughout the integration process, it is important to regularly monitor progress and evaluate the effectiveness of the new system. This may involve collecting feedback from stakeholders, monitoring key performance indicators, and making adjustments as needed. In conclusion, the integration process is a critical aspect of the implementation of the triple-entry method and blockchain in Ukraine's accounting system. In Table 1 below, we propose an indicative timeline for the implementation of such a system.

Data migration and system integration are key considerations when implementing the triple-entry method and blockchain technology in Ukraine's accounting system. This process involves transferring data from the existing system to the new system and integrating the new system with existing systems and processes. One of the key considerations for data migration is ensuring that data is accurately and completely transferred from the existing system to the new system. This may involve developing

Table 1 Indicative implementation timeline

Date	Activity	Description
Month 1	Planning	Develop a detailed plan for the implementation, including a clear understanding of the goals and objectives, resources required, and timeline for completion
Month 2–4	Training and awareness	Provide training and education for employees and stakeholders on the benefits and process of the implementation
Month 5–6	Technical integration	Begin the technical integration process by setting up the necessary systems, hardware, and software required to support the implementation
Month 7–8	Data migration	Begin the process of migrating existing data to the new system, ensuring that all data is accurately transferred and that the system is able to support the volume of data
Month 9–10	Testing and validation	Test the new system thoroughly to identify any potential issues and resolve them before the implementation is fully rolled out
Month 11	Launch	Launch the new system, with appropriate support in place for employees and stakeholders to ensure a smooth transition
Month 12 and beyond	Ongoing monitoring and support	Continuously monitor and maintain the system, providing support and making necessary improvements to ensure it operates optimally and delivers desired results

a data migration plan, verifying data accuracy, and testing the data migration process to ensure that all data is accurately transferred.

The integration process should be monitored and evaluated regularly to ensure that it is progressing smoothly and meeting the set goals. Any issues or challenges that arise should be addressed promptly to minimize disruption.

Also, the integration process should be continuously improved to ensure that it remains relevant and effective. This includes regular reviews and updates to the system to ensure that it meets the changing needs of the organization.

By following these best practices, organizations can ensure a seamless integration process and a successful implementation of the triple-entry method and blockchain technology in their accounting systems.

5 Conclusions and Further Research

Based on our research, we come to the following conclusions. A triple-entry accounting system based on blockchain technology can truly improve the security of accounting information, both in peacetime and in times of greater risk (military actions, climate disasters, epidemics, etc.). In the current circumstances, it is recommended to implement this two-step approach. In the first stage, the blockchain can be used as an additional technology in combination with the ERP system installed at the company. At the same time, the smart contracts necessary for all types of users should be developed and the complete methodology for using the new technology should be developed. And in the second stage, there is a transition to blockchain at the enterprise level and at the macro level.

For the first stage, it is recommended to use a permission blockchain with support for smart contracts. At the second stage, the blockchain must also support sharding technology. At this stage, we have made only general conclusions and suggestions; therefore, in the future, it is necessary to conduct the following studies:

(1) choose industries where the implementation of a triple accounting system using blockchain would be the least problematic for employees;
(2) develop protocols for accounting operations, on the basis of which appropriate smart contracts will be created;
(3) choose technological options for implementing the system (blockchain environment, type of consensus, type of stored information, number of accounts, etc.) in terms of minimizing the cost of material resources and maximizing the speed of operations.

It is also worth noting that implementation of the triple-entry method and blockchain solutions in Ukraine's accounting system requires a carefully planned and executed integration process. We provided an overview of the steps necessary for a successful integration, including considerations for data migration, system integration, and the development of blockchain-based solutions. The importance of addressing potential challenges and limitations was also emphasized, as well as the need for training and education of stakeholders on the new system. Overall, the conclusion highlights the significance of a seamless integration process in ensuring the success of the triple-entry method and blockchain in Ukraine's accounting system.

References

1. Grabchuk, I.: Orhanizatsiya zakhystu oblikovoyi informatsiyi v umovakh hibrydnoyi viyny (ukr.). Problemy teoriyi ta metodolohiyi bukhhalters'koho obliku, kontrolyu i analizu **3**(41), 20–24 (2018). https://doi.org/10.26642/pbo-2018-3(41)-20-24
2. Muravskyi, V.: Accounting and Cybersecurity: Monograph (2021). ISBN: 978-0-578-33183-6
3. Ministry of economy of Ukraine.: 558 relocated enterprises have already resumed operations in safe regions of the country (2022). (In Ukrainian) https://www.me.gov.ua/News/Detail?id=bd6e0089-4472-41b1-bf71-30fb261d87f0. Accessed 26 Sep 2022
4. Pacioli, L.: Paciolo on Accounting (Summa de Arithmetica, Geometria, Proportioni e Proportionalita: Distintio Nona—Tractatus XI, Particularis de Computis et Scripturis), translated by Brown, R. G., and K. S. Johnson. McGraw-Hill, New York (1514)
5. Sangster, A.: The genesis of double entry bookkeeping. Account. Rev. **91**(1), 299–315 (2016). https://doi.org/10.2308/accr-51115
6. Ijiri, Y.: A framework for triple-entry bookkeeping. Account. Rev. **61**(4), 745–759 (1986)
7. Grigg, I.: Triple Entry Accounting. Systemics, Inc. (2005). http://iang.org/papers/triple_entry.html. Accessed 26 Sep 2022

8. Tyra, M.J.: Triple Entry Bookkeeping with Bitcoin (2014). https://bitcoinmagazine.com/art icles/triple-entry-bookkeeping-bitcoin-1392069656. Accessed 26 Sep 2022

9. Deloitte.: Blockchain: Enigma, Paradox, Opportunity (2016). https://www2.deloitte.com/con tent/dam/Deloitte/uk/Documents/Innovation/deloitte-uk-blockchain-full-report.pdf. Accessed 26 Sep 2022

10. Grigg, I.: Is Bitcoin a Triple Entry System? (2011). http://financialcryptography.com/mt/arc hives/001325.html. Accessed 26 Sep 2022

11. Cai, C.: Triple-entry accounting with blockchain: How far have we come? Account. Financ. **61**(2) (2019). https://doi.org/10.1111/acfi.12556

12. Nakamoto, S.: Bitcoin: A Peer-To-Peer Electronic Cash System (2008). https://bitcoin.org/bit coin.pdf. Accessed 26 Sep 2022

13. Swan, M.: Blockchain: Blueprint for a New Economy. O'Reilly Media Inc., Boston, MA (2015)

14. Diffie, W.: The first ten years of public-key cryptography. Proc. IEEE **76**(5), 560–577 (1988). https://doi.org/10.1109/5.4442

15. Pilkington, M.: Blockchain Technology: Principles and Applications (2016). https://papers. ssrn.com/sol3/papers.cfm?abstract_id=2662660. Accessed 26 Sep 2022

16. Peters, G.W., Panayi, E.: Understanding modern banking ledgers through blockchain technolo-gies: Future of transaction processing and smart contracts on the Internet of Money. In: Banking Beyond Banks and Money, pp. 239–278. Springer International Publishing, New York (2016).

17. Kiviat, T.I.: Beyond bitcoin: issues in regulating blockchain transactions. Duke Law J. **65**, 569–608 (2015)

18. Szabo, N.: Smart Contracts (1994). http://www.fon.hum.uva.nl/rob/Courses/InformationInSp eech/CDROM/Literature/LOTwinterschool2006/szabo.best.vwh.net/smart.contracts.   html. Accessed 26 Sep 2022

19. Jacynycz, V., Calvo, A., Hassan, S., Sanchez-Ruiz, A.A.: Betfunding: A distributed bounty-based crowdfunding platform over Ethereum. In: 13th International Conference on Distributed Computing and Artificial Intelligence, pp. 403–411 (2016). https://doi.org/10.1007/978-3-319-40162-1_44

20. Wright, A., De Filippi, P.: Decentralized Blockchain Technology and the Rise of Lex Cryp-tographia. Working paper (2015). https://papers.ssrn.com/sol3/papers.cfm?abstract_id=258 0664. Accessed 26 Sep 2022

21. Corbett, J.C., Dean, J., Epstein, M., Fikes, A., Frost, C., Furman, J.J., Ghemawat, S., Gubarev, A., Heiser, C., Hochschild, P., Hsieh, W.C., Kanthak, S., Kogan, E., Li, H., Lloyd, A., Melnik, S., Mwaura, D., Nagle, D., Quinlan, S., Rao, R., Rolig, L., Saito, Y., Szymaniak, M., Taylor, C., Wang, R., Woodford, D.: Spanner: Google's globally distributed database. ACM Trans. Comput. Syst. **31**(3), 8:1–8:22 (2013). https://doi.org/10.1145/2491245. Accessed 26 Sep 2022

22. Luu, L., Narayanan, V., Zheng, C., Baweja, K., Gilbert, S., Saxena, P.: A secure sharding protocol for open blockchains. In: Proceedings of the 2016 ACM SIGSAC Conference on Computer and Communications Security, pp. 17–30. Vienna, Austria (2016). https://doi.org/10.1145/2976749.2978389

23. Kokoris-Kogias, E., Jovanovic, P., Gasser, L., Gailly, N., Syta, E., Ford, B.: Omniledger: A secure, scale-out, decentralized ledger via sharding. In: IEEE Symposium on Security and Privacy, SP 2018, Proceedings, 21–23, pp. 583–598. San Francisco, California, USA (2018). https://doi.org/10.1109/SP.2018.000-5

24. Zamani, M., Movahedi, M., Raykova, M.: Rapidchain: Scaling blockchain via full sharding. In: Proceedings of the 2018 ACM SIGSAC Conference on Computer and Communications Security, pp. 931–948. CCS 2018, Toronto, ON, Canada (2018). https://doi.org/10.1145/324 3734.3243853. Accessed 26 Sep 2022

25. Wang, J., Wang, H.: Monoxide: Scale out blockchains with asynchronous consensus zones. In: 16th USENIX Symposium on Networked Systems Design and Implementation, pp. 95–112. NSDI 2019, Boston, MA (2019). https://eprint.iacr.org/2019/263.pdf. Accessed 26 Sep 2022

26. Yizhong, L., Jianwei, L., Marcos, A., Zongyang, Z., Tong, L., Bin, H., Fritz, H., Rongxing, L.: Building blocks of sharding blockchain systems: concepts, approaches, and open problems (2021). https://doi.org/10.48550/arXiv.2102.13364

27. Yermack, D.: Corporate governance and blockchains. Rev. Financ. (forthcoming) (2017). https://doi.org/10.1093/rof/rfw074
28. Fanning, K., Centers, D.P.: Blockchain and its coming impact on financial services. J. Corp. Account. Financ. **27**(5), 53–57 (2016). 1002/jcaf.22179
29. Kuhn, J.R., Sutton, S.G.: Continuous auditing in ERP system environments: the current state and future directions. J. Inf. Syst. **24**(1), 91–112 (2010). https://doi.org/10.2308/jis.2010.24.1.91
30. Hitt, L.M., Wu, D.J., Zhou, X.: Investment in enterprise resource planning: Business impact and productivity measures. J. Manag. Inf. Syst. **19**(Summer), 71–98 (2002)
31. Dai, J., Vasarhelyi, M.: Toward blockchain-based accounting and assurance. J. Inf. Syst. **31**(3), 5–21 (2017). https://doi.org/10.2308/isys-51804
32. Swan, M.: Blockchain thinking: the brain as a DAC (Decentralized Autonomous Organization). In: Presented at the Texas Bitcoin Conference (2015). https://www.researchgate.net/publication/292391320_Blockchain_thinking_The_brain_as_a_dac_decentralized_autonomous_organization. Accessed 26 Sep 2022
33. Ahmed, W., Rasool, A., Javed, A.R., Kumar, N., Gadekallu, T.R., Jalil, Z., Kryvinska, N.: Security in next generation mobile payment systems: a comprehensive survey. IEEE Access **9**, 115932–115950 (2021). https://doi.org/10.1109/ACCESS.2021.3105450
34. Kryvinska, N., Strauss, C.: Conceptual model of business services availability versus interoperability on collaborative IoT-enabled eBusiness platforms. In: Internet of Things and Intercooperative Computational Technologies for Collective Intelligence, pp. 167–187 (2013). https://doi.org/10.1007/978-3-642-34952-2_7
35. Fauska, P., Kryvinska, N., Strauss, C.: The role of e-commerce in B2B markets of goods and services. IJSEM **5**, 41 (2013). https://doi.org/10.1504/IJSEM.2013.051872
36. Engelhardt-Nowitzki, C., Kryvinska, N., Strauss, C.: Strategic demands on information services in uncertain businesses: a layer-based framework from a value network perspective. In: 2011 International Conference on Emerging Intelligent Data and Web Technologies (2011). https://doi.org/10.1109/EIDWT.2011.28
37. Bagranoff, N., Simkin, M., Norman, C.: Core Concepts of Accounting Information Systems (2010)
38. ISO/IEC 27002:2022. Information security, cybersecurity and privacy protection—Information security controls. https://www.iso.org/ru/standard/75652.html. Accessed 26 Sep 2022
39. Law, C.C., Ngai, E.W.: ERP systems adoption: an exploratory study of the organizational factors and impacts of ERP success. Inf. Manag. **44**(4), 418–432 (2007). https://doi.org/10.1016/j.im.2007.03.004
40. Rainer, R., Kelly, P.B., Cegielski Casey, G.: Introduction to Information Systems, 5th edn. (2014). ISBN-10 1118988531

Modeling of the Strategy of Light Industry Enterprise Behavior under Crisis Conditions of Martial Law

Serhii Matiukh⑩, Yevhenii Rudnichenko⑩, and Nataliia Havlovska⑩

Abstract As the research result of peculiarities of industrial enterprise behavior strategy formation under crisis conditions, the model of the enterprise coalition cooperation is worked out. The model allows for forming rational plans (strategies) of behavior inside the enterprise coalition and maximizing profit for both the whole coalition and its members. The preconditions for such model development are determined and justified. The principal factors of a negative impact on the light industry enterprises are researched and their negative consequences for the functioning under conditions of martial law are detailed. The game theory is used as the application toolkit and transfer from an initial non-coalition game to a cooperative game with rethinking of the utility redistribution process on the basis of the study of the characteristic function of a non-coalition game is formed. The practical value of the model consists of the fact that after receiving the evaluated utility functions of all enterprises for broad coalition cases and incomplete coalition cases, each enterprise starts a corresponding game. That allows for determining the optimal strategy of enterprise behavior according to the actual conditions of its functioning.

Keywords Cooperative game · Mathematical model · Industry enterprise · Strategy

S. Matiukh · Y. Rudnichenko (✉) · N. Havlovska
Khmelnytskyi National University, 11, Instytuts'ka Str., Khmelnytskyi 29000, Ukraine
e-mail: rudnichenkoiem@khmnu.edu.ua

S. Matiukh
e-mail: matuh@khmnu.edu.ua

N. Havlovska
e-mail: havlovskant@khmnu.edu.ua

1   Introduction

The strategy of enterprise functioning and development differs essentially under conditions of ordinary functioning and of boundary aggravation of crisis phenomena. It is conditioned by the opportunity to reach determined strategic goals by traditional methods and approaches or an urgent search for new stabilization measures and business activity re-establishment under crisis conditions [1]. The methods of economic and mathematic modeling of economic phenomena and processes are rather often used in order to save resources and avert mistakes whose cost can be exceedingly high [2–9]. This enables to increase substantially the number of alternatives to managerial decision-making and also determines the final results of implementation of these decisions, taking into account the expended resources and gained profit.

Among a significant number of developed models and formed toolkits, only a few can satisfy enterprise management requirements and "give out" an adequate result on the basis of the actual state of the enterprise functioning environment. Traditionally, the requirements of the model are formed before its development, which gives the possibility to approach the selection of a respective modeling toolkit. In the process, the functioning specifics of corresponding economic entities, the development of information and communication components of the business, staffing, the employees' willingness to accept innovations, and the readiness of the owners to invest in the enterprise infrastructure improvement are taken into account.

2   Theoretical Basics of Modeling of Light Industry Enterprise Behavior Under Conditions of Their Functioning Environment Destabilization

Under current conditions of extremely tough competition and force-majeure circumstances (in particular, military actions and remaining threat of military strikes, shelling, and their all negative economic and political consequences), light industry enterprises, which produce similar production and compete toughly for consumers, resources and market segments, must change their approach to strengthening and backing up their own competitiveness at the proper level. In many regions of the country, there are communities where several light industry enterprises (usually, from two, three to five), which produce clothing (women's, men's, and children's), work. Production automation at such enterprises does not in any way correspond to automation of fabric production (materials) or, all the more, automation of raw material processing (for example, metal rolling at heavy industry plants). Garment production is impossible without the intensive involvement of human resources, and automation here constitutes not a large part of the manufacturing process. Thus, during the air raid serene, the garment sewing process (clothing production at the final stage) is interrupted with more negative consequences than any other manufacturing process, even in the same light industry (manufacturing of fabric, buttons, fasteners) and food industry (bakery production process is also much more automated). Hence, all light industry enterprises specializing in clothing production take these negative consequences.

Is there the possibility of at least partial elimination or reduction in the impact of these results? It is known that the participants of the manufacturing market always react collectively to common macroeconomic threats (for instance, raw material deficit), working out a collective deficit counteraction strategy. Sometimes, this strategy is initiated by the state itself, as it occurred during the fuel crisis in spring and in June 2022. Thus, the answer to the question is "yes"—the possibility of a decrease in the impact of negative consequences of air alerts and martial law, in general, ought to be sought in cooperation with those light industry subjects that suffer to some extent. What should this cooperation be? Certainly, we are not talking about a merger or a conglomerate. It is first and foremost about combining efforts both to overcome negative consequences and to prevent the effects of these negatives. For instance, if one enterprise is larger than another and has an equipped shelter at its disposal (maybe a cellar with a ventilation system), it could be temporal premises for manufacturing of both enterprises because, in this case, it would be necessary to relocate only equipment to the shelter, with the clothing industry equipment being usually compact. However, if an enterprise size does not allow setting equipment from both enterprises, the smaller enterprise could refocus on performing any other operation in the framework of this manufacture and temporally transfer employees to this operation. In particular, it could be sewing on buttons or cuffs, which can be done even at home (though, certainly, in case of the air alert, it is required for everybody without exception to descend into the nearest shelter). It is necessary to reach an agreement beforehand for performing such forced enterprise functioning. In future the, both enterprises are to share additional losses and profits connected with that for a fair return to the usual regime (if the regime of cooperation turns out less profitable than the ordinary one).

From the formal point of view, such cooperation may proceed in the framework of the coalition of the light industry enterprises that have the intention to cooperate. The intentions and everything that concerns them directly must be fixed in relevant legal documents. These documents are not only agreements that legally register all coalition cooperation subtleties. The important part of the coalition agreements is the algorithm of expense and income distribution while the agreement is in effect. However, before signing the agreements, each enterprise ought to analyze the possible results and consequences of its own investment in the coalition. It is entirely possible that these results will turn out to be not satisfactory after preliminary analysis, and the enterprise will suggest amendments (in particular, first of all, it concerns the expense and income distribution) [1, 10–14]. Otherwise, joining the coalition may not occur.

To make an in-depth analysis of the results and consequences of its own investment in the coalition, the enterprise must first study individually and collectively those factors that negatively influence manufacturing performance and sales of products. Since it is usually expected that each of the enterprises may work on a toll manufacturing basis with foreign partners, one ought to take into consideration the foreign factor impact.

On the basis of the current situation, the factors of the negative impact the light industry enterprises and, in particular, the clothing enterprises are the following:

1. Military actions and attacks on the region where the enterprise is located. As a result, one ought to consider unanticipated air alerts in time and duration, which have an extremely negative impact on manufacturing process dynamics.
2. Labor shortage. A part of the enterprise personnel, including managers, can leave the region or the country as a result of a military threat.
3. Resource deficit. Resource supply interruptions are characteristic of martial law. However, such interruptions may occur under normal social and economic conditions, though much less frequently compared with the military conditions.
4. Logistical problems. This factor is directly connected with martial law. The complexity of planning and implementing of logistic plans increases during military actions or constant threats.
5. Energy price growth. This factor is similarly characteristic of a conditionally normal social, economic and political state.
6. Difficulty in safe work environment creation. Not every enterprise has an equipped shelter with a ventilation system.
7. Competition in similar market segments. This factor is characteristic of any state because it is natural for economic and market processes.

The set of the above-enumerated factors is denoted by $\{c_i\}^7_{i=1}$, where the factor c_i refers to the number in the list of the factors. Let us mention that some factors are related to the other ones; in other words, they influence the other factors or are influenced by them. The fact that c_1 influences simultaneously five factors: c_2, c_3, c_4, c_5, c_6 is indisputable. Labor shortage and resource deficit may, to some extent, compensate for the corresponding negative consequences, though c_2 and c_3 are not correlated. However, c_6 makes an impact on c_2, because difficulty in safe work environment creation can become the demotivator for the employment or continuation of work under complicated conditions. Factors c_4 and c_5 are also correlated, though energy price growth does not directly aggravate logistics and vice versa. Only factor c_7 is more or less "independent", though it may become a significant negative factor for deficit development (c_2 and c_3), logistical problems (c_4), and difficulty in safe work environment creation (c_6).

Somewhat unexpectedly, a number of circumstances and processes that appeared to be favorable for light industry enterprises formed under the conditions of martial law before summer 2022. The following circumstances and processes turned out to be the most essential:

1. Decrease in volume of grey imports from China and neighboring countries, for example, Turkey. We are talking about both raw materials (that is, in this case, mainly fabric templates) and finished products. According to different estimations, the volume of grey raw material imports reduced most of all. For further analysis, this circumstance is denoted by *ZnyzSir*.
2. Potential support of domestic enterprises by foreign partner countries. These projects have not been implemented so far. They are directed to the support of

Ukraine at economic reconstruction both during the war and after its end and the establishment of the normal state. This process is denoted by *PotPidtr*.

3. A need for military uniforms. First of all, it is about government procurement. Temporal subsidizing for fast manufacturing of medium and large batches of military uniforms, kits, and camouflage covers. This factor will be further referred to as *FormaZSU*.

4. Implementation of regional support programs for industrial enterprises. These programs are a part of state support programs for the functioning of the domestic economy under conditions of martial law. Evidently, this item (let us denote it by *RegProg*), is partially related to the previous one—the factor *FormaZSU*, and also to the process *PotPidtr*.

Thus, the negative impact of factors $\{c_i\}^7_{i=1}$ may be compensated to some extent due to more or less favorable circumstances and processes.

$$SpryOP = \{ZnyzSir, PotPidtr, FormaZSU, Reg\,Propg\}. \qquad (1)$$

The set (1) is not constant as to the content and the impact on the enterprise functioning if its elements are evaluated according to a definite scale. It changes in time, depending more on domestic and foreign political circumstances and their development or slowing down than on season (autumn, winter, spring, summer). The scale of element evaluation *SpryOP* (i.e. minimum and maximum grades) will be considered further.

Certainly, the elements *SpryOP* are correlated in a particular way. There is the impact *PotPidtr* on *RegProg*. Partially, this impact is conditioned by *ZnyzSir*, decrease in the volume of grey import. Similarly, *RegProg* influences *FormaZSU*.

Are the factors $\{c_i\}^7_{i=1}$ and the set *SpryOP* correlated? It is likely that yes. Evidently, above all, c_1 influences (though in different ways) on all four processes in the set (1). Besides, extreme growth in need of uniforms for the army may cause deficit—negative growth will be observed c_2 i c_3.

3 The Model of the Coalition Cooperation for Light Industry Entities on the Basis of a Cooperative Game

The factors $\{c_i\}^7_{i=1}$ and the set *SpryOP* must be examined as a whole and from the point of view of their correlation. That gives the opportunity to predict more thoroughly and reasonably the results and consequences of the enterprise investment in the coalition (joining the coalition). It is entirely possible that two or more enterprises could provide overall resource savings, safe work environment and decrease in competition in some segments due to distinct specialization, working together in an equipped cellar shelter. However, as a result, profitability will not be equal—it must be distributed according to the individual investment of each enterprise. For instance, if two enterprises join manufacturing efforts and only one of them has an equipped shelter, the other enterprise will use this shelter de facto on the rights of a gratuitous agreement. However, such "gratuitous" rent must be paid after the termination of the coalition agreement (on cooperation). Certainly, the payment may be made monthly. Though, there will be plenty of likewise facilities (in particular, equipment, vehicles, labor resources) that will be shared during the term of the agreement, therefore, in case of three and more coalition members, it would be advisable to cover expenses not more often than quarterly.

Thus, the task consists in suggesting these enterprises such model of coalition interaction (cooperation) that will allow working out rational plans (strategies) of behavior inside the coalition and for the coalition as a whole. These strategies, in their turn, maximize the profit both of the whole coalition and its members.

Let us denote a set of players (agents), where $A = \{1, 2, \ldots, N\}$, $N \in \{2, 3, 4, \ldots\}$ by A. Each player personifies an enterprise. Let X_a—a set of strategies of the agent a, where $x_a \in X_a$—their arbitrary strategy, G_a—win function of the agent a, defined on the Cartesian product

$$\underset{a=1}{\overset{N}{\times}} X_a = X_1 \times X_2 \times \ldots \times X_{N-1} \times X_N.$$

In fact, this function depends on N variables, i.e.

$$G_a = G_a(x_1, x_2, \ldots, x_N), a \in A = \{1, 2, \ldots, N\}.$$

Then we have a non-coalition game for N individuals (personifications of N players or agents that are given corresponding opportunities in the form of the sets of the strategies $\{X_a\}_{a \in A}$) [7]

$$\Gamma = \left\langle A, \{X_a\}_{a \in A}, \{G_a(x_1, x_2, \ldots, x_N)\}_{a \in A}\right\rangle, \tag{2}$$

One of these agents is a (personified) state that puts fiscal pressure and sets formal (institutional) requirements (especially during martial law) that are manifested in the requirements for fire protection, lighting norms, rest breaks for employees, working time norms, etc. It should be mentioned that in current realities, these requirements have not been reconsidered for the period of martial law. To secure integrity, let us denote this agent by number N. The other agent is the personified external circumstances—both above enumerated negative influence factors and macroeconomic disruptions such as inflation processes, the unexpectedness of some national regulator decisions (including those, made as a result of our partners' foreign policy steps), the striking discrepancy between demand and supply as a consequence of military escalation or attacks on the region. This agent is denoted by number $N-1$. Thus, the first (according to the numbers) $N-2$ players are enterprises.

Let us consider the case when all enterprises preliminarily agree to contemplate the possibility of joining a coalition. Then, the largest coalition $K_{\max} \subset A$ will consist of $N-2$ players:

$$K_{\max} = \{1, \ 2, \ \ldots, \ N - 2\}. \tag{3}$$

The amalgamation of players into a coalition $K_{\max} \subset A$ means that they, from the viewpoint of game theory, turn into one player. All possible combinations of players' strategies from the coalition (3) are strategies of this player. Thus, these are the elements of the Cartesian production

$$\overset{N-2}{\underset{k=1}{\times}} X_k = X_1 \times X_2 \times \ldots \times X_{N-2} = X_{K_{\max} \subset A}. \tag{4}$$

These elements are denoted by $x_{K_{\max} \subset A} \in X_{K_{\max} \subset A}$. Similarly, the amalgamation of players into a coalition means that its win (if considered as conditional utility or conditional losses) is a sum of wins of the coalition agents [7, 10]:

$$G_{K_{\max} \subset A}\left(x_{K_{\max} \subset A}, \ x_{N-1}, \ x_N\right) = \sum_{k=1}^{N-2} G_k\left(x_{K_{\max} \subset A}, \ x_{N-1}, \ x_N\right). \tag{5}$$

Evidently, if the win functions in the game (2) correspond to conditional utility, then the aim of the coalition is maximization of the sum (5). Nevertheless, if the wins correspond to conditional losses, the coalition will strive to minimize the sum (5). Conditional economic utility can be obtained as a result of the evaluation of advantages from the viewpoint of profitability and economic security in all possible situations on the Cartesian product

$$\overset{N}{\underset{k=1}{\times}} X_k = X_1 \times X_2 \times \ldots \times X_{N-1} \times X_N = X_{K_{\max} \subset A} \times X_{N-1} \times X_N. \tag{6}$$

It is usually regarded as an entirely feasible process, thus, we will further consider the functions of wins in the game (2) to be conditional utility, according to which the player-coalition strives to maximize the function (5), and the state maximizes its function

$$G_N\left(x_{K_{\max} \subset A}, \ x_{N-1}, \ x_N\right). \tag{7}$$

The personified external circumstances in the cases close to the worst one are such that their function

$$G_{N-1}\left(x_{K_{\max} \subset A}, \ x_{N-1}, \ x_N\right), \tag{8}$$

obtains values that are either close to the maximum or the value of the function (7) decreases, or the utility function of the coalition (5) is close to its minimum values. Surely, it should be mentioned that the maximization of one utility function does not mean the minimization of the others in a non-coalition game. However, in our case, the personified external circumstances immediately mean unfavorable circumstances either for a state or for enterprises (one, some, all), or for both. Also, it should be stressed that the use of the term "coalition" and its derivatives is entirely reasonable because this game exists preliminarily, and coalition forming occurs only later, after thinking over the game content—analysis of the utility functions (5), (8) and (7).

Let us continue considering the worst cases. Let us assume that the state and personified external circumstances can similarly contextually "amalgamate" and prevent the coalition from obtaining the largest utility. Then, they are also a collective player. The set of its strategies is denoted by $Y_{A \backslash K_{max}} = X_{N-1} \times X_N$. The strategies of such a virtual agent are virtually denoted by $y_{A \backslash K_{max}} \in Y_{A \backslash K_{max}}$. Hence, taking into account the potentially possible worst cases of inflation processes, deficit, and problems with logistics, one can proceed from a non-coalition game (2) to a game with a zero-sum [15]

$$\Gamma_{K_{max} \subset A} = \left\langle X_{K_{max} \subset A}, \ Y_{A \backslash K_{max}}, \ G_{K_{max} \subset A} \left(x_{K_{max} \subset A}, \ y_{A \backslash K_{max}} \right) \right\rangle. \tag{9}$$

In a "new" game (9) that is simply a logical "instant" development of the game (2), the first player is an enterprise coalition, and the second—"the coalition" of a state and personified external circumstances which is further called disruptions to simplify the things. The set of coalition strategies is $X_{K_{max} \subset A}$, and the set of disruption coalition strategies—$Y_{A \backslash K_{max}}$. . In addition, as in each game with a zero sum, only the coalition utility function (5) is considered and the utility function of disruption "interests" is not taken into account. However, it should be mentioned that the set of the situations for the game (9) is the same as for the source game (2)—that is the set (6).

It is a known fact that while transferring from a source non-coalition game to a corporative one, rethinking the utility redistribution process (sometimes of the coalition common profit) occurs on the basis of the study of the characteristic function of a non-coalition game. In the case under research, this function is defined not by the set of all subsets of the set of the agents $A = \{1, 2, \ldots, N\}$, but by the set of all subsets of the set

$$K \subseteq \{1, 2, \ldots, N - 2\} \subset \{1, 2, \ldots, N\},$$

because the state and personified external circumstances and their "coalition"- disruption cannot be the members of the coalition with enterprises. Thus, the characteristic function $H_\Gamma(K)$ of the non-coalition game (2) may obtain 2^{N-2} possible variants of values, in particular:

Variant 1. The value of the characteristic function $H_\Gamma(\emptyset)$—if nobody joined the coalition; evidently that $H_\Gamma(\emptyset) = 0$;

Variants 2, 3, ..., N–1. The value of the characteristic function $H_\Gamma(K_1)$, where $K_1 = \{a_1\} \in \{1, 2, \ldots, N - 2\}$,—when only one enterprise joined the coalition a_1; according to the definition, similar to (5) (the coalition utility is the sum of utility of its agents), in this case, the value $H_\Gamma(K_1)$ depends only on the function $G_{a_1}(x_1, x_2, \ldots, x_N)$.

Variants $N, N + 1, \ldots, N^*$, where

$$N^* = \frac{(N-2)!}{2! \cdot (N-2-2)!} + N - 1 = \frac{(N-2)!}{2 \cdot (N-4)!} + N - 1$$
$$= \frac{(N-2) \cdot (N-3)}{2} + N - 1.$$

The value of the characteristic function $H_{\Gamma}(K_2)$, where

$$K_2 = \{a_1, a_2\} \subset \{1, 2, \ldots, N-2\},$$

two enterprises a_1 and a_2 joined the coalition; In this case the value $H_{\Gamma}(K_2)$ depends on the sum

$$G_{a_1}(x_1, x_2, \ldots, x_N) + G_{a_2}(x_1, x_2, \ldots, x_N);$$

Variants $2^{N-2}-N+2$, $2^{N-2}-N+3$, ..., $2^{N-2}-1$. The value of the characteristic function $H_{\Gamma}(K_{N-3})$, where

$$K_{N-3} = \{a_1, a_2, \ldots, a_{N-3}\} \subset \{1, 2, \ldots, N-2\},$$

all enterprises except one joined the coalition;
Variant 2^{N-2}. The value of the characteristic function $H_{\Gamma}(K_{N-2}) = H_{\Gamma}(K_{max})$, where

$$K_{N-2} = K_{max} = \{1, 2, \ldots, N-2\},$$

all enterprises joined the coalition; this variant is considered now, hence, the value of the characteristic function by such coalition is the value of the game according to the function (5).

If there are, for example, five enterprises, then we have totally 32 variants, among which there are both the unavailability of the coalition (variant 1) and variants 2, 3, 4, 5, 6 (because here $N = 7$), which mean a formal coalition that consists only of one enterprise, what is equivalent to the unavailability of coalitions. It is necessary to examine the rest of 26 variants. There is its own coalition agreement for each of the variants. Hence, the best agreement ought to be selected from 26 ones. But now it is not about the characteristic function for a non-coalition game (2), because we consider variant 2^{N-2}. It is done because the enterprises, first of all, study the opportunity for amalgamating in general, without exceptions (considering only the variant of forming a "broad coalition").

As it was mentioned above, the set of all strategies of the broad coalition $K_{max} \subset A$ is a result of the Cartesian product (4). In the simplest case, if each of N–2 coalition members has only two strategies, the result means that the broad coalition receives at its disposal 2^{N-2} potential strategies, and each of these strategies contains N–2 components (a component from every coalition member). In addition, one should mention that the initial enterprise strategies can consist of some components. For instance, each strategy of a light industry enterprise producing clothing would rather contain such components: (1) the volume of raw material purchase (fabric, templates); (2)

the level of (average) salary; (3) expenses on promotion (variations—promotion of information on the enterprise goods in Facebook, Instagram, YouTube, other social networks, etc.).

All these components are, at least formally, correlated. Depending on the market and other circumstances, negative impact factors, favorable circumstances, and processes *SpryOP* according to the formula (1), each of the enumerated components can be regulated to some extent. In the simplest case that is a minimal regulation during a time unit, in other terms, a simple decrease or increase. Thus, the example of such a three-component strategy is an increase in the volume of purchases simultaneously with a decrease in the average salary and an increase in the expenses for promotion. If some component must remain without changes at all, then one can conditionally use a pair of strategies where this component initially decreases or increases and then—visa versa (for compensation). Correspondingly, eight possible strategies for one enterprise contain the full list of all variants. However, if the enterprise plans to increase the volume of raw material purchase considerably (10…20% and more), then this simply means the gradual repeated use of strategies where the first component is a minimum increase in the volume of purchase (two other components may have any values—they do not depend on the volume of purchase).

Since the enterprises intend to function from the viewpoint of cooperation, it is natural to equal all their strategies with the coalition strategies $x_{K_{max}CA} \in X_{K_{max}CA}$. It simplifies considerably the construction and analysis of the utility function of the coalition (5). Thus, according to the above-mentioned enterprise three-component strategy (now—coalition), the set of all strategies is the following

$$
\begin{aligned}
&x_{K_{max}CA}(1) = \{-1, \ -1, \ -1\}, x_{K_{max}CA}(2) = \{1, \ -1, \ -1\}, \\
&x_{K_{max}CA}(3) = \{-1, \ 1, \ -1\}, x_{K_{max}CA}(4) = \{1, \ 1, \ -1\}, \\
&x_{K_{max}CA}(5) = \{-1, \ -1, \ 1\}, x_{K_{max}CA}(6) = \{1, \ -1, \ 1\}, \\
&x_{K_{max}CA}(7) = \{-1, \ 1, \ 1\}, x_{K_{max}CA}(8) = \{1, \ 1, \ 1\},
\end{aligned}
\tag{10}
$$

where value –1 corresponds to the component decrease, and value 1—component increase. In particular, for instance, the strategy of the coalition

$$
x_{K_{max}CA}(6) = \{1, \ -1, \ 1\},
$$

means an increase in the volume of raw material purchases and expenses on promotion with a simultaneous decrease in the average salary level (certainly, this measure is only temporary in practice). However, if nobody is going to change their salary, then the coalition strategy is the pair of strategies

$$
\{x_{K_{max}CA}(6), \ x_{K_{max}CA}(8)\} = \{\{1, \ -1, \ 1\}, \{1, \ 1, \ 1\}\},
$$

that ought to be implemented during the same period as any other strategy. It is equivalent to the implementation of the strategy $\{1, 0, 1\}$, where the salary is not changed (or this component is unavailable), but such a "truncated" strategy is absent from the above list. On the basis of such principle, one can generate any other "truncated" strategies, where, two or even three components are unavailable. To clarify, the latter means that the strategy $\{0, 0, 0\}$, which corresponds to a steady development of the coalition functioning process, ought to be formally generated on the basis of the implementation of the pair of strategies

$$\left\{x_{K_{max} \subset A}(1), \ x_{K_{max} \subset A}(8)\right\} = \{\{-1, \ -1, \ -1\}, \ \{1, \ 1, \ 1\}\}. \tag{11}$$

In reality, this means the unavailability of any changes for the volume of raw material purchase, for the average salary level, promotion expenses; and enterprises (coalition members) do not use either the strategy $x_{K_{max} \subset A}(1) = \{-1, \ -1, \ -1\}$ or the strategy $x_{K_{max} \subset A}(8) = \{1, \ 1, \ 1\}$. In fact, it may seem that the coalition utility function will not change. Nevertheless, this is not the case because in reality, even unavailability of any actions or changes in a short-term policy of business activity does not for sure means the absence of changes in the results of this activity compared with the previous (short-term) period. It should be mentioned that for a formal imitation of the strategy $\{0, 0, 0\}$, one can, except (11), use other pairs of strategies, for example:

$$\left\{x_{K_{max} \subset A}(2), \ x_{K_{max} \subset A}(7)\right\} = \{\{1, \ -1, \ -1\}, \ \{-1, \ 1, \ 1\}\},$$
$$\left\{x_{K_{max} \subset A}(3), \ x_{K_{max} \subset A}(6)\right\} = \{\{-1, \ 1, \ -1\}, \ \{1, \ -1, \ 1\}\},$$
$$\left\{x_{K_{max} \subset A}(4), \ x_{K_{max} \subset A}(5)\right\} = \{\{1, \ 1, \ -1\}, \ \{-1, \ -1, \ 1\}\}.$$

Thus, the utility function of the enterprise coalition must be evaluated on the set (10) in the way, so that combining of strategies from (10) for excluding one or more components from consideration could correspond to the adequate change of this function. This requirement means that utility function of the k enterprise of the broad coalition $G_k\left(x_{K_{max} \subset A}, \ x_{N-1}, \ x_N\right)$ with sufficient accuracy for practice must meet the condition of steady functioning

$$G_k(\{0, \ 0, \ 0\}, \ x_{N-1}, \ x_N)$$
$$\approx \frac{G_k\left(x_{K_{max} \subset A}(1), \ x_{N-1}, \ x_N\right) + G_k\left(x_{K_{max} \subset A}(8), \ x_{N-1}, \ x_N\right)}{2}, \tag{12}$$

for arbitrary strategies x_{N-1} and x_N. Evidently, the condition (12) may be violated in the case when the coalition is not broad. In fact, in this case, at least one enterprise that has not joined the coalition, evidently, from the viewpoint of the reasoning behind profitability, will try to implement the combination (sequence) of its strategies from the same set (10). For a certain period of time, this combination or at least its part will not coincide with the corresponding coalition combination. Then the formal use of the combination (11) in fact will mean at first real decrease and then increase in the level of the components, and the enterprise that has not joined the coalition will react to it. Its reaction consists in triggering usual market mechanisms, which are certain to work in a competitive environment. Then the condition (12) may be violated.

The condition (12) is in fact a convention for a broad coalition because in this case no enterprise will object to this approach. This condition does not always work for smaller coalitions, and the set of potential strategies for such coalitions must be supplemented with the strategies where the change of one or some (possibly—all) components does not occur.

Since the set of disruption strategies $Y_{A \backslash K_{max}} = X_{N-1} \times X_N$ consists of the binary combination of the strategies of the state and (in particular, unfavorable) external circumstances, each disruption strategy is a two-component one:

$$y_{A \backslash K_{max}} = \begin{bmatrix} x_{N-1} & x_N \end{bmatrix} \in Y_{A \backslash K_{max}},$$

where $x_{N-1} \in X_{N-1}$—a state strategy, and $x_N \in X_N$—a strategy of external circumstances. However, similar to initial enterprise strategies, the state strategy is a multicomponent one. Thus, it must contain:

(1) the level of fiscal pressure that can be increased or decreased;
(2) the use of donations and subsidization to support the small business entities that are at the initial stage of their functioning;
(3) the use of the simplified tax reporting procedures;
(4) temporary suspension or simplification of the mandatory planned inspections regarding compliance with sanitary and epidemiological conditions.

The latter three points include binary management—to provide donations (+1) or not to provide (−1), to enable simplification of the tax reporting procedures (+1) or not to enable (−1), simplify the inspections (+1) or remain the inspections unchanged (−1). Increase (−1) or decrease (+1) in the level of fiscal pressure can be caused by minimal regulations for the time unit, though sometimes these changes are not insignificant as a result of making political decisions instead of decisions based on economic calculations.

Evidently, external circumstances cannot have any "strategies" because they do not plan them. Each conditional strategy of external circumstances forms on the basis of taking into account both the negative impact factors $\{c_i\}_{i=1}^{7}$ and favorable circumstances and processes (1). Then, formally, the conditional strategy of external circumstances consists of these 11 components, though some of them are more or less closely correlated. However, the consideration of even the set of only binary combinations of 11 components is inappropriate from the practical point of view because, in this case, it is required to evaluate $2^{11} = 2048$ different strategies, which is too many.

Thus, for necessary reduction, the impact of external circumstances may be presented in a binary way:

0—unavailability of the negative impact or there is even little progress as the result of the strengthening of the positive impact of state strategies;

1—availability of the negative impact or the strengthening of negative consequences as the result of implementation by the state of its definite strategies.

In addition, the level of fiscal pressure and the use of simplified tax reporting procedures may be combined into one strategy, which will be the first component.

Let us give this component "the right" to get the value 0 in the case there are no significant changes in the tax legislation. The second component is the provision of donations or subsidization. The third is temporary suspension or simplification of the mandatory planned inspections. The fourth component of the disruption strategy is the flag (0 or 1) of the external circumstance impact. Then, the set of disruption strategies is shown in this way:

$$
\begin{aligned}
&y_{A\backslash K_{max}}(1) = \{0, -1, -1, 0\}, y_{A\backslash K_{max}}(2) = \{0, 1, -1, 0\}, \\
&y_{A\backslash K_{max}}(3) = \{0, -1, 1, 0\}, y_{A\backslash K_{max}}(4) = \{0, 1, 1, 0\}, \\
&y_{A\backslash K_{max}}(5) = \{1, -1, -1, 0\}, y_{A\backslash K_{max}}(6) = \{1, 1, -1, 0\}, \\
&y_{A\backslash K_{max}}(7) = \{1, -1, 1, 0\}, y_{A\backslash K_{max}}(8) = \{1, 1, 1, 0\}, \\
&y_{A\backslash K_{max}}(9) = \{-1, -1, -1, 0\}, y_{A\backslash K_{max}}(10) = \{-1, 1, -1, 0\}, \\
&y_{A\backslash K_{max}}(11) = \{-1, -1, 1, 0\}, y_{A\backslash K_{max}}(12) = \{-1, 1, 1, 0\}, \\
&y_{A\backslash K_{max}}(13) = \{0, -1, -1, 1\}, y_{A\backslash K_{max}}(14) = \{0, 1, -1, 1\}. \\
&y_{A\backslash K_{max}}(15) = \{0, -1, 1, 1\}, y_{A\backslash K_{max}}(16) = \{0, 1, 1, 1\}, \\
&y_{A\backslash K_{max}}(17) = \{1, -1, -1, 1\}, y_{A\backslash K_{max}}(18) = \{1, 1, -1, 1\}, \\
&y_{A\backslash K_{max}}(19) = \{1, -1, 1, 1\}, y_{A\backslash K_{max}}(20) = \{1, 1, 1, 1\}, \\
&y_{A\backslash K_{max}}(21) = \{-1, -1, -1, 1\}, y_{A\backslash K_{max}}(22) = \{-1, 1, -1, 1\}, \\
&y_{A\backslash K_{max}}(23) = \{-1, -1, 1, 1\}, y_{A\backslash K_{max}}(24) = \{-1, 1, 1, 1\}.
\end{aligned}
\tag{13}
$$

For instance, the strategy

$$
y_{A\backslash K_{max}}(11) = \{-1, -1, 1, 0\},
$$

means increase in fiscal pressure on condition of subsidization unavailability and the simultaneous simplification of the mandatory planned inspections without the growth of the negative impact of the increased taxes and unavailability of donations. Instead, the strategy

$$
y_{A\backslash K_{max}}(23) = \{-1, -1, 1, 1\},
$$

means the same, but includes the strengthening of the negative consequences of tax increase and unavailability of donations. Such strengthening does not depend on the business entities or on the public authorities. Nevertheless, it must be taken into account as a part of reasonable fluctuations in micro- and macroeconomic environments (similar to fluctuations on the currency and security markets, etc.), which are impossible to be predicted.

Each enterprise must obtain its utility function, on whose basis the utility function of the coalition (5) can be gained as

$$G_{K_{\max} \subset A}\left(x_{K_{\max} \subset A}, \ y_{A \backslash K_{\max}}\right) = \sum_{k=1}^{N-2} G_k\left(x_{K_{\max} \subset A}, \ x_{N-1}, \ x_N\right). \tag{14}$$

Then, it will give the possibility to start the game (9) and to work out the best collective strategy for the coalition.

Taking as a basis the set of strategies of each enterprise (10) and the set of disruption strategies (13), it can be seen that this is the game with the matrix of wins (utility) of the size 8×24. In fact, there are N–2 such matrices, which then combine (strictly saying, they are added) according to the rule (14).

The utility matrix $\mathbf{G}_k = [g_{kij}]_{8 \times 24}$ for the k enterprise can be obtained with the expert evaluation of the utility of the game situations (9). Let us denote by $\mathbf{G}_k^{<u>} = [g_{kij}^{<u>}]_{8 \times 24}$ the variant of such a matrix, provided by the u respondent (expert). The value $g_{kij}^{<u>}$ is the grade of the conditional utility obtained by the k enterprise, using its strategy $x_{K_{\max} \subset A}(i)$ on condition that the disruption strategy is $y_{A \backslash K_{\max}}(j)$. The set of all possible grades ought to be done discrete in order to simplify and speed up the experts' answers. Let it be a five-point scale with grades from 1 to 5, in other terms, from minimum utility (or, to be exact, its unavailability) to maximum:

$$g_{kij}^{<u>} \in \{1, \ 2, \ 3, \ 4, \ 5\}.$$

If the experts are given a continuous scale instead of a discrete one, where only the lower and the higher values are marked, they will still try to discretize such a scale for themselves in some way in the process of evaluation.

If there are in total U experts, taking part in the expert survey, then the median utility matrix for the k enterprise $\tilde{\mathbf{G}}_k = [\tilde{g}_{kij}]_{8 \times 24}$ is preliminarily defined as

$$\tilde{\mathbf{G}}_k = \frac{1}{U} \cdot \sum_{u=1}^{U} \mathbf{G}_k^{<u>}, \tag{15}$$

where

$$\tilde{g}_{kij} = \frac{1}{U} \cdot \sum_{u=1}^{U} g_{kij}^{<u>} \ \left(\text{for each } i = i = \overline{1, \ 8} \text{ and}, \ j = \overline{1, \ 24} \, k = \overline{1, \ N-2}\right). \tag{16}$$

The variant of the questionnaire is given in Table 1. The whole set with 192 situations is divided into two parts. The left part corresponds to the unavailability of the strengthening of negative consequences, and the right one – to the situation when such strengthening occurs. Since one person is unable to give equally clear evaluations of all 192 situations, the expert can suggest the following variant. The expert is supposed to give not fewer than four grades for each enterprise strategy in the left part. Five or more grades (maximum—12) may be given at will, but such maximum is allowed only for three arbitrary enterprise strategies. The experts must do the same for the right part.

The restrictions on giving all 192 grades are justified by the fact that the more grades the expert gives (no matter what the level of their proficiency is), in other terms, the longer the evaluation process lasts, the more probable making mistakes is. Hence, it is worthwhile to suggest some "middle ground", according to which the minimal volume of evaluation accounts for 64 grades. Then, one person is able to provide qualitative evaluation during a work day or several hours. In order not to overstrain the experts, the restriction on giving 12 ratings is accepted as a recommendation not to glut a matrix questionnaire with filled-in squares because, in this case, their credibility drops rapidly. But how to determine the rest of the grades in the not filled-in squares? The answer to this question is connected with the coherence of expert grades.

If the grade $g_{kij}^{<u>}$ is not given, then we temporarily accept $g_{kij}^{<u>} = 0$. If

$$g_{kij}^{<u>} = 0 \text{ for each } u = \overline{1, \; U}, \tag{17}$$

then the situation $\{x_{K_{max} \subset A}(i), \; y_{A \setminus K_{max}}(j)\}$ does not obtain any grades (what, in fact, is entirely possible if the number of experts is restricted). In this case, all experts are suggested to evaluate all these situations and the situations that have received fewer than three grades. Thus, the matrix may contain zero elements $\mathbf{G}_k^{<u>}$ even after the second stage of the evaluation, but an obligatory condition of proceeding to the verification of the grade coherence must be fulfilled:

$$\begin{aligned} g_{kij}^{<u_*>} &\neq 0 \text{ for } u_* \in \mho_* \subset \{\overline{1, \; U}\} \text{ by } 3 \leq |\mho_*| \leq U \text{ and} \\ g_{kij}^{<u_0>} &= 0 \text{ for } u_0 \in \mho_0 \subset \{\overline{1, \; U}\} \setminus \mho_* \text{ for each } i = \overline{1, \; 8} \text{ and } j = \overline{1, \; 24}. \end{aligned} \tag{18}$$

Before starting coherence verification, the zero elements of each matrix $\mathbf{G}_k^{<u>}$ must be replaced by the corresponding average not zero grades. This is done with the help of the following algorithm:

$$g_{kij}^{<u_0>} = \frac{1}{|\mho_*|} \cdot \sum_{u_* \in \mho_* \subset \{\overline{1, \; U}\}} g_{kij}^{<u_*>} \tag{19}$$

for $u_0 \in \mho_0 \subset \{\overline{1, \; U}\} \setminus \mho_*$ for each $i = \overline{1, \; 8}$ and $j = \overline{1, \; 24}$.

After that, all matrices $\{\mathbf{G}_k^{<u>}\}_{u=1}^{U}$ for each enterprise ($k = \overline{1, \; N-2}$) will contain some (not zero) grades. Then, determination of the median utility matrix for the k enterprise $\tilde{\mathbf{G}}_k$ according to the formulae (15), (16) will be valid, because by adding according to (16) there will not be invalid (zero) values that are not grades.

Coherence validation is of great importance here. One can work with utility matrix evaluation (15) only if the grades of all its 192 elements are coherent. Otherwise, the game solutions (9) will be incompatible with reality. Then, on what condition can the grades of a certain situation be considered as coherent, taking into account that some situations, according to the condition (18) may have only three grades? Evidently,

Table 1 Matrix questionnaire for evaluation of the utility of the broad coalition situations

j	No strengthening of negative consequences												The strengthening of negative consequences is available												
i	1	2	3	4	5	6	7	8	9	10	11	12	13	14	15	16	17	18	19	20	21	22	23	24
1																								
2																								
3																								
4																								
5																								
6																								
7																								
8																								

one must start with consideration of the average spread of all expert grades of the matrix (15). It is commonly accepted that this spread must not exceed 25% of the maximum average spread.

The grades 1 and 5 being the minimum and maximum possible values, the average one is equal to 3. It follows that dispersion in this case equals 4. The corrected maximum dispersion is equal to

$$Dysp_{max} = \frac{4U}{U-1}.$$

Then, the condition of grade coherence in matrices $\left\{\mathbf{G}_k^{<u>}\right\}_{u=1}^{U}$ is the following one:

$$\frac{1}{192} \cdot \sum_{i=1}^{8} \sum_{j=1}^{24} \sum_{u=1}^{U} \frac{\left(g_{kij}^{<u>} - \tilde{g}_{kij}\right)^2}{U-1} \le \frac{U}{U-1}.$$

This condition can be simplified to:

$$\frac{1}{192} \cdot \sum_{i=1}^{8} \sum_{j=1}^{24} \sum_{u=1}^{U} \left(g_{kij}^{<u>} - \tilde{g}_{kij}\right)^2 \le U. \tag{20}$$

If the condition (20) is fulfilled for all enterprises ($k = \overline{1, \ N-2}$), then we find the evaluated utility function of the coalition (5) with the formula (14), by substituting the median utility matrices of all enterprises into this matrix:

$$G_{K_{max}CA}\left(x_{K_{max}CA}, \ y_{A\backslash K_{max}}\right) = \sum_{k=1}^{N-2} \tilde{G}_k. \tag{21}$$

Then we start the game (9), which acquires the following form for the broad coalition

$$\Gamma_{K_{max}CA} = \left\langle X_{K_{max}CA}, \ Y_{A\backslash K_{max}}, \sum_{k=1}^{N-2} \tilde{G}_k \right\rangle. \tag{22}$$

For the broad coalition, the solution of the game (22) is the maximin strategy [13]

$$x_{K_{max}CA}^{*} \in X_{K_{max}CA}^{*} = \left\{ \arg \max_{i=\overline{1, \ 8}} \min_{j=\overline{1, \ 24}} \sum_{k=1}^{N-2} \tilde{g}_{kij} \right\} \subset X_{K_{max}CA}, \tag{23}$$

on condition that

$$\max_{i=\overline{1,\,8}} \min_{j=\overline{1,\,24}} \sum_{k=1}^{N-2} \tilde{g}_{kij} = \min_{j=\overline{1,\,24}} \max_{i=\overline{1,\,8}} \sum_{k=1}^{N-2} \tilde{g}_{kij}. \qquad (24)$$

In this case, the strategy (23) is optimal for the broad coalition. However, if the condition (24) is not fulfilled, then the condition is considered [15]

$$\max_{\mathbf{M}\in M_8} \min_{\mathbf{L}\in L_{24}} \sum_{i=1}^{8}\sum_{j=1}^{24} \mu_i\lambda_j \sum_{k=1}^{N-2} \tilde{g}_{kij} = \min_{\mathbf{L}\in L_{24}} \max_{\mathbf{M}\in M_8} \sum_{i=1}^{8}\sum_{j=1}^{24} \mu_i\lambda_j \sum_{k=1}^{N-2} \tilde{g}_{kij} = H_\Gamma(K_{\max}),$$

$$\qquad (25)$$

where

$$\mathbf{M}\in M_8 = \left\{ \mathbf{M} = [\mu_i]_{1\times 8} : \mu_i \geq 0,\ \sum_{i=1}^{8}\mu_i = 1 \right\},$$

$$\mathbf{L}\in L_{24} = \left\{ \mathbf{L} = [\lambda_j]_{1\times 24} : \lambda_j \geq 0,\ \sum_{j=1}^{24}\lambda_j = 1 \right\},$$

which is always obligatorily fulfilled. In this case, the optimal strategy of a broad coalition is such a set of probabilities $\mathbf{M}^* = [\mu_i^*]_{1\times 8} \in M_8$, which meets the condition (25) [11, 13]:

$$H_\Gamma(K_{\max}) = \sum_{i=1}^{8}\sum_{j=1}^{24} \mu_i^*\lambda_j^* \sum_{k=1}^{N-2} \tilde{g}_{kij} = \mathbf{M}^* \cdot \left(\sum_{k=1}^{N-2} \tilde{\mathbf{G}}_k\right) \cdot (\mathbf{L}^*)^{\mathrm{T}}. \qquad (26)$$

Thus, each enterprise must obtain its utility function on the basis of coherent expert grades. Using this as a starting point, a utility function of the broad coalition can be found (21). This allows starting the game (22) and working out the best collective strategy of the broad coalition $\mathbf{M}^*$. However, in reality other variants of the coalition may occur.

As we have mentioned above, variants $2, 3, \ldots, N{-}1$, when the coalition consists of one enterprise with the value of the characteristic function $H_\Gamma(K_1)$, are not considered because, evidently, it is inappropriate to oppose the interests of the enterprise to the interests of all other ones. In any way, all enterprises will consider the case of a broad coalition, but the variant of an incomplete coalition, where at least one enterprise will not join, will also be considered. The exception here is the variant with two enterprises which consider only the variant of their (broad) coalition and then compare their utility with the variant of the unavailability of the coalition. In the case of three enterprises, the following variants are to be considered: the broad coalition of all three enterprises and the coalition of two enterprises that contains three subvariants of the value $H_\Gamma(K_2)$. Thus, four corresponding optimal variants of utility will be compared and then they will be compared with the optimal utility $H_r(K_1)$.

The set of the incomplete coalition strategies must be supplemented with strategies where the steadiness of one, two or even of all three components is expected. Hence, the set of the incomplete coalition strategies is the following one: to the summand to the subset of the strategies (10), it is

$$
\begin{aligned}
x_{KCA}(9) &= \{0,\ 0,\ 0\},\ x_{KCA}(10) = \{1,\ 0,\ 0\},\ x_{KCA}(11) = \{0,\ 1,\ 0\}, \\
&x_{KCA}(12) = \{0,\ 0,\ 1\},\ x_{KCA}(13) = \{1,\ 1,\ 0\}, \\
x_{KCA}(14) &= \{1,\ 0,\ 1\},\ x_{KCA}(15) = \{0,\ 1,\ 1\},\ x_{KCA}(16) = \{-1,\ 0,\ 0\}, \\
&x_{KCA}(17) = \{0,\ -1,\ 0\},\ x_{KCA}(18) = \{0,\ 0,\ -1\}, \\
&x_{KCA}(19) = \{-1,\ -1,\ 0\}, \\
x_{KCA}(20) &= \{-1,\ 0,\ -1\},\ x_{KCA}(21) = \{0,\ -1,\ -1\}, \\
x_{KCA}(22) &= \{-1,\ 1,\ 0\},\ x_{KCA}(23) = \{1,\ -1,\ 0\}, \\
&x_{KCA}(24) = \{-1,\ 0,\ 1\}, \\
x_{KCA}(25) &= \{1,\ 0,\ -1\},\ x_{KCA}(26) = \{0,\ -1,\ 1\}, \\
&x_{KCA}(27) = \{0,\ 1,\ -1\}.
\end{aligned}
\tag{27}
$$

In addition, the set of the disruption strategies may not be supplemented with the strategies of the enterpirses that have not joined the coalition because there "accession" to disruptions influences only the intensiveness of the latter ones, which now will "try" competing more fiercely with the coalition. Therefore, for the incomplete coalition K, the set of the disruption strategies $Y_{A\backslash K}$ is formally equivalent to the set (13): $Y_{A\backslash K} = Y_{A\backslash K_{max}}$. Thus, the games for incomplete coalitions have the size 27×24, in other terms, one game has 648 situations. Accordingly, the variant of the questionnaire for the evaluation of utility of the incomplete coalition situations presented in Table 2 is three times larger. Similarly to the case of the broad coalition grades by the layout in Table 1, the expert is not to provide all 648 grades, but only a third of them: arbitrary four grades for the left and right parts. However, in the case shown in Table 2, it is not a suggestion any longer but an obligatory condition— only eight grades, not more, for each line. It is done with the aim of restricting the duration of the evaluation process and improving its reliability. In order to help to orient, a column with the prompts, what changes are caused by the strategies (27), is introduced to Table 2: the marks "+1", "+2", "−1", "−2", "±1", as well as "no changes" for $x_{KCA}(9)$.

The processing of the grades for an incomplete coalition proceeds in the same way as for a broad coalition: the utility function of the incomplete coalition is defined

$$
G_{KCA}(x_{KCA},\ y_{A\backslash K}) = \sum_{k \in K \subset \{\overline{1,\ N-2}\}} \tilde{G}_k,
\tag{28}
$$

where the utility matrix for the $k$enterprise $\tilde{G}_k = \left[\tilde{g}_{kqj}\right]_{27 \times 24}$ is defined as (15) with

Table 2 Matrix questionnaire for evaluation of the utility of the incomplete coalition situations

j q	1	2	3	4	5	6	7	8	9	10	11	12	13	14	15	16	17	18	19	20	21	22	23	24
	No strengthening of negative impacts												The strengthening of negative impacts is available											
1	These																							
2	estimates																							
3	may differ from those																							
4	for the																							
5	broad																							
6	coalition																							
7																								
8																								
9	No changes																							
10																								
11	+1																							
12																								
13																								
14	+2																							
15																								
16																								
17	−1																							
18																								
19																								
20	−2																							
21																								

(continued)

Table 2 (continued)

j	No strengthening of negative impacts	The strengthening of negative impacts is available
22		
23		
24	±1	
25		
26		
27		

Notes

+1: one component increases and two others are without changes

+2: two components increase and one is without changes

−1: one component decreases and two others are without changes

−2: two components decrease and one is without changes

±1: one component increases, the other one—decreases and one component remains without changes

$$\tilde{g}_{kqj} = \frac{1}{U} \cdot \sum_{u=1}^{U} g_{kqj}^{<u>}$$

$$\left(\text{for each } q = \overline{1, \ 27} \text{ and } j = \overline{1, \ 24}, \ k \in K \subset \left\{\overline{1, \ N-2}\right\}\right). \tag{29}$$

Before that an obligatory condition of proceeding to the grade coherence verification must be fulfilled:

$$g_{kqj}^{<u_*>} \neq 0 \text{ for } u_* \in \mathfrak{V}_* \subset \left\{\overline{1, \ U}\right\} \text{ with } 3 \leq |\mathfrak{V}_*| \leq U \text{ and}$$
$$g_{kqj}^{<u_0>} = 0 \text{ for } u_0 \in \mathfrak{V}_0 \subset \left\{\overline{1, \ U}\right\}\backslash\mathfrak{V}_* \text{ for each } q = \overline{1, \ 27} \text{ and } j = \overline{1, \ 24}. \tag{30}$$

Before starting the process of coherence verification we replace the zero elements of each matrix $\mathbf{G}_k^{<u>}$ with the corresponding average not zero grades using the same algorithm as in (19):

$$g_{kqj}^{<u_0>} = \frac{1}{|\mathfrak{V}_*|} \cdot \sum_{u_* \in \mathfrak{V}_* \subset \{\overline{1, \ U}\}} g_{kqj}^{<u_*>}$$

$$\text{for } u_0 \in \mathfrak{V}_0 \subset \left\{\overline{1, \ U}\right\}\backslash\mathfrak{V}_* \text{ for each } q = \overline{1, \ 27} \text{ and } j = \overline{1, \ 24}. \tag{31}$$

After that all the matrices $\left\{\mathbf{G}_k^{<u>}\right\}_{u=1}^{U}$ for each enterprise $k \in K \subset \left\{\overline{1, \ N-2}\right\}$ of an incomplete coalition will have no zero grades.

As soon as the condition of coherence is fulfilled

$$\frac{1}{648} \cdot \sum_{q=1}^{27} \sum_{j=1}^{24} \sum_{u=1}^{U} \left(g_{kqj}^{<u>} - \tilde{g}_{kqj}\right)^2 \leq U \text{ for all }, k \in K \subset \left\{\overline{1, \ N-2}\right\}, \tag{32}$$

we find the utility function of the incomplete coalition (28) and (29). Transfer from the non-coalition game (2) to the game with a zero sum

$$\Gamma_{KCA} = \left\langle X_{KCA}, \ Y_{A\backslash K}, \ \sum_{k \in K \subset \{\overline{1, \ N-2}\}} \tilde{\mathbf{G}}_k \right\rangle, \tag{33}$$

for the incomplete coalition is the development of the game (2). For an incomplete coalition, the solution of the game (33), which is more difficult than the game (22) for each of these coalitions, is a maximin strategy

$$x_{KCA}^* \in X_{KCA}^* = \left\{\arg \max_{q=\overline{1, \ 27}} \min_{j=\overline{1, \ 24}} \sum_{k \in K \subset \{\overline{1, \ N-2}\}} \tilde{g}_{kqj}\right\} \subset X_{KCA}, \tag{34}$$

on condition that

$$\max_{q=\overline{1, \ 27}} \min_{j=\overline{1, \ 24}} \sum_{k \in K \subset \{\overline{1, \ N-2}\}} \tilde{g}_{kqj} = \min_{j=\overline{1, \ 24}} \max_{q=\overline{1, \ 27}} \sum_{k \in K \subset \{\overline{1, \ N-2}\}} \tilde{g}_{kqj}. \tag{35}$$

In this case, the strategy (34) is optimal for an incomplete coalition. However, the probability of the condition (35) is lower than the probability of the condition (24) because there are 27 variants of the strategy x_{KCA}—(10) and (27). Thus, we should not count on the optimality of the type (34). On the contrary, the condition (35) is the partial case of the condition

$$
\max_{\mathbf{N} \in N_{27}} \min_{\mathbf{L} \in L_{24}} \sum_{q=1}^{27} \sum_{j=1}^{24} \eta_q \lambda_j \sum_{k \in K \subset \{\overline{1, N-2}\}} \tilde{g}_{kqj} =
$$
$$
= \min_{\mathbf{L} \in L_{24}} \max_{\mathbf{N} \in N_{27}} \sum_{q=1}^{27} \sum_{j=1}^{24} \eta_q \lambda_j \sum_{k \in K \subset \{\overline{1, N-2}\}} \tilde{g}_{kqj} = H_\Gamma(K),
$$

(36)

where

$$
\mathbf{N} \in N_{27} = \left\{ \mathbf{N} = [\eta_q]_{1 \times 27} : \eta_q \geq 0, \sum_{q=1}^{27} \eta_q = 1 \right\},
$$

and $\mathbf{L} \in L_{24}$, which is always fulfilled. In this case, the optimal strategy of the incomplete coalition is the following set of probabilities $\mathbf{N}^* = [\eta_q^*]_{1 \times 27} \in N_{27}$, which meets the condition (36)

$$
H_\Gamma(K) = \sum_{q=1}^{27} \sum_{j=1}^{24} \eta_q^* \lambda_j^* \sum_{k \in K \subset \{\overline{1, N-2}\}} \tilde{g}_{kqj} = \mathbf{N}^* \cdot \left(\sum_{k \in K \subset \{\overline{1, N-2}\}} \tilde{\mathbf{G}}_k \right) \cdot (\mathbf{L}^*)^{\mathrm{T}}.
$$

(37)

Each variant of the incomplete coalition has its own optimal strategy $\mathbf{N}^*$, the partial case of which is one strategy (37), and the corresponding optimal utility $H_\Gamma(K)$. Let us mention that utility (37) is a mathematical expectation, in other terms, expected utility which is gained by an incomplete coalition only after practical implementation (realization) of its optimal strategy $\mathbf{N}^*$ during a definite period of time. The same concerns the utility (26) and the optimal strategy of the broad coalition $\mathbf{M}^*$, whose implementation can theoretically last shorter. However, it is not about the three times shorter implementation period. We will return to this peculiarity below. Certainly, if the strategy $\mathbf{N}^*$ contains only one not zero element (which is equal to 1), i.e. it is equivalent to the optimal strategy (34), then the practical implementation is carried out and the corresponding result (37) is obtained for a very short period of time, during which the enterprise coalition (though, in fact, each enterprise separately) gets the reaction of the economic environment to the changes made according to the strategy (34).

If there is no coalition at all, then each enterprise considers the simplified variant of the game (33). For the k enterprise, it is the game

$$
\Gamma_{K_1CA} = \langle X_{K_1CA}, \, Y_{A \setminus K_1}, \, \tilde{\mathbf{G}}_k \rangle = \langle X_k, \, Y_{A \setminus \{k\}}, \, \tilde{\mathbf{G}}_k \rangle,
$$

(38)

where $Y_{A \setminus \{k\}}$ is the set of strategies (13). For the game (38), the equation always holds

$$\max_{N \in N_{27}} \min_{L \in L_{24}} \sum_{q=1}^{27} \sum_{j=1}^{24} \eta_q \lambda_j \tilde{g}_{kqj} = \min_{L \in L_{24}} \max_{N \in N_{27}} \sum_{q=1}^{27} \sum_{j=1}^{24} \eta_q \lambda_j \tilde{g}_{kqj} = H_\Gamma(K_1). \tag{39}$$

Due to this fact the optimal strategy of the k enterprise is the set of probabilities

$$\mathbf{N}_k^* = \left[\eta_q^*(k)\right]_{1 \times 27} \in N_{27},$$

which meets the condition (39)

$$H_\Gamma(K_1) = \sum_{q=1}^{27} \sum_{j=1}^{24} \eta_q^*(k) \lambda_j^* \tilde{g}_{kqj} = \mathbf{N}_k^* \cdot \tilde{\mathbf{G}}_k \cdot \left(\mathbf{L}^*\right)^{\mathrm{T}}. \tag{40}$$

In the partial case, the optimal strategy of the k enterprise is a strategy

$$x_k^* \in X_k^* = \arg \max_{q=\overline{1,27}} \min_{j=\overline{1,24}} \tilde{g}_{kqj} \subset X_k, \tag{41}$$

if the equation holds

$$\max_{q=\overline{1,27}} \min_{j=\overline{1,24}} \tilde{g}_{kqj} = \min_{j=\overline{1,24}} \max_{q=\overline{1,27}} \tilde{g}_{kqj}. \tag{42}$$

Though, for the incomplete coalitions consisting of two and more enterprises, it is not worthwhile to count on the condition (42); in the cases when the Eq. (42) does not hold, the maximin strategy (41) is not optimal.

4   Conclusions

As the result of the use of the worked out model, the enterprise management may make decisions regarding the forming of their behavior strategy with determination of the peculiarities of the coalition building of the corresponding enterprises on the basis of the profit maximization criterion. In addition, the decision on the coalition joining is made on the basis of the following considerations. After obtaining the evaluated utility functions of all enterprises for the cases of a broad and an incomplete coalition, each enterprise starts the corresponding game and determines its optimal utility according to this game. These are the following functions:

$H_\Gamma(K_{\max})$—for the case of a broad coalition (there is only one such case);

$H_\Gamma(K)$—for the cases of an incomplete coalition (there are at least two such cases—for $N = 5$, when there are three enterprises, and the first enterprise considers the variant of its incomplete coalition with the second one and the variant of its incomplete coalition with the third one; the first enterprise does not consider the incomplete coalition of the second and the third enterprises; the cases of incomplete coalitions, containing three enterprises and more for $N \geq 6$ also must be taken into account);

$H_\Gamma (K_2)$—for the cases of an incomplete coalition, containing two enterprises (if $N \geq 5$);

$H_\Gamma (K_1)$—for one case, when the enterprise remained alone (in fact, not having joined the coalition, though this case is formally considered as a coalition which this enterprise joined).

For instance, if there are three enterprises, these functions are the following: one function $H_\Gamma (K_{max})$, two functions $H_\Gamma (K_2)$, and one function $H_\Gamma (K_1)$. In a general case, the values of all these functions are sorted in descending order. The coalition, either broad or incomplete, forms on condition that utility (profit) of the enterprises that are its members are maximum for them, and higher enterprise utility is available only in other variants of the coalition that are unfavorable for one or more enterprises. The broad coalition forms if each enterprise gains less (or not more than in a complete coalition, the complete coalition being in priority) in any other variants of an incomplete coalition. Then, all N–2 enterprises agree to form a broad coalition. The incomplete coalition forms in the case when one or more enterprises do not receive some part of their own profit (certainly, after profit distribution) in a complete coalition.

References

1. Aleksanyan, L., Huiban, J.-P.: Economic and financial determinants of firm bankruptcy: evidence from the French food industry. Rev. Agric. Food Environ. Stud. **97**(2), 89–108 (2016). https://doi.org/10.1007/s41130-016-0020-7
2. Amilon, H.: A neural network versus Black-Scholes: a comparison of pricing and hedging performances. J. Forecast. **22**(6–7), 317–335 (2003). https://doi.org/10.1002/for.867
3. Aumann, R.J.: Agreeing to disagree. Ann. Stat. Inst. Math. Stat. **4**(6), 1236–1239 (1976)
4. Bruin, B.: Game theory in philosophy. Topoi **24**, 197–208 (2005). https://doi.org/10.1007/s11 245-005-5055-3
5. Esmaeili, M., Allameh, G., Tajvidi, T.: Using game theory for analysing pricing models in closed-loop supply chain from short- and long-term perspectives. Int. J. Prod. Res. **54**(7), 2152–2169 (2016). https://doi.org/10.1080/00207543.2015.1115907
6. Nash, J.F., Nagel, R., Ockenfels, A., Selten, R.: The agencies method for coalition formation in experimental games. Proc. Natl. Acad. Scie. **109**(50), 20358–20363 (2012). https://doi.org/10.1073/pnas.121636110
7. Nisan, N., Roughgarden, T., Tardos, É., Vazirani, V.V.: Algorithmic Game Theory, p. 778. Cambridge University Press, Cambridge, UK (2007)
8. Rudnichenko, Y., Korchevska, L., Mykolaichuk, V., Berezhniuk, I., Havlovska, N., Nagorichna, O.: Customs qualitative impact on the system of enterprise economic security: modeling and evaluating the results. Tem J. Technol. Educ. Manag. Inf. **8**(4), 1176–1184. https://doi.org/10.18421/TEM84-10
9. Rudnichenko, Y., Havlovska, N., Matiukh, S., Lopatovskyi, V., Yadukha, S.: Optimization of the interaction of industrial enterprises and foreign counterparties using pure player strategies

in a non-cooperative game. Tem J. Technol. Educ.. Manag. Inf. **8**(1), 182–188 (2019). https://doi.org/10.18421/TEM81-25

10. Bogutska, O.: Financial and economic mechanism of ensuring investment activity of enterprises within institutipnal models of financing the real sector of economics. Innovat. Technol. Sci. Solutions Ind **3**(5), 79–86 (2018). https://doi.org/10.30837/2522-9818.2018.5

11. Hua, M., Lai, I.K.W., Tang, H.: Analysis of advertising and a points-exchange incentive in a reverse supply chain for unwanted medications in households based on game theory. Int J Prod Econ **217**(C), 259–268 (2019). https://doi.org/10.1016/j.ijpe.2019.02.004

12. Fairfield, P.M., Yohn, T.L.: Using asset turnover and profit margin to forecast changes in profitability. Rev. Acc. Stud. **6**, 371–385 (2001). https://doi.org/10.1023/A:1012430513430

13. Gilad-Bachrach, R., Navot, A., Tishby, N.: Margin based feature selection: theory and algorithms. In: Greiner, R., Schuurmans, D. (eds.) Proceedings of the Twenty-First International Conference on Machine Learning, pp. 337–344. ACM Press, Banff, Alberta, Canada, New York (2004)

14. Horváthová, J., Mokrišová, M.: Risk of bankruptcy, its determinants and models. Risks **6**(4), 1-22 (2018). https://doi.org/10.3390/risks6040117

15. Leyton-Brown, K., Shoham, Y.: Essentials of Game Theory: A Concise, Multidisciplinary 3. Introduction, p. 104. Morgan & Claypool Publishers (2008)

Innovative Technologies to Make Effective Business Decisions at Every Stage of a Mining Company's Development

Nataliia Bariatska and Vadym Tarasov

Abstract The digital transformation and automation of all business processes and production operations, using innovative technology, allows for well-informed strategic decisions and the stabilisation of the business. The metals and mining industry is not a "digital leader" and is 30–40% less mature digitally than other similar industries, such as pharma, chemicals, logistics and others. Mining industry nevertheless has good potential for automation, digital transformation and advanced innovative technologies, taking into account a huge number of different influencing factors. Process automatisation and specialised innovative technologies can be applied at each stage of mining project development: geological and operational exploration; geological interpretation and modelling; block modelling and resourceestimation; economic assessment (pre- or feasibility study); optimisation and strategic planning, open pit or mine design, production scheduling and support; mining management and dispatching. Creating and evaluating multiple options at each stage makes each process iterative and allows different parameters and options to be changed and refined in real time. The company's high level of digital maturity, introduction of innovative technologies and automation of production processes ensure that strategic business decisions are better informed at every stage of the mining business' development.

Keywords Digital technology · Digital business · Digitalisation · Micromine · Geological structures · Optimisation · Software

N. Bariatska
Softmine LLC, 104 Raiduzhna Str., Kyiv 02218, Ukraine
e-mail: nbariatska@micromine.com

V. Tarasov (✉)
Volodymyr Dahl East Ukrainian National University, 17 John Paul II Street, Kyiv 01042, Ukraine
e-mail: tarasov@snu.edu.ua

A. Semenov et al. (eds.), *Data-Centric Business and Applications*, Lecture Notes on Data Engineering and Communications Technologies 194, https://doi.org/10.1007/978-3-031-53984-8_14

1 Introduction

The intensive development of information technology and the digitalisation of various areas of our lives is already a long-term global trend [1, 2]. A company's high level of digitalisation and the application of advanced, innovative technologies allow more better control and forecasting of business [3]. Digitalisation or digital transformation of business is the process of adoption and implementation of digital technology into various aspects of a business to optimise management of main business processes [4, 5]. Digital transformation is supposed to accelerate business growth and/or increase operational efficiency [6]. An indicator of a company's digital development, which describes the extent and success of its digital transformation, is the level of digital maturity [7].

Digital acceleration is a business strategy concerning digitalisation. It is a move from process digitisation to adopting the latest technologies such as business process automation, data visualisation, machine learning and artificial intelligence. Carefully selected and implemented technologies can have a positive impact on the cost of running the business. It can already be seen in many industries including manufacturing, logistics and banking. Among the companies using digital technology to advance their business, around 30% are achieving "extreme" efficiency. In some cases, "digital leaders" are able to achieve such outcomes [8–10]:

- 4 times better at operational efficiency;
- 3 times better at customer experience;
- 3.5 times better at increasing revenue;
- >10% higer market share grows;
- >15% higer ROI (return on investment);
- 7 times more efficient data management.

The mining business is unique compared to other industries in which many factors can affect its success. Each process and each operation can vary depending on a lot of aspects. Therefore, consideration of multiple factors, correct resource distribution and process optimisation help to minimise losses and stable company operations. Digital transformation has the potential to help mining companies reduce their costs and improve profits by streamlining their work processes and incorporating greater data insights to drive strategic decisions. According to the Boston Consulting Group research, the metals and mining industry is 30–40% less mature digitally than other similar industries [11, 12]. Research is based on the use of the Digital Acceleration Index (DAI), a specific indicator for analyzing and understanding the digital maturity of companies, industries and countries. Most metals and mining companies have ambitious digital strategies, but the gap between strategy and execution (Fig. 1). As can be seen, this gap between planned and achieved digitalisation levels averages around 30%, which is significantly higher compared to other industries, such as pharma, chemicals, logistics and others. In the mining industry, digitalisation is most important for production processes such as preparation and extraction of ore, processing and enrichment. But also important is the application of digital innovative technologies for procurement, supply, sales and marketing, finance and planning. Although the mining industry is not a "digital leader", it nevertheless has good potential for automation, digital transformation and advanced innovative technologies. The level of potential return of investment of such implementations is also positively assessed [8].

Digitalisation of general business processes such as finance, supply chain, marketing are not the focus of this study. At the same time, the mining industry has a number of specific areas with high potential for digitalisation and the introduction of innovative technologies, such as:

- Geological exploration, including additional reservoir and operational exploration;
- Geological modelling using geological data interpretation and wireframing;

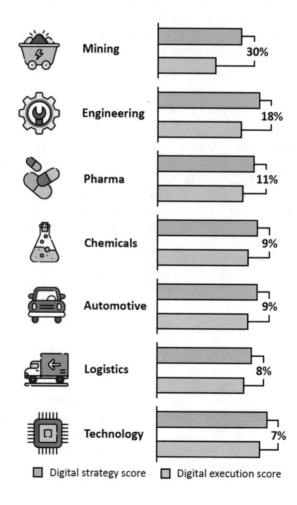

Fig. 1 The strategy-execution gap in metals and mining, according to [11]

- Resource estimation of commercial components using three-dimensional block modelling,
- (Pre)feasibility study, including ore reserve estimation, considering economic, processing and mining constraints;
- Mining optimisation and strategic planning to minimise costs, maximise revenues over Life of Mine;
- Design of open pit and underground mines, taking into account the ore reserve distribution, mining-engineering, hydrogeological and other parameters;
- Production scheduling to achieve optimal sequencing of operations and efficient allocation of production resources;
- Surveying and geological support of mining operations;
- Mining management with automation and control of all production processes.

In fact, the areas of a mining operation which require automatisation and digital transformation may differ depending on the type of mineralisation, mining and processing method, etc. Let's consider some examples of the application of innovative technologies at each main stage of deposit (mining company) development.

2 Geological Exploration

The world's largest geological and mining companies understand that geological information can be a company asset and affect value. The collection and analysis of geological information, whether historical or recent, is one of the most important. The completeness and reliability of the raw data crucially determines the confidence of the result—a resource model of the deposit, a mine design or a mining schedule.

Modern technology allows to combine all types of necessary information in one software (Fig. 2).

This can be geological, geochemical, geophysical, hydrogeological data; surface, drill hole and underground data; topographic surfaces, satellite imagery, laser and LIDAR surveys; geotechnical, technological, economic information etc.

Specific tools and functions provide a wide range of capabilities for data analysis, processing and management [13, 14]. These can include classic statistics, variographic analysis, a variety of statistical plots and charts, top-cut grade determination, QA/QC analysis and many others (Fig. 3). Drilling planning capabilities make the process faster, easier and more accurate.

Collecting, managing and analysing huge volumes of data allows us, on the one hand, to provide a basis for a reliable resource estimate and, on the other hand, to better inform decision-making during the geological exploration stage of a project. Collecting, managing and analysing huge volumes of data allows us, on the one hand, to provide a basis for a reliable resource estimate and, on the other hand, to better inform decision-making during the geological exploration stage of a project. For example, when drilling programmes are created, we can more accurately estimate

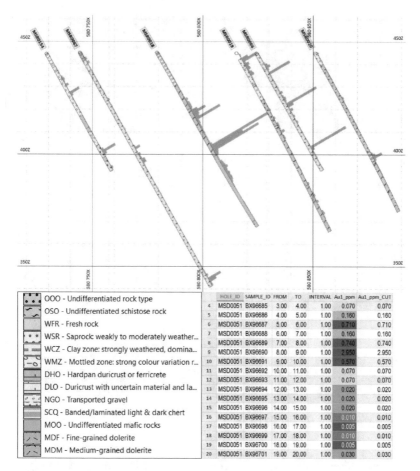

Fig. 2 Drillhole database (visualisation on the top, geological legend on the bottom left and assay table on the bottom right)

	OOO - Undifferentiated rock type
	OSO - Undifferentiated schistose rock
	WFR - Fresh rock
	WSR - Saprock: weakly to moderately weather...
	WCZ - Clay zone: strongly weathered, domina...
	WMZ - Mottled zone: strong colour variation r...
	DHO - Hardpan duricrust or ferricrete
	DLO - Duricrust with uncertain material and la...
	NGO - Transported gravel
	SCQ - Banded/laminated light & dark chert
	MOO - Undifferentiated mafic rocks
	MDF - Fine-grained dolerite
	MDM - Medium-grained dolerite

	HOLE_ID	SAMPLE_ID	FROM	TO	INTERVAL	Au1_ppm	Au1_ppm_CUT
4	MSD0051	BX96685	3.00	4.00	1.00	0.070	0.070
5	MSD0051	BX96686	4.00	5.00	1.00	0.160	0.160
6	MSD0051	BX96687	5.00	6.00	1.00	0.710	0.710
7	MSD0051	BX96688	6.00	7.00	1.00	0.160	0.160
8	MSD0051	BX96689	7.00	8.00	1.00	0.740	0.740
9	MSD0051	BX96690	8.00	9.00	1.00	2.950	2.950
10	MSD0051	BX96691	9.00	10.00	1.00	0.570	0.570
11	MSD0051	BX96692	10.00	11.00	1.00	0.070	0.070
12	MSD0051	BX96693	11.00	12.00	1.00	0.070	0.070
13	MSD0051	BX96694	12.00	13.00	1.00	0.020	0.020
14	MSD0051	BX96695	13.00	14.00	1.00	0.020	0.020
15	MSD0051	BX96696	14.00	15.00	1.00	0.020	0.020
16	MSD0051	BX96697	15.00	16.00	1.00	0.010	0.010
17	MSD0051	BX96698	16.00	17.00	1.00	0.005	0.005
18	MSD0051	BX96699	17.00	18.00	1.00	0.010	0.010
19	MSD0051	BX96700	18.00	19.00	1.00	0.005	0.005
20	MSD0051	BX96701	19.00	20.00	1.00	0.030	0.030

the drilling and associated scope and predict the results, especially if the drilling programme is based on a preliminary model [15].

3 Geological or Grade Modelling of Deposit

A geological model can consist of lithologies (rocks), geological structures, faults, contacts, represented by wireframes and surfaces. The geological modelling process includes data interpretation and wireframing. Such model provides a geological understanding and is the basis for block resource modelling.

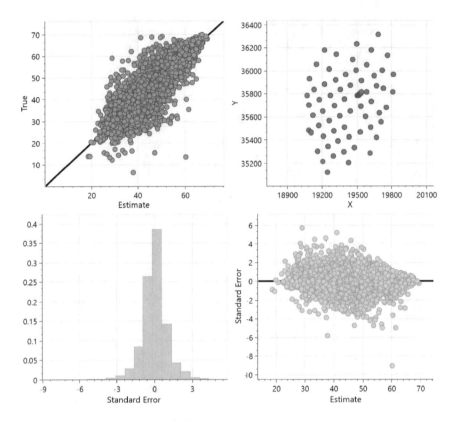

Fig. 3 Chart, graph and plot examples

Another option of wireframe modelling of a deposit is grade modelling. This model is typically used in cases where mineralisation is not distinct lithologically controlled and is also used for further block modelling and resource estimation.

Advanced implicit modelling tools allow to create wireframe models using the radial basis function algorithm (Fig. 4). When selecting the correct modelling parameters, wireframing is performed correctly automatically. This provides an opportunity to quickly update the model based on changed or supplemented source data.

Unlike manual modelling, automatic functions allow you to create "complete" geological models that include all geological elements (lithological varieties, veins, faults, contacts, etc.), spatially consistent with each other. In manual mode, it requires a lot of time.

Fig. 4 Three-dimensional wireframe model created using Micromine implicit modelling tools

4   3-D Block Modelling and Resource Estimation

A block model of the deposit is created based on geological wireframes and available sampling from drill holes or other workings. The contents of elements of interest and other parameters such as density are interpolated in each block using special geostatistical tools. Block modelling tools allow you to create, validate, re-block, combine, optimise block models (Fig. 5). Interpolation of the grades and other parameters can use a wide range of methods: inverse distance method, simple, ordinal, ranking, multiple indicator kriging, etc. Cross validation functions, Swath plots, Quantitative Kriging Neighbourhood Analysis (Fig. 6), etc., allow assessment of the quality of interpolation and resource estimation [16].

Modern block resource modelling techniques estimate resources according to many different parameters and restrictions, such as cut-off grades, ore types, categories, etc. (Fig. 7). Maximum accurate local grade distributions subsequently allow for more accurate mining design and planning.

The resource model takes into account the natural characteristics of the project and is the basis for applying economic, technical and other restrictions. In general, the more accurate the resource model, the more predictable the future production.

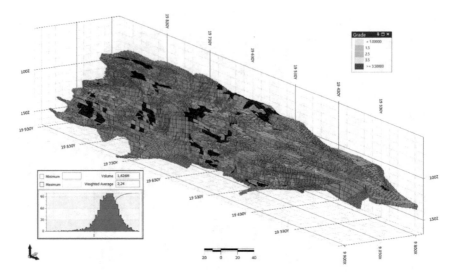

Fig. 5 Block resource model

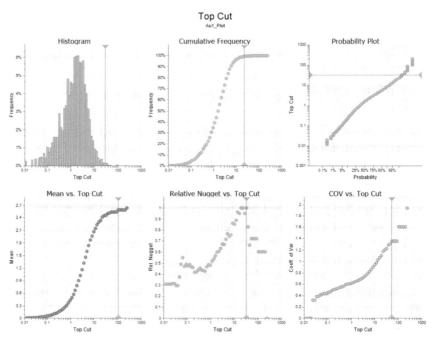

Fig. 6 Determine the top-cut value by different methods in Micromine

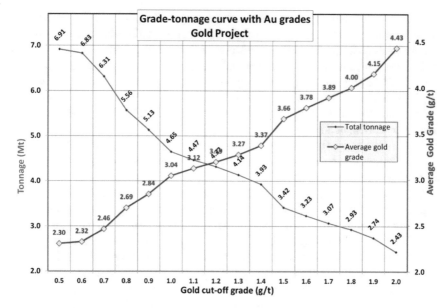

Fig. 7 Grade-tonnage curve based on the block model

5 Economic Assessment and Optimization

Optimisation process of a mining operation depends on the corporate interests of the company developing it. Actually, the strategic development plan for a mining project may vary among different companies depending on their distinct corporate goals and targets.

One of the most common overarching objectives for most companies is to increase shareholder value in the long-term by maximizing NPV. Net Present Value (NPV) is a financial metric used to evaluate the profitability of an investment project by comparing the present value of its expected cash inflows to the present value of its expected cash outflows.

A mining company may have various objectives apart from just maximizing profits, such as extending the life of the mine, reducing production costs, improving the efficiency of operations, increasing the recovery of metals, etc.

Open pit optimization in open mining or stope optimization in underground mining is the process of determining the optimal contour or shell to be mined to maximize profit while meeting operational requirements.

The main basis is the block resource model, which in addition to grades may contain other parameters such as ore type, process and geotechnical properties, etc. In addition to the block model, a number of mining, process and economic parameters are required for optimisation.

For open mining, the main geotechnical parameters are the slope angle for different areas or rocks, the spatial restrictions of the open pit, which may be determined by protective zones, buildings, etc. For underground mining, we have to specify the stope parameters: size, position (azimuth, dip) and boundaries (restrictions). Mining dilution and recovery also to be considered for both mining methods.

Economic parameters such as ore and waste rock mining costs, rehabilitation costs, finished product price, administration costs, discount rates by period are also required. Some of these indicators can be regulated by means Cost Adjustment Factors. So, MCAF (mining cost adjustment factor) is particularly used to adjust mining costs with depth.

Process parameters typically include processing costs, recovery and dilution of all commercial components for each processing method, as well as additional processing costs, if required.

Each unit in the model receives an optimisation value, which is the income from the sale of all elements minus all costs. If the optimisation value is positive, including the block in a solution increases the value of that solution; conversely, if the optimisation value is negative, including the block decreases the value of the solution. Unfortunately, it is often necessary for negatively valued blocks to be included to ensure that the requirements of the design are met. The role of the optimiser is to maximise the number of positively valued blocks in the solution whilst minimising the number of negatively valued blocks that must be included to satisfy the requirements of the design.

As a result, the optimiser generates nested pit shells (stope outlines) for different target profit levels corresponding to the specified design and economic parameters. The NPV is sequentially calculated for each optimisation (Figs. 8 and 9).

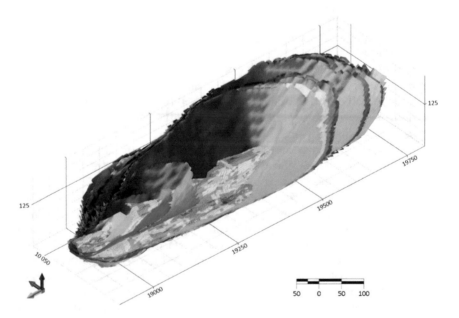

Fig. 8 Nested shells generated by pit optimization

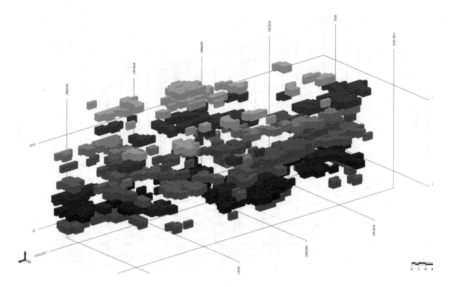

Fig. 9 Stope optimisation outcomes

If we consider the pit shells (outlines of excavation units) in order from the highest to the lowest value of mined tonnes (from the highest to the lowest profit), this sequence can represent the optimum mining dynamics.

The optimisation results chart helps to select the optimum ultimate pit (stope) with the maximum economic effect (Fig. 10). In most cases, a company chooses the option with the highest NPV, but the volume of mined ore, waste rock and other indicators should also be taken into account [17].

A sensitivity analysis of economic and technological parameters can be completed at this stage. You can see how the project is sensitive to changes in prices and costs, processing recovery, etc. Once the final optimum ore volume has been selected, the next step is to create a detailed design that can be safely and efficiently mined [18, 19].

6 Open Pit or Mine Design

After optimization, it is necessary to develop a detailed mining project that deviates as little as possible from the economically justified shell/stope (Fig. 11).

The design should:

- take into account the boundaries of mine allotment, exclusion zones and access to the open pit,
- minimise transportation distances,
- maximise ore extraction,

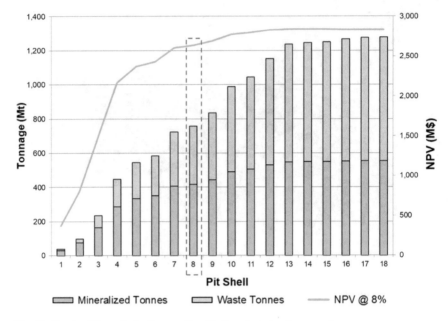

Fig. 10 Graph of the optimization results with the selected optimum option

- maintain compliance with production parameters,
- consider geotechnical and safety parameters.

It is common for there to be a loss of value between the optimised open pet or stope and the mine design due to compliance with various mining and technical requirements.

For example, for open mining, the addition of ramps and berms results in extra waste rock and/or ore loss.

Modern design techniques allow for different parameters to be used for different areas or zones of a mining project. We can also design stockpiles and waste dumps, plan blast and development drilling (Fig. 12). These additional features allow for a more accurate assessment of upcoming costs and forecasting results.

The design is typically iterative process. If necessary, the optimization step can be repeated, taking into account the refined parameters obtained from the design.

Manual design was a rather time-consuming, labor- and resource-intensive process. Modern advanced technologies allow designers and engineers to quickly create and test multiple design options and analyze their performance, cost, and reliability. This helps to make more informed and considered decisions about the final project.

For example, by designing several options for the mine profiles and equipment corresponding to different projected annual production rates, we can choose the most feasible and economically efficient option.

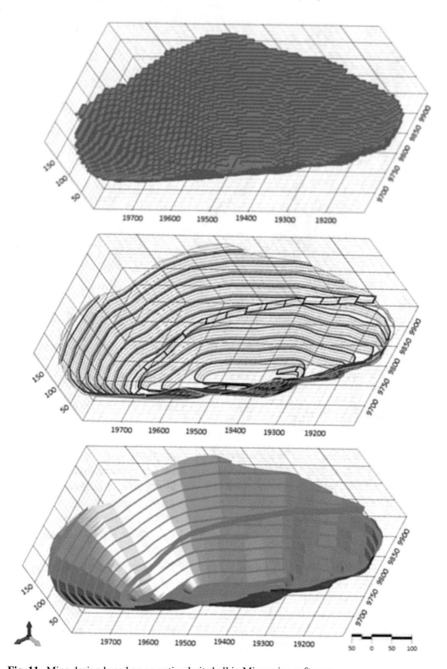

Fig. 11 Mine design based on an optimal pit shell in Micromine software

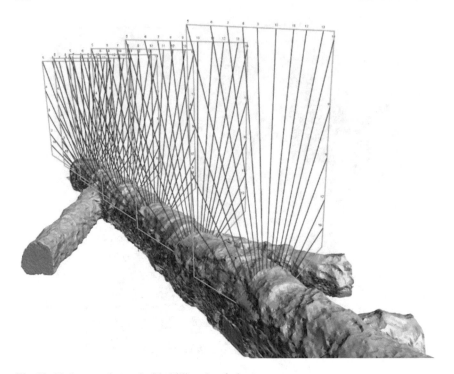

Fig. 12 Underground tunnel with drilling ring design

7 Mining Scheduling

Production scheduling is an integral part of the mine planning process. Scheduling is undertaken to plan and coordinate the development and production activities of a mine. The scheduling workflow can generally be described as a sequence of steps: creating mining blocks, determining mining tasks and preparing a schedule [20, 21].

During the mining block creation, blocks are assigned some attributes, such as volume, elevation level, block index etc. (Fig. 13). Other attributes are assigned based on the block model—tonnage, average grade, etc.

The next planning step is to create production objectives, such as stripping, drilling, mining, etc. Each task should have a set of parameters, such as volume, tonnage, rate, productivity, etc. All tasks are visualised in a Gantt chart according to the specified parameters (Fig. 14).

The Gantt chart is linked to each task and its parameters on the one hand and to the 3D mine blocks and their attributes on the other. So any changes to the original block design wireframes, for example updated face positions, can be swiftly applied to tasks in the Gantt (Fig. 15).

Dependencies between production tasks, created by defining horizontal, vertical, inter-stage links, allow the development of a production sequence corresponding

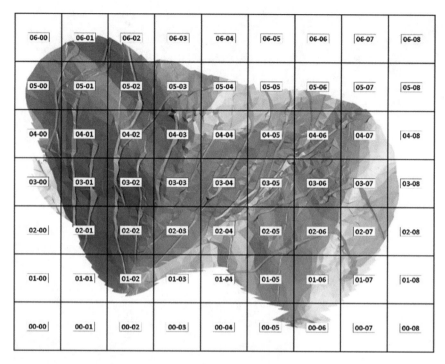

Fig. 13 Automatic creation of mining blocks for the open pit

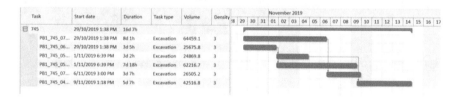

Fig. 14 Gantt chart for mine extraction with linked task list

to the main operational tasks. Creating dependencies between operational tasks by defining horizontal, vertical, inter-stage links allows the development of a production sequence that corresponds to the main operational objectives. In addition, modern technology allows to optimise the schedules by maximising the defined parameters (block cost or others) using advanced mathematical algorithms, taking into account the restrictions.

Mine scheduling typically involves the preparation of several schedules with different levels of detail [23, 24]:

1. strategic or Life of mine (LOM) plan—a conceptual development plan, which may include consideration of resource categories, different scenarios and mining strategies;

Fig. 15 Mining and resource blocks correlations with the corresponding parameters [22]

2. medium-term plan, focuses on part of the mine Life of mine plan, e.g. a 5-year plan;
3. the short-term plan should reflect a detailed operational schedule, such as a rolling monthly, weekly or shift plan, and describe the day-to-day operational activities.

8 Surveying and Geological Support of Mining Operations

Modern technology greatly simplifies the mining support. There are many tools available to monitor grade (quality) and "payback" of reserves during production. A number of special tools allow to monitor grade (quality) and update reserves in real time.

In addition, a special functionality allows to predict the parameters of extracted ore based on geological and operational exploration. This helps to adjust in real time the parameters of ore entering the processing plant. Special interactive grade control tools allow you to control the ratios and parameters of the different ore types within a mining block (Fig. 16). Quickly generating and updating relevant reports helps to monitor the current condition [25].

A wide range of modern equipment such as laser scanners, drones, GPS stations, electronic tacheometers and others are available to carry out surveying work in open pits or mines. Advanced mining software products allow data to be imported directly from recorders and processed in the same software (Fig. 17).

The use of advanced 3D scanning and mapping technology allows mining companies to accurately map mines and analyse data in real time, leading to improved mining efficiency [26].

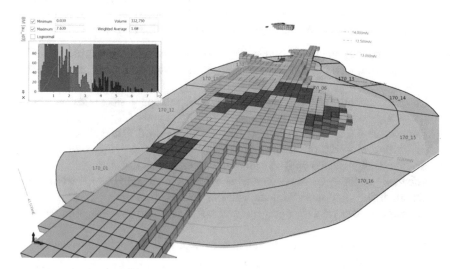

Fig. 16 Control of grades in mining blocks using a resource block model (Micromine)

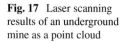

Fig. 17 Laser scanning results of an underground mine as a point cloud

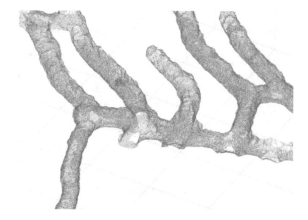

9 Dispatching and Management of Mining Operations

By leveraging advanced sensors, Internet devices, and data analytics platforms, mining companies can gather real-time data on various parameters, such as ventilation, equipment performance, geological conditions etc.

Comprehensive mine control and fleet management solutions are used for capturing, managing, analysing and optimising holistic mine activity. Core operational asset data (equipment, material, locations and people) are brought together to deliver data-driven insights in real or near-real-time, improving operational outputs and outcomes.

GPS sensors and trackers are widely used on mining and truck equipment to automate production operations. Dispatch fleet management systems help mine optimize its haulage cycle and dramatically reduce truck idle times [27].

Advanced Integrated Mine Control Systems provide the ability to:

- Record, monitor and report on fleet, location and personnel productivity;
- Measure equipment availability, equipment utilisation and report on Utilisation of Available;
- Record, reconcile and manage material flow, including stockpiles and grade (Fig. 18);
- Plan, report and optimise operational activity including drill and blast, short-term scheduling, and shift planning (Fig. 19);
- Automate, digitise and optimise time-intensive, manual tasks;
- Enhance site safety across personnel activities to understand which workers are behind digital signs describing high-risk or restricted areas;
- Interrogate, report and optimise advanced multidimensional mining data analytics;
- Visualise 3D real-time equipment and personnel locations.

This data-driven approach enables better decision-making, predictive maintenance, and optimised resource allocation.

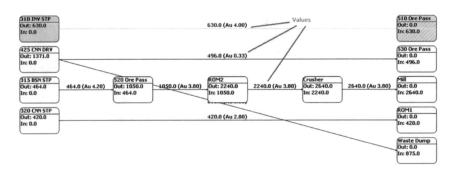

Fig. 18 Material movement in a mining operation

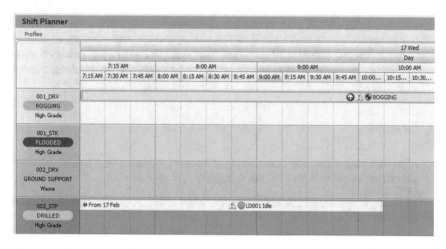

Fig. 19 Shift planner window

The automatisation and digitalisation of a mining company's production operations allows:

- Make more informed and accurate decisions from shift-level to long-term.
- Improve response time and course-correct sooner to stay on-task and on-target.
- Undertake ad-hoc analysis of your mine to streamline and optimise operational activities across departments.
- Gain greater control and visibility of workforce and assets.
- Identify and act on opportunities faster.

It should also be noted that the automatisation and digitalisation of operational processes is also widely used in other areas of the mining business. In land rehabilitation or environmental monitoring, for example, it is necessary to collect and analyse information in real time, to assess the current situation objectively and to react promptly to extraordinary events.

10   Conclusions

At a time of intense artificial intelligence and robotics development, the digitalisation of business is inevitable. For the mining industry, the effect is very significant due to its complexity and multifactorial nature.

The main advantage is the creation and evaluation of a lot of options for project development at each stage. Innovative technologies allow each process to be made iterative. We can compare results with actual data, identify reasons for discrepancies, and refine the initial parameters at each iteration. The cycle is repeated until the desired outcome is achieved.

Already during the mining operation, given the mining results, we can adjust various parameters and make changes to existing models and designs, as well as adjust the timing and direction of activities. In general, innovative technology ensures more informed decision-making at every stage of the mining business' development.

References

1. Al-Maitah, M., Semenova, O., Semenov, A., Kulakov, P., Kucheruk, V.: A hybrid approach to call admission control in 5G networks. Adv. Fuzzy Syst. **2018**, 1–7 (2018). https://doi.org/10.1155/2018/2535127

2. Semenova, O., Semenov, A., Voitsekhovska, O.: Neuro-fuzzy controller for handover operation in 5G heterogeneous networks. In: 2019 3rd International Conference on Advanced Information and Communications Technologies (AICT) (2019). https://doi.org/10.1109/AIACT.2019.8847898

3. Kryvinska, N., Bickel, L.: Scenario-based analysis of IT enterprises servitization as a part of digital transformation of modern economy. Appl. Sci. **10**, 1076 (2020). https://doi.org/10.3390/app10031076

4. Fauska, P., Kryvinska, N., Strauss, C.: The role of e-commerce in B2B markets of goods and services. IJSEM **5**, 41 (2013). https://doi.org/10.1504/IJSEM.2013.051872

5. Kryvinska, N., Kaczor, S., Strauss, C., Greguš, M.: Servitization—its raise through information and communication technologies. In: Lecture Notes in Business Information Processing, pp. 72–81 (2014). https://doi.org/10.1007/978-3-319-04810-9_6

6. Engelhardt-Nowitzki, C., Kryvinska, N., Strauss, C.: Strategic demands on information services in uncertain businesses: a layer-based framework from a value network perspective. In: 2011 International Conference on Emerging Intelligent Data and Web Technologies (2011). https://doi.org/10.1109/EIDWT.2011.28

7. Stoshikj, M., Kryvinska, N., Strauss, C.: Efficient managing of complex programs with project management services. Glob. J. Flex. Syst. Manag. **15**, 25–38 (2013). https://doi.org/10.1007/s40171-013-0051-8

8. Serheichuk, N.: Digital acceleration: What, why, and how (2021). https://www.n-ix.com/digital-acceleration/

9. Young, A., Rogers, P.: A review of digital transformation in mining. Min. Metall. Explor. **36**(4), 683–699 (2019)

10. Tavana, M., et al.: A review of digital transformation on supply chain process management using text mining. Processes **10**(5), 842 (2022)

11. Ganeriwalla, A., Harnathka, S, Voigt, N.: Racing Toward a Digital Future in Metals and Mining (2021). https://www.bcg.com/publications/2021/adopting-a-digital-strategy-in-the-metals-and-mining-industry

12. Gao, S., et al.: Digital transformation in asset-intensive businesses: Lessons learned from the metals and mining industry (2019)

13. Semenov, A., Baraban, S., Semenova, O., Voznyak, O., Vydmysh, A., Yaroshenko, L.: Statistical express control of the peak values of the differential-thermal analysis of solid materials. SSP **291**, 28–41 (2019). https://doi.org/10.4028/www.scientific.net/SSP.291.28

14. Semenov, A.O., Baraban, S.V., Osadchuk, O.V., Semenova, O.O., Koval, K.O., Savytskyi, A.Y.: Microelectronic pyroelectric measuring transducers. In: IFMBE Proceedings, pp. 393–397 (2019). https://doi.org/10.1007/978-3-030-31866-6_72

15. Bariatska, N.: Three-dimensional model as the basis for exploration planning. In: Geoinformatics: Theoretical and Applied Aspects, pp. 1–5 (2020). https://doi.org/10.3997/2214-4609.2020geo044

16. Bariatska, N., Safronova, N.: Some aspects of resource model validation. In: 18th International Conference on Geoinformatics, Theoretical and Applied Aspects, pp. 1–5 (2019). https://doi.org/10.3997/2214-4609.201902115

17. Breed, M.F., Heerden, D.: Post-pit optimization strategic alignment. J. South. Afr. Inst. Min. Metall. **116**(2), 109–114 (2016). http://www.scielo.org.za/pdf/jsaimm/v116n2/04.pdf

18. Hustrulid, W., Kuchta, M.: Open pit mine planning and design (1995)

19. Newman, A.M., et al.: A review of operations research in mine planning. Interfaces **40**(3), 222–245 (2010)

20. Esmaili. A.: Open Pit Mine Planning and Design–A guide to using Open Pit data in Micromine's Scheduler (2016). https://www.micromine.com/open-pit-mine-planning-design-guide-using-open-pit-data-micromines-scheduler-part-1/

21. Kaiser, P.K., et al.: Innovations in mine planning and design utilizing collaborative immersive virtual reality (CIRV). In: Proceedings of the 104th CIM Annual General Meeting (2002)

22. Haselgrove, S.: Micromine steps up mine design offering with Micromine 2020, Australian Mining (2019). https://www.australianmining.com.au/micromine-steps-up-mine-design-offering-with-micromine-2020/

23. Alonso, T.: Digital Transformation in Mining: Why it's a Necessity, https://www.cascade.app/blog/digital-transformation-mining. Last accessed 21 June 2023

24. Sganzerla, C.: Constantino, seixas, and alexander, conti: disruptive innovation in digital mining. Procedia Eng. **138**, 64–71 (2016)

25. Dagdelen, K.: Open pit optimization-strategies for improving economics of mining projects through mine planning. In: 17th International Mining Congress and Exhibition of Turkey, p. 117 (2001)

26. Dowd, P.A., Xu, C., Coward, S.: Strategic mine planning and design: some challenges and strategies for addressing them. Min. Technol. **125**(1), 22–34 (2016)

27. Wu, Q., Xu, H.: Three-dimensional geological modeling and its application in digital mine. Sci. China Earth Sci. **57**, 491–502 (2014)

Innovative Method of Forecasting the Manifestation of Dangerous Properties of Coal Seams

Yevhen Rudniev, **Vitalii Popovych**, **Rostyslav Brozhko**, and **Vadym Tarasov**

Abstract One of the main directions for reducing accidents is the reliability of the forecast for the manifestation of hazardous properties of coal seams. The results of the conducted studies revealed the predominant occurrence of accidents due to the manifestation of the hazardous properties of coal seams. Improvement of normative documents for the safe conduct of mining operations in terms of the formation of hazardous properties of coal seams during geological processes, their reliable forecast, and the development of preventive measures are very relevant for all coal-mining countries of the world. Based on the results of the research, the existing problems of establishing a critical combination of the parameters of the influencing factors of the three blocks for the occurrence of an emergency are reflected. The formation and manifestation of hazardous properties of coal seams according to the requirements of current regulatory documents are due only to the degree of metamorphic transformations of the organic (combustible) mass, and other stages of coal formation (sedimentation, peat, and brown coal) are not taken into account. As a result, the content of mineral impurities and some fluids are not considered as factors influencing the manifestation of hazardous properties of coal seams.

Keywords Digital technology · Digital business · Digitalization · Coal formation · Accidents · Coal seams · Metamorphism · Hazardous properties · Geological processes · Safety · Mining engineering · Regulatory framework · Software

Y. Rudniev · V. Popovych · R. Brozhko · V. Tarasov (✉)
Volodymyr Dahl East Ukrainian National University, 17 John Paul II Street, Kyiv 01042, Ukraine
e-mail: tarasov@snu.edu.ua

Y. Rudniev
e-mail: rudnev_es@snu.edu.ua

V. Popovych
e-mail: asp-263-22-431@snu.edu.ua

R. Brozhko
e-mail: brozhko@snu.edu.ua

1 Introduction

An analysis of accidents in coal mines [1, 2] over the past few decades shows that the root cause of their occurrence, in the general case, maybe the combined influence of factors from three blocks [3]. The factors of the first block were formed under natural conditions during the geological processes of the formation of coal seams. As a result of the implementation of these processes, individual coal seams have a tendency to exhibit certain dangerous properties during mining operations. The most dangerous properties leading to accidents with severe consequences include increased gas release into mine workings and flammable gases, sudden emissions of coal and gas, the occurrence of spontaneous combustion of coal, the release of dust and its tendency to form explosive mixtures with methane and some other features that complicate mining coal seams during mining operations. The factors of the second block include the mining and geological conditions of the occurrence of coal deposits. They characterize the thickness of the developed and adjacent seams, the strength properties of the host rocks, their water content, and some other indicators that ultimately determine the adoption of technological decisions for the effective exploitation of a coal deposit [4].

The technological parameters [5–7] of the operation of a coal enterprise are characterized by the factors of the third block. These include schemes for opening and preparing the deposit, schemes for ventilation, drainage, and transport of minerals, materials, and equipment, delivery to employee workplaces and some other operations that ensure the life of the mine. For an emergency to occur in underground conditions, some critical individual combination of the factor parameters of each block or a complex combination of the factor values of all three blocks is necessary. The difficulty in establishing the critical combination of factors leading to an accident lies in the absence of the possibility, as a rule, of reproducing it again. For this reason, during an individual investigation of an accident, according to the official version, in most cases its occurrence is associated with a human factor or a natural phenomenon. In all investigated cases of the causes of accidents, the critical combination between the factors of all three blocks, which differ from each other in the nature of their occurrence, remains essentially unestablished. The factors related to the second and third blocks are most reliably established. The mining and geological conditions of the deposit are quite accurately determined at the exploration stage and then refined during the mining process. The technological parameters of the third operating unit of the coal enterprise are laid down in its construction projects.

The most difficult to establish is the factors of the first block, which determine the conditions for the formation of dangerous properties of coal seams in the process of their geological transformations. Available information about accidents that have occurred over the past thirty years, mainly in coal-producing countries of the world [1, 2, 8–24], can be divided into two categories (Table 1). One of them includes accidents that occurred under the influence of the natural properties of mine seams (gas content, tendency to release coal and gas, occurrence of spontaneous combustion, explosiveness of coal dust, etc.). The second includes all other accidents which were not associated with the hazardous properties of coal seams. In all cases, accidents caused by the manifestation of dangerous properties of coal seams led to the most severe consequences with human casualties. In all major coal-mining countries, the number of victims from the manifestation of the dangerous properties of coal seams is several, or even several dozen times, higher than the number of victims in other accidents (Table 1).

Periodically recurring accidents due to the hazardous properties of coal seams with severe consequences in all coal-mining countries of the world indicate the existing variable problems in their timely prevention. One of the most important indicators of reducing accident rates is the reliability of the forecast for the manifestation of hazardous properties of coal seams and the advanced development of preventive measures in accordance with the requirements of regulatory documents for the safe conduct of mining operations.

The results of the analysis of the causes of accidents show that scientific research to improve these regulatory documents is very relevant for all coal-mining countries of the world.

In the above analysis (Table 1), in most cases, accidents with more than five casualties are taken into consideration. If we take into account accidents with fewer casualties, the relevance of research work to improve the regulatory framework for safe mining operations increases many times over.

In addition to human casualties, accidents caused by the dangerous properties of coal seams lead to significant material losses. Cases of endogenous fires were not fully included in the analysis, since, in most such cases, the number of victims was less than five. Along with this, material damage in some cases was catastrophically large, since the operation of both individual excavation sites and enterprises as a whole ceased.

Manifestation of hazardous properties of coal seams during mining operations most researchers associate them with processes of metamorphic transformations. Taking into account the possible diverse and versatile nature of the influence of these processes on the occurrence of dangerous properties, it is necessary to consider modern interpretations of the concept of metamorphism.

2 The Stages of Coal Formation and the Formation of Dangerous Properties of Coal Seams

The formation and manifestation of dangerous properties of coal seams, judging by the requirements of some current regulatory documents [25], are determined by the degree of metamorphic transformation of fossil coals.

According to the officially accepted definition [26], metamorphism means the transformation of brown coal successively into hard coal and anthracite as a result of changes in the chemical composition, structure, and physical properties of coal

Table 1 Information on accidents that occurred in the mines of the main coal-mining countries at the end of the 20th century and at the beginning of the 21st century

No.	Number of victims		Number of victims in one accident, people	Types of accidents	
	On factors of hazardous properties of coal mine seams	Other accidents		On factors of hazardous properties of coal seams	By other factors
China (1991 ÷ 2009) [8–11]					
1	1612	607	More than 100	Gas explosion, gas, and coal dust explosion, sudden release of coal and gas	Flooding
Ukraine (1992 ÷ 2021) [1, 12, 13]					
2	863	64	5 or more	Coal dust explosion, gas mixture explosion, methane and coal dust explosion, sudden release of coal and gas, methane release followed by combustion	Roof collapse, cage breakage, headframe collapse, water breakthrough, rope breakage
Türkiye (1983 ÷ 2014) [14–16]					
3	_[a]	_[a]	_[a]	Explosion of a gas mixture, fire	Collapse, flooding, transformer explosion
Kazakhstan (1993 ÷ 2021) [1, 2]					
4	149	6	5 or more	Explosion of a gas mixture, sudden release of coal and gas	Fall of the cage
USA (2006 ÷ 2010) [1, 17–20]					
5	34	_[a]	5 or more	Explosion of a gas mixture	Rock collapse
Mexico (2002 ÷ 2006) [1, 2]					
6	65	13	More than 5	Explosion of a gas mixture	Mine flooding

Note [a]No data

in the bowels of the Earth, mainly under the influence of increased pressure and temperature.

In a more general case, such transformations apply not only to coal seams but also to all metamorphic rocks. They consist of a change in the mineral composition or the size and structure of grain aggregates, without a significant change in the chemical composition (except for the H_2O and CO_2 content) under the influence of fluids, temperature, and pressure [27].

Metamorphic rocks were formed due to the transformation of igneous or sedimentary rocks under the influence of high pressure, temperature, and hot gas–water solutions. As physical bodies, they are characterized by density, elasticity, strength, thermal, electrical, magnetic, radiation, and other properties, which primarily depend on their mineral composition and macrostructure (structural-textural feature) [28, 29].

Metamorphism, according to [30], is the process of changing the structure, mineralogical, and sometimes chemical composition of rocks in the earth's crust under the influence of temperature, pressure, and chemical activity of deep solutions.

A similar definition of the process of metamorphism is given in the encyclopedic dictionary [28–32]; it is considered a significant change in the texture, structure, and mineralogical composition of rocks under the influence of temperature, pressure, and chemical activity of deep solutions (fluids). According to another encyclopedic dictionary, metamorphism refers to a set of deep natural processes that cause changes in the chemical and mineralogical composition and structure of rocks in the bowels of the Earth. Metamorphism is caused by changes in pressure and temperature and the activity of deep solutions. Metamorphism is associated with the formation of metamorphic rocks.

Information on changes in fossil coals and rocks under the influence of metamorphic processes according to [28–32] is given in Table 2.

According to generally accepted ideas [25–27, 33], the degree of metamorphic transformations should be judged by changes in the elemental composition of coals and their physical and mechanical properties. In contrast to this definition of metamorphism in regulatory documents, completely different criteria have been adopted as indicators of metamorphic transformations of coals to predict the hazardous properties of coal seams. They do not directly characterize the change in the elemental composition and properties of coals during the process of metamorphic transformations of coal seams.

Metamorphic transformations of coal seams, according to developed by Academician Ammosov I. I., the general scheme of coal formation is a logical continuation of the peat and lignite stages. At these stages of coal formation, changes in the elemental composition and properties of coals also occur. Such changes are not considered in a logical sequence in the regulatory framework for the safe conduct of mining operations, which affects the reliability of the forecast of the hazardous properties of coal seams during mining operations.

According to the general scheme of coal formation, metamorphic processes relate only to the coal and anthracite stages, which must be considered as a continuation of the peat and lignite stages [34–36]. The possible reliability of the prediction of

Table 2 Information on changes in the composition and properties of fossil coals and rocks under the influence of metamorphic processes

Objects that have undergone metamorphic transformations	Factors determining metamorphic processes	Changes due to metamorphic processes		
		In chemical and mineralogical compositions	In the structure	In physical and mechanical properties
Brown coal, hard coal, anthracite	Increased pressure and temperature	Change in chemical composition	Change of structure	Change In physical properties
Sedimentary and igneous rocks	Fluids, temperature, pressure	Change in mineralogical composition without significant change in chemical composition (with the exception of H_2O and CO_2)	Recrystallization	Varies widely
Sedimentary and igneous rocks	High pressure, temperature, hot gas–water solutions	Change in mineral composition	Change in macrostructure	Changes in density, elasticity, strength, thermal, electrical, magnetic, radiation properties
Rocks	Temperature, pressure, chemically active deep solutions	Changes in mineralogical and sometimes chemical compositions	Change of structure	–
Mineralogical composition of rocks	Temperature, pressure, chemically active deep solutions (fluids)	Significant changes in mineralogical composition	Significant change in texture and structure	–
Composition and structure of rocks	Pressure and temperature changes, activity of deep solutions	Changes in mineralogical and chemical composition	Change of structure	–

the hazardous properties of mine seams can be judged by the degree of scientific validity of the coal conversion indicators used for these purposes in regulatory documents and their compliance with the definition of the generally accepted concept of metamorphism.

3   Modern Regulatory Framework for Safe Mining Operations and Applied Indicators of Metamorphic Transformations of Coal Seams

In modern regulatory documents [27, 34–36] for assessing the degree of metamorphic transformations of coal seams and their hazardous properties, the yield of volatile substances during thermal decomposition of fuel without air access is accepted as the main one. According to methods for determining the mass (V^{daf}) and volumetric (V_V^{daf}) yield of volatile substances during thermal destruction [35, 37], these indicators, taking into account the methods of their determination, cannot directly and simultaneously characterize changes in the elemental composition and properties of coals during the process of metamorphic transformations of coal seams [38, 39]. In addition, coal metamorphism and thermal decomposition are different stages of their transformation. The quantitative and qualitative composition of the resulting volatile substances during thermal decomposition is not directly related to previously occurring metamorphic processes under natural conditions, under which some of the gaseous products and moisture had already been removed. Thermal destruction is the result of a new (next) artificial stage of transformation of the original organic matter raised to the earth's surface.

Indicators of thermal decomposition of coals without air access are used in current regulatory documents to predict gas release into mine workings, gas-dynamic phenomena, prevention and extinguishing of endogenous fires [40], as well as to assess the dust-forming ability of mine seams.

The yield of volatile substances, taking into account the determination methods, does not correspond to the classical (generally accepted) characteristic of metamorphism as a change in the composition and properties of coals in the process of geological transformations of coal seams. One indicator, even the most universal, cannot reliably and comprehensively characterize the change in the ratio of all components of organic mass and its properties under the influence of metamorphic processes.

Methods for predicting the hazardous properties of coal seams have not changed over the past few decades. In addition to mass V^{daf} and V_V^{daf} in different combinations, two more auxiliary indicators of metamorphism are used—the thickness of the coal seam (y) and the logarithm of the electrical resistivity of anthracite ($\lg \rho$). Their scientific substantiation for predicting the outburst hazard of mine seams has not yet been carried out.

All indicators of metamorphic transformations of coal seams, used in the modern regulatory framework and some other documents, are borrowed from industrial classifications of consumer qualities of coal, including those officially in force. It is based on the brand of coal, which is reflected in the definition of the concept of metamorphism. It consists of the sequential transformation of brown coal into hard coal and anthracite. In turn, a coal grade is a symbol for varieties of coal that are similar in genetic characteristics and basic energy and technological characteristics. In some cases, in regulatory documents on safe mining operations, a conventional grade of coal is used to characterize the hazardous properties of coal seams, but it is, to a greater extent, developed and reflects the consumer qualities of coal.

Proving the possibility of using indicators of the consumer qualities of coal to predict the hazardous properties of mine seams requires a more detailed study of the principles of constructing a modern industrial classification [27].

4 Modern Classification Indicators for Establishing the Consumer Qualities of Coals

To establish the hazardous properties of coal seams during mining operations, the regulatory framework of Ukraine uses a limited number of classification indicators of fossil coals. Classification indicators, according to the logic of their use, should characterize changes in the composition and properties of coals in the process of their metamorphic transformations. For these purposes, the industrial classification [25] uses 10 indicators. According to them, all coals are ranked according to the degree of metamorphism. They are systematized into three types, 49 classes, 8 categories, 31 types (of which 6 are for brown coal, 21 for hard coal, 4 for anthracite), and 33 subtypes (brown—4, stone—23, anthracite—6). Based on this division, 17 grades, 27 groups, 44 subgroups, and 81 types and subtypes have been established for all fossil coals. In total, more than 20 classification indicators are known, with the help of which it is possible to comprehensively characterize changes in the composition and properties of coals in the process of their geological transformations.

It is now a generally accepted fact that some accidents in mines are related to the properties of fossil coals that appeared as a result of geological processes and metamorphic transformations of the original organic matter. This is confirmed by the use of the modern regulatory framework of Ukraine to predict gas release, gas-dynamic phenomena, endogenous fires, and the dust-forming ability of coal seams, indicators of the degree of metamorphism of coals. One of the main signs of increased metamorphic transformation of coals is an increase in the elemental content of carbon (C_o) and a decrease in other components in the original organic matter. This, one of the basic principles of metamorphic transformations, is not observed when predicting the dangerous properties of coal seams. Changes in the elemental composition of the original organic matter, as the main indicator of the degree of metamorphism of coals, are not considered at all in regulatory documents. Methods for establishing the degree of metamorphism of coals in the documents under consideration were developed based on the experience of using industrial classifications. They, first of all, provided for the determination of the technological properties of coals associated with their tendency to coking and calorific value.

The degree of metamorphism was initially characterized by the yield of coke to organic matter (K) and moisture content (W). It is these indicators, for the most part, that directly characterize the consumer properties of coal. In the process of improving industrial classifications, more than 30 indicators were additionally studied, which in their totality, more differentiated, compared to C_o and W, determine the techno-logical properties of coals. For this reason, the significance of the original Co and W indicators, as classification indicators for establishing technological properties, was somewhat lost. Modern industrial classification provides for the use of 10 indicators. Carbon content (C_o), as one of the main indicators of metamorphic transformations, is not considered. The maximum moisture capacity (W_{max}^{af}) for the ash-free state is used to a limited extent only for dividing brown coals into types. One of the main indicators of the stages of metamorphism in industrial classifications and regulatory documents began to be the volumetric yield of volatile substances during the thermal decom-position of coals into a dry, ash-free state V_V^{daf} for anthracites. At its core, thermal decomposition is an artificial continuation of the stages of coal transformation at temperatures (900 °C) significantly higher than the transformation temperature of hard coals (300–500 °C) and anthracite (500–650 °C). Thermal decomposition prod-ucts do not directly characterize changes in the composition and properties of coals in past geological periods of time. The feasibility of using V^{daf} and V_V^{daf} indicators to establish the consumer properties of coal is confirmed by the successful experi-ence of using industrial classification. This possibility of experimental verification is absent when predicting the hazardous properties of coal seams.

Periodically recurring accidents in coal mines indicate the relevance of improving the regulatory framework in terms of predicting the manifestation of hazardous prop-erties of coal seams. Its main disadvantage is the use of indicators that do not directly characterize the change in the elemental composition and properties of coals in the process of their metamorphic transformations. The use of V^{daf} and V_V^{daf} indicators, without proper scientific justification, is copied from industrial classifications. Indus-trial classification provides for the systematization of coals according to indicators characterizing their suitability for industrial use [34]. According to them, coal grades were established. They conventionally designate varieties of coal that are similar in genetic characteristics and basic energy and technological characteristics. Industrial classifications do not provide for the prediction of the manifestation of hazardous properties of coal seams during mining operations.

Establishing the propensity of coal seams to manifest their dangerous properties based on genetic characteristics should take into account the definition of the concept of metamorphism according to [34]. It consists of the transformation of brown coal successively into hard coal and anthracite as a result of changes in the chemical composition, structure, and physical properties of coal in the depths, mainly under the influence of elevated temperature and pressure. None of the indicators used in regulatory documents fully corresponds to this definition. A direct change in the composition of the original organic matter is not considered when metamorphism increases as an increase in carbon content and a decrease in other components.

The research methodology was developed on the basis of modern knowledge in the field of geological science, chemical analysis of coals, experience in mining oper-ations and the results of statistical processing, and the available extensive material obtained in different coal basins by many researchers. When conducting research, the basic principles were observed, according to which the metamorphism of fossil coals manifests itself in a change in the elemental composition of the original organic matter and its properties. Classification indicators should directly characterize the change in carbon content in an individual ratio with each component of the organic mass. These basic principles of the manifestation of metamorphism are not taken into account by the modern regulatory framework and the current industrial classification.

The stages of metamorphic transformations of coal seams according to known industrial classifications cannot be determined for several reasons:

- the main indicators of the degree of metamorphism of coals are based on determining the products of thermal decomposition of coals. This process is an artificial continuation of the stages of transformation of organic matter that occurred under natural conditions in past geological periods of time;
- most classification indicators are determined for dry ash-free mass, which does not correspond to the conditions for the presence of coal in mine seams. In all cases, there is the presence of moisture and mineral impurities;
- conventional grades of coal are determined for their industrial use. They are artificially established by a certain combination of several indicators that are not directly related to changes in the elemental composition of organic matter and the manifestation of hazardous properties during mining operations;
- formation moisture is not considered part of organic matter, which excludes its influence on the physical state of coal seams.

The technique assumed that the main components of organic matter are carbon (C_o), hydrogen (H_o), nitrogen (N_o), sulfur (S_o), oxygen (O_o), and moisture (W). Their total content at all stages of strata metamorphism is 97,5% or more. The average compositions of coals from the Donetsk basin at different stages of seam metamorphism [30] based on the results of processing more than a thousand samples were taken into account for the analysis. Their results completely coincide with the average values of the elemental composition of organic matter obtained in other coal basins when processing about three thousand more pairs of data [28–33]. The obtained results of statistical processing of such a quantity of experimental data do not raise doubts about their reliability.

Changes in the content of components of organic matter (C_o, O_o, H_o, S_o, N_o) and moisture (W) at different stages of metamorphism of layers are given in Table. 3. The initial criterion for determining the stages of metamorphism was the release of coke into organic matter (K). According to K values from 52 to 100%, the transformation of layers from young hard coals to anthracites was evenly divided into 10 stages. The average values of coke yield ($\overline{K}$) at each stage are set to the average values of organic matter components ($\overline{C}_o$, $\overline{H}_o$, $\overline{N}_o$, $\overline{S}_o$, $\overline{O}_o$) and moisture content ($\overline{W}$) in the initial samples. The final member of the coal metamorphism series is graphite ($C_o \approx 100\%$). The main component of organic matter at all stages of transformation is carbon. Its minimum value for hard coals is about 70%. The average values of carbon content at the 10 stages of formation metamorphism under consideration varied in the range of $80,19 \div 93,65\%$ (Table 3).

The average amount of removed fluids ($\overline{V}_K$) formed at different stages of formation transformation is determined from the relationship:

Table 3 Information on the average composition of coals of the Donetsk basin at different stages of seam metamorphism

Indicators, %	Stages of metamorphic transformations of layers and composition of the original organic matter									
	I	II	III	IV	V	VI	VII	VIII	IX	X
K	52 ÷ 55	55 ÷ 60	60 ÷ 65	65 ÷ 70	70 ÷ 75	75 ÷ 80	80 ÷ 85	85 ÷ 90	90 ÷ 95	95 ÷ 100
$\overline{K}$	53,5	57,5	62,5	67,5	72,5	77,5	82,5	87,5	92,5	97,5
V_K	48 ÷ 45	45 ÷ 40	40 ÷ 35	35 ÷ 30	30 ÷ 25	25 ÷ 20	20 ÷ 15	15 ÷ 10	10 ÷ 5	5 ÷ 0
$\overline{V}_K$	46,5	42,5	37,5	32,5	27,5	22,5	17,5	12,5	7,5	2,5
$\overline{W}$	7,34	6,44	2,59	1,59	1,15	0,99	0,88	0,78	1,29	3,32
$\overline{C}_o$	80,19	81,57	84,29	86,43	88,33	89,53	90,43	91,46	92,67	93,65
$\overline{H}_o$	5,34	5,31	5,31	5,21	5,10	4,81	4,60	4,30	3,75	1,93
$\overline{N}_o$	1,43	1,44	1,44	1,46	1,52	1,51	1,51	1,38	1,32	1,05
$\overline{S}_{o,}$	2,28	1,83	1,42	1,24	1,10	1,04	1,06	1,03	1,00	0,74
$\overline{O}_o$	10,76	9,85	7,54	5,66	3,95	3,11	2,40	1,83	1,26	0,63

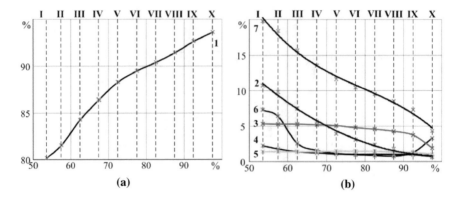

Fig. 1 Changes in the average content of organic matter components from the yield of coke at different stages of metamorphic transformations of coal seams 1, 2, 3, 4, 5, 6—curves of average carbon content in organic matter $(\overline{C}_o)$, oxygen $(\overline{O}_o)$, hydrogen $(\overline{H}_o)$, sulfur $(\overline{S}_o)$, nitrogen $(\overline{N}_o)$ and moisture (W); 7—the curve of changes in the sum of components of organic mass $(\overline{H}_o, \overline{N}_o, \overline{S}_o, \text{and } \overline{O}_o)$. I–X—stages of metamorphic transformations of seams based on coke output

$$\overline{V}_K = 100 - \overline{K} \tag{1}$$

With this definition $\overline{V}_K$, its values will slightly exceed the average yield of volatile substances during the thermal decomposition of coals $(\overline{V}^{daf})$. In composition $\overline{V}^{daf}$, only gaseous decomposition products are taken into account, and the presence of pyrogenetic moisture and coal tar cannot be excluded in the removed fluids. This once again confirms the discrepancy between the $\overline{V}^{daf}$ index and the metamorphic transformations of the strata. As metamorphic processes intensified (growth $\overline{K}$) a unilateral increase in carbon content occurred (Fig. 1a). Simultaneously with these processes, a decrease in the amount of other components $(\sum \overline{H}_o, \overline{N}_o, \overline{S}_o, \overline{O}_o)$ of organic matter was also observed. (Fig. 1 b curve 7).

The individual change in the content of organic matter components was not so clear. The most intense decrease in oxygen content was observed (stages I–IV), then a more gradual decrease was observed (stages V–X) to almost zero (Fig. 1b curve 2). The hydrogen content remained practically unchanged at stages I–VI and then decreased according to a nonlinear dependence (Fig. 1b curve 3). Nitrogen and sulfur contents underwent minor changes (Fig. 1b curves 5 and 4). In addition to the considered components of organic matter $(\overline{C}_o, \overline{H}_o, \overline{N}_o, \overline{S}_o, \overline{O}_o)$ its composition includes formation moisture $(\overline{W})$. Determination methods $\overline{W}$ do not allow it to be considered together with other components $(\overline{C}_o, \overline{H}_o, \overline{N}_o, \overline{S}_o, \overline{O}_o)$ in a 100% organic matter composition. Percentage content $\overline{W}$ is related to the original coal sample.

The reduction in moisture content occurred intensively at the initial stages (I–III) of the transformation of coal seams (Fig. 1b, curve 6). Then its reduction was insignificant (stages IV–VII, Table 3), and an increase was observed at stages VIII–X. Such an ambiguous change in the indicator $\overline{W}$ certainly has an impact on the processes of metamorphism and the manifestation of the dangerous properties of coal seams. This is evidenced by a change in the position $\overline{W}$ in the ranking of indicators for coke yield. As metamorphism processes intensify at stages I–VI, the role $\overline{W}$ decreases compared to other components. In the ranking series, the $\overline{W}$ value moves from third to sixth position.

Then, at the final stages IX–X, the importance of $\overline{W}$ in the ranking series increases again.

The unconditional leading position in the ranking at all stages of formation transformation is occupied by carbon content. The presence of other components ($\overline{H}_o$, $\overline{N}_o$, $\overline{S}_o$, $\overline{O}_o$) and moisture ($\overline{W}$) in organic matter directly depends on the $\overline{C}_o$ content. This gives grounds for using the change in the ratio between the carbon content and other components to characterize the stages of metamorphism of mine seams instead of the release of coke. The ratio $\overline{C}_o$ to the sum of hydrogen ($\overline{H}_o$) and oxygen $\overline{O}_o$ content was defined as an indicator of carbonation. Using only two components to determine the carbonization index (C_n) does not provide a complete picture of the influence of nitrogen ($\overline{N}_o$), sulfur ($\overline{S}_o$) and moisture ($\overline{W}$) content on the increase in carbon (c) with intensification of metamorphic processes. It is more appropriate when determining the C_n index to consider all the main components of organic matter ($\overline{H}_o$, $\overline{N}_o$, $\overline{S}_o$, $\overline{O}_o$) and moisture ($\overline{W}$).

The carbonization index C_n^{Σ} corresponds to the sum of the components of organic matter, according to the data in Table 4, was calculated using the equation:

$$C_n^{\Sigma} = \frac{\overline{C}o}{\overline{H}o + \overline{N}o + \overline{S}o + \overline{O}o} \tag{2}$$

Similarly, approximately, since ($\overline{W}$) is not part of the 100% organic matter component content, the carbonation index for moisture content was calculated:

$$C_n^{W} = \frac{\overline{C}o}{\overline{W}} \tag{3}$$

The results of determining indicators C_n^{Σ} and C_n^{W} for different stages of metamorphism of layers are given in Table 4.

The total carbonation index (C_n) is related to C_n^{Σ} and C_n^{W} by the expression:

$$\frac{1}{C_n} = \frac{1}{C_n^{\Sigma}} + \frac{1}{C_n^{W}} \tag{4}$$

Taking the value C_n^{-1} as one for each stage of metamorphism of layers, we determined the participation shares of the sum of components of organic matter $\left(\Delta C_n^{\Sigma}\right)$ and moisture $\left(\Delta C_n^{W}\right)$, respectively:

Table 4 Results of determining carbonization indicators for components of organic matter at different stages of metamorphic transformations of coal seams

Indicators	Carbon content in organic matter, $\overline{C}o\%$									
	80,19	81,57	84,29	86,43	88,33	89,53	90,43	91,46	92,67	93,65
C_n^Σ	4,05	4,43	5,37	6,37	7,57	8,55	9,45	10,71	12,64	21,53
C_n^W	10,93	12,67	32,5	54,36	76,81	90,43	102,76	117,26	71,84	28,21
$(C_n^\Sigma)^{-1}$	0,25	0,23	0,19	0,16	0,13	0,12	0,11	0,09	0,08	0,05
$(C_n^W)^{-1}$	0,09	0,08	0,03	0,02	0,013	0,011	0,01	0,009	0,014	0,035
$(C_n^\Sigma)^{-1} + (C_n^W)^{-1}$	0,339	0,31	0,22	0,18	0,143	0,131	0,120	0,099	0,094	0,085
ΔC_n^Σ	0,730	0,74	0,86	0,89	0,910	0,92	0,92	0,91	0,85	0,59
ΔC_n^W	0,270	0,26	0,14	0,11	0,090	0,08	0,08	0,09	0,15	0,41
$\overline{H}_o$	5,34	5,31	5,31	5,21	5,10	4,81	4,60	4,30	3,75	1,93
$\overline{H}_o \Delta C_n^\Sigma$	3,90	3,93	4,56	4,64	4,64	4,81	4,23	3,91	3,19	1,14
$\overline{N}_o$	1,43	1,44	1,44	1,46	1,52	1,51	1,51	1,38	1,32	1,05
$\overline{N}_o \Delta C_n^\Sigma$	1,04	1,07	1,24	1,30	1,38	1,39	1,39	1,26	1,12	0,62
$\overline{S}_o$	2,28	1,83	1,42	1,24	1,10	1,04	1,06	1,03	1,00	0,74
$\overline{S}_o \Delta C_n^\Sigma$	1,66	1,35	1,22	1,10	1,00	0,96	0,98	0,94	0,85	0,44
$\overline{O}_o$	10,76	9,85	7,54	5,66	3,95	3,11	2,40	1,83	1,26	0,63
$\overline{O}_o \Delta C_n^\Sigma$	7,96	7,30	6,47	5,04	3,60	2,86	2,21	1,67	1,07	0,38
$\overline{W}$	7,34	6,44	2,59	1,59	1,15	0,99	0,88	0,78	1,29	3,32
$\overline{W} \Delta C_n^W$	1,98	1,67	0,67	0,41	0,30	0,26	0,23	0,20	0,34	0,86
C_n^H	20,30	20,77	18,44	18,63	19,04	20,21	21,38	23,39	29,05	82,15

(continued)

Table 4 (continued)

Indicators	Carbon content in organic matter, $\overline{C}o\%$									
	80,19	81,57	84,29	86,43	88,33	89,53	90,43	91,46	92,67	93,65
C_n^N	75,65	76,23	67,98	66,48	64,01	64,41	65,06	72,59	82,74	151,05
C_n^S	47,45	60,42	69,1	78,57	88,33	93,26	92,28	97,30	109,02	212,84
C_n^O	10,07	11,2	13,01	17,15	24,6	31,30	40,92	54,77	86,61	246,44
C_n^W	41,98	48,84	125,81	210,8	294,43	344,35	393,17	457,3	272,56	108,9
$(C_n^H)^{-1}$	0,049	0,048	0,054	0,054	0,053	0,049	0,047	0,043	0,034	0,012
$(C_n^N)^{-1}$	0,013	0,013	0,015	0,015	0,016	0,016	0,015	0,014	0,012	0,007
$(C_n^S)^{-1}$	0,021	0,017	0,014	0,013	0,011	0,011	0,011	0,010	0,009	0,005
$(C_n^O)^{-1}$	0,098	0,089	0,077	0,058	0,041	0,032	0,024	0,018	0,012	0,004
$(C_n^W)^{-1}$	0,024	0,020	0,008	0,005	0,003	0,003	0,003	0,002	0,004	0,009
$\sum(C_n^i)^{-1}$	0,206	0,187	0,168	0,145	0,124	0,111	0,10	0,087	0,071	0,037
ΔC_n^H	0,237	0,257	0,32	0,37	0,43	0,44	0,47	0,49	0,48	0,32
ΔC_n^N	0,063	0,070	0,09	0,10	0,13	0,14	0,15	0,160	0,17	0,19
ΔC_n^S	0,101	0,09	0,08	0,09	0,09	0,10	0,110	0,11	0,13	0,14
ΔC_n^O	0,478	0,47	0,46	0,40	0,33	0,29	0,24	0,21	0,17	0,11
ΔC_n^W	0,121	0,11	0,05	0,03	0,02	0,03	0,03	0,02	0,05	0,24

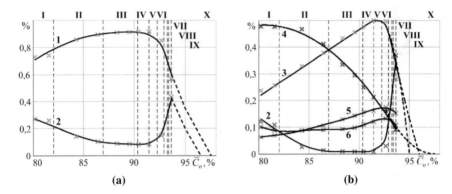

Fig. 2 Dependence of changes in the share of participation of organic matter components in carbonization of seams on the average carbon content ($\overline{C}_o$) according to [28]. 1, 2—curves of changes in the share of participation in carbonization, respectively, of the sum of components (C_n^{Σ}) of organic matter ($\overline{H}_o$, $\overline{N}_o$, $\overline{S}_o$, $\overline{O}_o$) and moisture (ΔC_n^W); 3, 4, 5, 6—curves of changes in the share of participation in carbonization, respectively, of hydrogen (ΔC_n^H), oxygen (ΔC_n^O), nitrogen (ΔC_n^N) and sulfur (ΔC_n^S); I, II, III, IV, V, VI, VII, VIII, IX, X—stages of metamorphism of seams

$$\Delta C_n^{\Sigma} = \frac{1}{C_n^{\Sigma}} \cdot \frac{1}{C_n} \tag{5}$$

The calculated values for each stage of metamorphism are given in Table 4. The share of participation of the sum of the components of organic matter $\left(\Delta C_n^{\Sigma}\right)$ and moisture $\left(\Delta C_n^W\right)$ in carbonization changes significantly as the carbon content increases (Fig. 2a). For this reason, we adjusted the initial values of $\overline{H}o$, $\overline{N}o$, $\overline{S}o$, $\overline{O}o$ and $\overline{W}$ for each stage, multiplying them by ΔC_n^{Σ} and ΔC_n^W, respectively (Table 4).

Adjusting the values of $\overline{H}_o$, $\overline{N}_o$, $\overline{S}_o$, $\overline{O}_o$ and $\overline{W}$ for the shares of their participation in carbonization (ΔC_n^{Σ} and ΔC_n^W) had virtually no effect on changes in the ranking series of components compared to the ranking series by coke yield. In the early stages of metamorphism (I, II), the main components influencing carbonization were the oxygen content ($\overline{O}_o$) and moisture ($\overline{W}$). At the following stages, oxygen and hydrogen (III, IV), hydrogen and oxygen (V–VIII), hydrogen and moisture (IX), and moisture and hydrogen (X) occupied the leading position successively. The compared series of ranking the components of organic matter by coke yield and carbonization index practically do not differ from each other in the location of the remaining members of these series.

For example, at the final stage (X), they are arranged in the same order—$\overline{W}$, $\overline{H}_o$, $\overline{N}_o$, $\overline{S}_o$, $\overline{O}_o$.

This indicates that when establishing the stages of metamorphism of coal seams, taking into account their classical definition, it is possible, instead of the coking index ($\overline{K}$) to use the value of carbon content ($\overline{C}_o$,) in organic matter. An additional argument for accepting carbon content as the main criterion for dividing at the stage

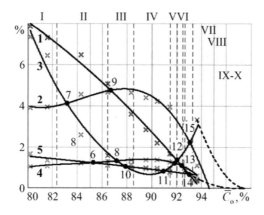

Fig. 3 Dependence of average values of component content in organic substances on the environment [28] 1, 2, 3, 4, 5—curves of the average content of oxygen ($\overline{O}_o$), hydrogen ($\overline{H}_o$), moisture ($\overline{W}$), nitrogen ($\overline{N}_o$), and sulfur ($\overline{S}_o$) in organic matter, respectively; 6, 7, 8, 9, 10, 11, 12, 13, 14—characteristic points of intersection of curves that define the boundaries of the stages of metamorphism of coal seams; I, II, III, IV, V, VI, VII, VIII, IX, X—stages of metamorphism of coal seams, determined by characteristic points of intersection of curves

of metamorphism of coal seams is provided by individual graphs of mutual changes in the components of organic matter (Fig. 3).

The intersection points of curves characterizing individual changes in the components of organic matter indicate not only a change in the chemical composition but also a change in physical and mechanical properties. With this assumption, the initial stage of metamorphism of layers (I) corresponds to a carbon content of less than 83%. The upper limit of 83% is determined by points 6 and 7 of the intersection of curves 4, 5 and 2, 3, respectively.

Curves 4 and 5 characterize, respectively, the change in nitrogen ($\overline{N}_o$) and sulfur ($\overline{S}_o$) content, and curves 2 and 3—the content of hydrogen ($\overline{H}_o$) and oxygen ($\overline{O}_o$) and moisture ($\overline{W}$). The upper limit ($\overline{C}_o = 87\%$) of stage II was determined in a similar way at points 8 and 9. They correspond to the intersection of curves 3, 4, and 1, 2, which determine, respectively, the content of moisture ($\overline{W}$) and nitrogen ($\overline{N}_o$), and sulfur ($\overline{S}_o$), and curves 2 and 3—the content of oxygen ($\overline{O}_o$) and hydrogen ($\overline{H}_o$). The upper limit of stage III ($\overline{C}_o = 89\%$) is determined by point 10 of the intersection of curves 3 and 5. The intersection points 11 and 12 correspond to the end of stage IV (curves 3, 5 and 1, 4). Point 13 corresponds to the intersection of curves 3, and 4 (stage V). The intersection at point 14 of curves 1 and 5 determines stage VI. Stages VII and VIII differ by tenths of the percentage carbon content, and the boundaries of stages IX-X cannot be established based on available data (Table 5), since, in the case under consideration, the carbon content is less than 93.6%.

Stages of metamorphism of layers at $\overline{C}_o > 93.6\%$ can differ significantly in properties due to the unpredictability of the relationship between the components of organic matter [40, 41]. At these stages, even a minimal difference between the

Table 5 Change in elemental carbon content in organic matter at different stages of coal seam transformation

The criterion for determining stages	Stages of coal seams metamorphism and carbon content, %									
	I	II	III	IV	V	VI	VII	VII	IX	X
Coke output, $\overline{K}$, % [6]	80,19	81,57	84,29	86,43	88,33	89,53	90,43	91,46	92,67	93,65
Carbonization by individual content of organic matter and moisture components	80,0–83,0	83,0–87,0	87,0–89,0	89,0–92,0	92,0–92,7	92,7–93,0	93,0–93,2	93,2–93,6	More than 93,6	
Individual share of participation of organic matter components in carbonization	80,0–83,4	83,4–86,0	86,0–90,4	90,4–91,8	91,8–92,5	92,5–93,2	93,2–93,4	93,4–93,6	More than 93,6	

components can cause the emergence of new properties of coal seams. As the influence of metamorphism processes increases, the boundaries of the stages determined by the percentage of carbon become narrower. Using the method described above (Eqs. 2–5), we calculated the adjusted shares of individual participation in carbonation of hydrogen (ΔC_n^H), oxygen (ΔC_n^O), nitrogen (ΔC_n^N) and sulfur (ΔC_n^S) and moisture (ΔC_n^W). Their values (Table 4) changed significantly in the process of metamorphic transformations of the coal seams. There are also differences in the ranking series by the share of components in carbonization, compared to the series ranked by the elemental composition of organic matter. The presence of moisture plays a significant role in the early stages (I, II) of coal seam transformation [42, 43]. At the following stages (III–IX), the share of its influence on carbonation is minimal compared to other components. At the last stage (X), the role of moisture increases again, and it returns to one of the leading places in the ranking series. After reaching a carbon content of more than 93.6%, a sharp decrease in the share of participation in carbonization (Fig. 2b) of all other components ($\overline{H}_o$, $\overline{N}_o$, $\overline{S}_o$, $\overline{O}_o$. and $\overline{W}$) is predicted. Their total share in organic matter, in this case, will not exceed 6.4%, which undoubtedly affects the properties of coal seams during mining operations.

The established stages of formation transformation according to the individual share of the components in carbonization practically do not differ from the boundaries of the stages established by the elemental composition of organic matter (Fig. 2b, Table 5). It should also be noted that the average carbon content at the stages of metamorphism of coal seams, determined by coke yield, in most cases does not coincide with the ranges of variation ($\overline{C}_o$) established by the individual content of components or the shares of their participation in carbonization (Table 5).

The instability of the boundaries of the stages of metamorphism of coal seams makes it unacceptable to use the coke yield indicator as the main criterion for assessing the transformation of coal, and, even more so, the manifestation of dangerous properties of the coal seams.

5   Economic Assessment and Optimization

According to KD 12.01.402–2000, the propensity to self-ignition of coal seam under development must be determined by research organizations as necessary, but not less often than once every 5 years. Samples for determining the fire hazard of the coal seam are taken according to the methodology regulated [44, 45]. For the method of determining the susceptibility of coal seam to endogenous fire hazard based on the indicators of carbon, hydrogen, oxygen, sulfur, moisture, and ash content of samples, the software has been developed for personal computers, which, by the method of information/data collection and analysis, will allow workers of mines and associations to make an express assessment of the fire hazard of the developed coal seams. The developed program will warn about the danger of carrying out work on this site, which, in combination with other enterprise management systems, will allow to exclude the chance of harming the health of employees, as well as to introduce preventive measures to reduce damage to the mine's infrastructure and equipment. The software is implemented in a high-level programming language Python (Fig. 4). The principle of the method for determining the fire hazard of mining plastic is the recalculation of the hydrogen content with corrections for the content of mineral impurities and general (coal seam) moisture in accordance with generally accepted methods [46–48]. Assigning the mine plastic to one of the zones (see items 7 and 8), after which the program generates a message (result) regarding the tendency of the mine plastic to self-ignite (according to item 9). To test the developed method, we will calculate the probability of occurrence of endogenous fires for mines, which are given in КD 12.01.401–96 [45]. All of the reviewed mining plastics fell into zone I with the maximum H_π^r hydrogen content value (zone I—mining plastics are most prone to self-ignition) and zone III with an average content value (zone III—mining plastics are prone to the manifestation of an endogenous fire hazard). Not a single coal seam specified in КD 12.01.401–96 reached the II zone with the minimum value of hydrogen content (II zone—mine plastic is not prone to the occurrence of self-ignition fires), which are installed in the working condition of the samples. This confirms the correctness (adequacy, correctness) of the results obtained with the help of the developed method of collecting and analyzing information/data (see point 9), which will make it possible to improve the management of the enterprise.

Endogenous fires are one of the most difficult types of accidents in coal mining enterprises, the elimination of which is long, expensive, and labor-intensive. In terms of number and caused damage, endogenous fires prevail over other types of accidents.

The amounts of material damages, which reach 40% of all costs, are due to the results of accidents at coal enterprises in Ukraine, which directly affect the financial mechanisms of the economic development of enterprises. The analysis of statistical data shows that the endogenous fire hazard is constantly increasing, which can be explained by the complication of mining and geological conditions and the increase in the temperature of rocks [49]. On average, endogenous fires make up about 20% of the total number of fires in the mines of the Donetsk basin. Costs for measures to improve working conditions in coal mines of Ukraine are usually several times higher than at enterprises in other industries, and also make the field less attractive for investments. However, in conditions of shortage of financial resources, they are often insufficient, and measures often turn out to be ineffective, and sometimes economically impractical. One of the reasons for this is the imperfection of the known methods of assessing the effectiveness of costs aimed at improving conditions and increasing the safety of miners.

From the point of view of the economy, costs for improving working conditions are investments [50], since they are carried out with the aim of obtaining long-term benefits due to increased labor productivity, improved product quality, preservation of labor potential, etc. Investments have the most important role in restoring and increasing production resources, and, accordingly, in ensuring stable rates of economic growth.

Strategies for the development of investment activities in the economy of the enterprise are one of the main factors in ensuring financial stability because they form the material basis of its development. According to the economic dictionary [50],

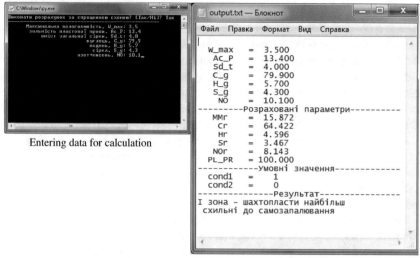

Entering data for calculation

Calculation result

Fig. 4 Data analysis method for determining the susceptibility of coal seam to endogenous fires based on hydrogen content (software implemented in Python)

investments are long-term investments of capital in one's own country or abroad in enterprises of various industries, entrepreneurial projects, socio-economic programs, and innovative projects [51]. These can be monetary, property, and intellectual values. The main difference between investments and other expenses is that they give a return after a significant period of investment. Investments can be made in fixed (buildings, machines, equipment, etc.) and circulating (inventories of commodity values, etc.) funds, in intangible resources and assets (securities, patents, licenses, etc.). Investments also include spending on education, scientific research, and personnel training, which represent long-term investments in «human capital», which are gaining more and more importance today, because the main factor of modern economic development is increasingly the intellectual product that affects innovative processes, and also determines the economic position of the country in the world hierarchy of states and the competitiveness of its economy.

The effectiveness of investments in improving working conditions can be expressed in social, socio-economic, and economic effects. Calculation of the social and economic effectiveness of investments is recommended to be carried out both in operating enterprises and in the process of designing, constructing, and reconstructing of enterprises.

The social effect of investments (increasing work capacity, reducing injuries and illnesses, reducing staff turnover, increasing the prestige of the profession, the degree of job satisfaction, etc.) should always be positive. As a rule, it leads to an increase in labor productivity, improvement in the quality of manufactured products, preservation of labor resources, and other positive consequences.

In cases where the social effect can be expressed in monetary terms (for example, when increasing labor productivity), the socio-economic efficiency of investments is considered. That is, the economic efficiency of investments in improving working conditions is nothing more than an economic expression of a social effect its inevitable consequence.

The economic results of investment are expressed in the form of the economic effect of the implementation of labor protection measures and indicators of efficiency or comparative efficiency, if necessary, a comparative assessment of several directions of investment in improving working conditions. Assessment of both social and economic effects, which are closely related, can be carried out independently using individual indicators, as well as based on their aggregate.

The occurrence of accidents and emergency situations, as well as accidents that cause injuries, has a probabilistic nature. Therefore, when determining the amount of predicted damage from accidents 3_{ab}, it is necessary to take into account both possible expected economic losses in the event of the occurrence of a certain type of accidents $3(A)_i$, as well as the probable nature of their occurrence. In this case, the maximum value of averted damage as a result of the implementation of the measure, which excludes the possibility of the i-th type of accident at the mine, will have the form [52]

$$3_{abi} = p_i \, 3(A)_i \tag{6}$$

where p_i—the probability of an accident occurring at an object of the i-th type.

In connection with the fact that such types of accidents as gas and dust explosions, sudden emissions, fires destroy or damage material assets, lead to the death and injury of people, cause long-term disruptions in the production process, the expected economic losses $3(A)_i$ consist of direct losses, costs for repair and restoration of facilities, expenses related to disruption of the production and economic activity of the mine, material damage from injuries for the entire period of payments in accordance with current legislation.

The economic effect of this measure depends on how accurately the forecast of expected economic losses $3(A)_i$ and the probability of occurrence of an accident p_i is carried out. Obtaining reliable predictive data is the most important and, at the same time, the most complex stage of calculations, which confirms the need for the introduction of enterprise information systems and innovative models of management for more accurate forecasting of accidents and their prevention.

The maximum value of averted damage (6) is achieved in the case when, as a result of the implementation of the measure, the possibility of the occurrence of the i-th type of accident at the mine is completely excluded. In reality, no measure, no matter how expensive it is, is unable to completely eliminate the possibility of an accident. Therefore, when determining the effectiveness of measures aimed at preventing accidents, we can only talk about reducing the probability of their occurrence. In this case, the amount of averted loss due to the implementation 3_{asi} of the measure is defined as [52, 53]:

$$3_{asi} = (p_i - p_i^B) \, 3(A)_i \tag{7}$$

where p_i, p_i^B is the probability of an accident occurring at the facility of the ith type, respectively, before and after the implementation of measures. $3(A)_i$ is the amount of expected damage from accidents of the i-th type.

We denote the probability difference as [52, 53]:

$$\Delta p_i = (p_i - p_i^B), \tag{8}$$

In this case, the amount of averted damage due to the implementation of measure 3_{asi} is determined as:

$$3_{asi} = \Delta p_i \, 3(A)_i \tag{9}$$

It follows from (9) that in order to determine the amount of averted damage, it is necessary to have reliable information about reducing the probability of an accident at the facility, while the limited amount of statistical information about accidents, which affects the reliability of forecasting the probability of a certain type of accident at the mine, will be significant less influence on the determined amount of averted loss. In turn, the determination of value Δp_i can be carried out using known methods of mathematical modeling and physical models of the occurrence and development of an accident.

The occurrence of accidents and emergency situations, as well as accidents that cause injuries, has a probabilistic nature, therefore, when determining the amount of predicted damage from accidents, it is necessary to take into account both the possible expected economic losses in the event of a certain type of accidents, and the probable nature of their occurrence. Predictive assessment of material damage as a result of accidents at coal mines should be differentiated by the type of accident, the place and time of the accident, the state and time of operation of the equipment, the preparedness of mine workers to eliminate accidents, the equipment and condition of the equipment intended for its elimination.

The average statistical value of losses during the liquidation of the accident, determined based on the results of the analysis of accidents performed by the State Militarized Mining Rescue Service (SMMRS), can be taken as the basic predicted value of the loss of a coal enterprise from a certain type of accident. The calculated value of the predicted loss of a coal enterprise takes into account its basic value, adjusted depending on the mining-geological and mining technical conditions. The correction is carried out in the case of establishing a dependence between the amount of actual losses and the factors that characterize the mining-geological and mining-technical conditions at the mine.

According to the information of the Ministry of Energy and Coal Industry of Ukraine regarding fires and emergency situations that occurred at enterprises of the coal-industrial complex [6], the economic loss amounted to: in 2017—losses from fires are estimated; in 2018—$1.9 million; in 2019—$3.10 million; in 2020—$1.9 million; in 2021—$0.2 million. On average, the amount of losses was $1.76 million per year. According to statistical data, the amount of damage caused by endogenous fires is 40% of all costs caused by accidents at coal enterprises of Ukraine. So, the average amount of damage caused by endogenous fires is $0.75 million per year.

It is clear that no method of predicting the occurrence of endogenous fires in mine linings and even an exhaustive list of all possible preventive measures to prevent the occurrence of endogenous fires, taking into account their stochastic nature, are not able to completely eliminate the probability of the occurrence of this phenomenon.

The probability of an endogenous fire at these mines is within $0.45 \div 0.75$. Let's conditionally accept this indicator as 0.7. According to the method of determining the fire hazard based on the elemental composition of coal, this mine belongs to the increased danger due to spontaneous combustion of coal, the probability of the occurrence of endogenous fires is 0.9. According to (6), the difference between the probability of fire occurrence before and after the implementation of fire hazard forecasting measures will be 0.2. According to formula (9), the averted loss due to the implementation of measure A, taking into account the average amount of losses of $0.75 million per year, will be $0.14 million per year. Thus, as a result of the introduction of the method of forecasting the endogenous fire hazard of coal seams according to the hydrogen content, the value of the expected averted damage caused by the occurrence of fires in the coal-industrial complex is $0.14 million per year.

The amount of averted damage takes into account the expected economic losses in the event of a certain type of accident and the probability of their occurrence. In the simplest case, the probability of the occurrence of a certain type of accident is determined by studying the available statistical data on such accidents for a certain observation period and extrapolating this data to the future period of time.

A set of innovative enterprise management technologies [54], modern IT tools and a digital information-theoretical base [55] will allow modern and future enterprises of the coal industry to simulate possible accidents in an automated manner in order to reduce the probability of their occurrence or reduce damages from accidents, if they cannot be prevented.

6 Conclusions

The main conclusions obtained as a result of the conducted research are summarized as follows:

- The difficulty of studying the influence of the complex of factors of the three blocks on the occurrence of emergency situations lies in the impossibility of reproducing them. The only exception is the dust-forming ability of coal seams. Practice has proven the connection between the dust-generating capacity and other dangerous properties of mine plastics (endogenous fire hazard, occurrence of fire-explosive situations). Having studied the complex of factors that determine the processes of dust formation and, having established the relationship between the indicators of different blocks, it is possible to develop, by analogy, a methodology for choosing the factors of three blocks, the ratio of which can lead to the manifestation of any dangerous property of mine plastic during mining operations.
- On the basis of the conducted research, the peculiarities of the selection of indicators of metamorphic transformations of coal are established, which are used in parallel, respectively, to establish the quality of fuel and forecast the dangerous properties of coal seams. Inconsistencies in the indicators of the degree of metamorphism of the state of fuel during mining operations in underground conditions, which are used in the current regulatory framework for the safe conduct of mining operations, were revealed. In order to improve normative documents on the safe conduct of mining operations, it is necessary to consider indicators that characterize not only the organic (combustible) part of the fuel, but also the presence of mineral impurities and moisture in coal in the area of mining operations.
- Software was developed for the method of determining the propensity of coal seams to endogenous fire hazard based on hydrogen content, which will allow mine workers and associations to assess the fire hazard of the developed coal seam. The integration of the information system and the enterprise management system will reduce the risk of accidents using the developed methodology.
- According to the results of calculations, the averted loss from the implementation of the proposed measures will decrease by 5 times.

References

1. World Health Organization. [Last Access date 11 September 2023]. http://www.who.int/en/ [Internet]
2. Safety and Health at Work [Internet] [Access date 30.08.2023 https://www.ilo.org/global/top ics/safety-and-health-at-work/lang--en/index.htm
3. Antoshchenko, M., Tarasov, V., Nedbailo, O., Zakharova, O., Rudniev, Y.: On the possibilities to apply indices of industrial coal-rank classification to determine hazardous characteristics of workable beds. Min. Miner. Deposits **15**(2), 1–8 (2021). https://doi.org/10.33271/mining15. 02.001
4. Bunko, T.V., Shevchenko, V.G., Yashchenko, I.A., Kokoulin, I.E.: Vdoskonalennia systemy upravlinnia vyrobnytstvom ta okhoronoiu pratsi. J. Mizhvid. zb. nauk. prats Heotekhnichna mekhanika **127**, 3–17 (2016)
5. Bondarenko, V.I., Kuzmenko, O.M, Hriadushchyi, Y.B, Haiduk, V.A., Kolokolov, O.V.: Tekhnolohiia pidzemnoi rozrobky plastovykh rodovyshch korysnykh kopalyn Natsionalnyi hirnychyi universytet (2005)
6. Liashok, Y., Iordanov, I., Chepiga, D., Podkopaiev, S.: Experimental studies of the seam openings competence in different methods of protection under pitch and steep coal seams development. Min. Miner. Deposits **12**(4), 9–19 (2018)
7. Podkopaiev, S., Iordanov, I., Chepiha, D.: Stability of the coal seam roof during the sudden collapse of lateral rocks. Min. Miner. Deposits **11**(3), 101–110 (2017)
8. Zhang, Y., Shao, W., Zhang, M., Li, H., Yin, S., Xu, Y.: Analysis 320 coal mine accidents using structural equation modeling with unsafe conditions of the rules and regulations as exogenous variables. Accid. Anal. Prev.. Anal. Prev. **92**, 189–201 (2016). https://doi.org/10.1016/j.aap. 2016.02.021
9. Wright, T.: The political economy of coal mine disasters in China: "Your Rice Bowl or Your Life." China Q. **179**, 629–646 (2004). https://doi.org/10.1017/S0305741004000517
10. Wu, L., et al.: Major accident analysis and prevention of coal mines in China from the year of 1949 to 2009. Min. Sci. Technol. (China) **21**(5), 693–699 (2011)
11. Han, S., Chen, H., Stemn, E., Owen, J.: Interactions between organizational roles and environmental hazards: the case of safety in the Chinese coal industry. Resour. Policy. Policy **60**, 36–46 (2019). https://doi.org/10.1016/j.resourpol.2018.11.021
12. The State Emergency Service of Ukraine. [Last Access date 11 September 2023]. https://dsns. gov.ua/en/news/nadzvicaini-podiyi [Internet]
13. Skipochka, S., Palamarchuk, T., Prokhorets, L.: The concept of risk-based technical solutions for the protection of coal mine workings. In: IOP Conference Series: Earth and Environmental Science, vol. 1156, No. 1, p. 012017. IOP Publishing (2023)
14. Küçük, F.Ç.U., Ilgaz, A.: Causes of coal mine accidents in the world and Turkey. Turk. Thorac. J. **16**(Suppl), S9–S14 (2015Apr). https://doi.org/10.5152/ttd.2015.003
15. Akgün, M.: Coal mine accidents. Turk. Thorac. J. **16**(Suppl 1), S1 (2015)
16. Sari, M., Duzgun, H.S.B., Karpuz, C., Selcuk, A.S.: Accident analysis of two Turkish underground coal mines. Saf. Sci.. Sci. **42**(8), 675–690 (2004). https://doi.org/10.1016/j.ssci.2003. 11.002
17. Blanch, J., Freijo Álvarez, M., Alfonso, P., Sanmiquel Pera, L., Vintró Sánchez, C.: Occupational injuries in the mining sector (2000–2010). Comparison with the construction sector. DYNA **81**(186), 153–158 (2014). https://doi.org/10.15446/dyna.v81n186.39771
18. Blanch, A., Torrelles, B., Aluja, A., Salinas, J.A.: Age and lost working days as a result of an occupational accident: a study in a shift work rotation system. Saf. Sci.. Sci. **47**(10), 1359–1363 (2009). https://doi.org/10.1016/j.ssci.2009.03.001
19. Mine Safety and Health Administration (MSHA).: Historical Data on Mine Disasters in the United States (2013)
20. Asfaw, A., Mark, C., Pana-Cryan, R.: Profitability and occupational injuries in U.S. underground coal mines. Accid. Anal. Prev.. Anal. Prev. **50**, 778–786 (2013). https://doi.org/10. 1016/j.aap.2012.07.002

21. Ivaz, J., Stojadinović, S., Petrović, D., Stojković, P.: Analysis of fatal injuries in Serbian underground coal mines–50 years review. Int. J. Inj. Contr. Saf. Promot.Saf. Promot. **27**(3), 362–377 (2020)
22. Sanmiquel, L., Freijo, M., Edo, J., Rossell, J.M.: Analysis of work related accidents in the Spanish mining sector from 1982–2006. J. Saf. Res. **41**(1), 1–7 (2010). https://doi.org/10.1016/j.jsr.2009.09.008
23. Government of India, Ministry of Labour and Employment, Directorate-General of Mines Safety.: Statistics of mines in India (p. 41). Vol. I (Coal). Government of India, Ministry of Labour and Employment, Directorate-General of Mines Safety (2014)
24. Mining Review.: Coal mining sector in South Africa applauded for safety improvement (2016). Accessed 12 July 2019, from https://www.miningreview.com/southern-africa/sa-coal-mining-sector-applauded-for-safety-improvement/
25. Antoshchenko, M., Tarasov, V., Liubymova-Zinchenko, O., An, H., Kononenko, A.: About possibility to use industrial coal-rank classification to reveal coal layers hazardous characteristics. Civil Eng. Architect. **2**, 507–511 (2021). https://doi.org/10.13189/cea.2021.090223
26. Rudniev, Y., Galchenko, A., Antoshchenko, M., Tarasov, V.: Moisture as an indicator of the manifestation of hazardous properties of coal seams. Geofizicheskii Zhurnal (Geophysical Journal) **44**(3), 66–79 (2022). https://doi.org/10.24028/gj.v44i3.261969
27. Antoshchenko, M., Tarasov, V., Rudniev, Y., Zolotarova, O.: On the issue of establishing the stages of coal metamorphism for predicting the hazardous properties of coal seams. Nat. Environ. Pollut. Technol.Pollut. Technol. **20**(4), 1495–1503 (2021). https://doi.org/10.46488/NEPT.2021.v20i04.011
28. Mironov, K.V.: Spravochnik geologa-ugolshhika Moskva Nedra, p. 311 (1982)
29. Zheldakov, M.E., Ivanova, E.I.: Spravochnik po kachestvu antracitov Sovetskogo Sojuza Nedra (1980)
30. DonUGI.: Geologo-uglehimicheskaja karta Doneckogo bassejna VIII Ugletehizdat, p. 430 (1954)
31. Anciferov, A.V, Golubev, A.A., et al.: Gazonosnost i resursy metana ugolnyh bassejnov Ukrainy: v 3-h tomah. t.1. Geologija i gazonosnost zapadnogo, jugo-zapadnogo i juzhnogo Donbassa Veber Doneckoe otdelenie, p. 451 (2009)
32. Chernousov, J.M.: Geologija ugolnyh mestorozhdenij Vishha shkola, p. 176 (1977)
33. Ghosh, S., Ojha, A., Varma, A.K.: Spectral manifestations of coal metamorphism: insights from coal microstructural framework. Int. J. Coal Geol. **228**, 103549 (2020)
34. DSTU 3472:2015.: Vuhillia bure, kamiane ta antratsyt. Klasyfikatsiia, DP "UkrNDNTs" (2016)
35. DSTU 9220:2023 Palyvo tverde mineralne. Metody vyznachennia vykhodu letkykh rechovyn, DP "UkrNDNTs"
36. GOST 3168–93 (ISO 647–74) Mezhgosudarstvennyj standart. Toplivo tverdoe mineralnoe. Metody opredelenija vyhoda produktov polukoksovanija, Minsk 17
37. Zhukov, P.P., Lashhenko, V.I.: Ob ocenke vyhoda letuchih veshhestv iz uglja. J. Ugol Ukrainy **9**, 39–40 (1985)
38. Medvedev, E.N., Saranchuk, V.I., Kachan, V.N.: Ocenka pyleobrazujushhej sposobnosti uglej v rjadu metamorfizma. J. Ugol Ukrainy **8**, 32–33 (1984)
39. Ochkur, N.N., Kozyrskaja, V.F.: Pokazatel svobodnogo vspuchivanija dlja ocenki spekaemosti i marochnogo sostava Doneckih uglej. J. Ugol Ukrainy **2**, 60–64 (1992)
40. Semenov, A., Baraban, S., Semenova, O., Voznyak, O., Vydmysh, A., Yaroshenko, L.: Statistical express control of the peak values of the differential-thermal analysis of solid materials. SSP **291**, 28–41 (2019). https://doi.org/10.4028/www.scientific.net/SSP.291.28
41. Semenov, A.O., Baraban, S.V., Osadchuk, O.V., Semenova, O.O., Koval, K.O., Savytskyi, A.Y.: Microelectronic pyroelectric measuring transducers. In: IFMBE Proceedings, pp. 393–397 (2019).. https://doi.org/10.1007/978-3-030-31866-6_72
42. Semenov, A., Zviahin, O., Kryvinska, N., Semenova, O., Rudyk, A.: Device for measurement and control of humidity in crude oil and petroleum products. Metrol. Meas. Syst. (2023). https://doi.org/10.24425/mms.2023.144865

43. Osadchuk, O.V., Semenov, A.O., Zviahin, O.S., Semenova, O.O., Rudyk, A.V.: Increasing the sensitivity of measurement of a moisture content in crude oil. Nauk. visn. nat. hirn. univ. 49–53 (2020). https://doi.org/10.33271/nvngu/2021-5/049

44. Pashkovskij, P.S. Kostenko, V.K. Zaslavskij, V.P. Horolskij, A.T. Zabolotnyj, A.G. et al.: KD 12.01.401–96. Endogenous res in the coal mines of Donbass. Prevention and suppression. Instructions (1997)

45. Derzhavnyi normatyvnyi akt pro okhoronu pratsi.: Skhemy ta sposoby keruvannia hazovy-dilenniam na vyimkovykh dilnytsiakh vuhilnykh shakht. Derzhavnyi departament promyslovoi bezpeky, okhorony pratsi y hirnychoho nahliadu (2006)

46. Cashdollar, K.L.: Coal dust explosibility. J. Loss Prev. Process Ind. 9(1), 65–76 (1996)

47. Rice, D.D.: Composition and origins of coalbed gas. Hydrocarbons Coal: AAPG Stud. Geol. 38(1), 159–184 (1993)

48. Standart Minvuhlepromu Ukrainy.: SOU-P 10.1.00174088.016:2009. Pravyla vyznachennia efektyvnosti vyperedzhalnoho zakhystu plastiv, skhylnykh do hazody-namichnykh yavyshch (2009)

49. Friedman, J.P.: Dictionary of Business and Economic Terms. Simon and Schuster (2012)

50. Vahonova, O.H., Romaniuk, N.M.: Economic justification of stages of the investment project of a mining and processing enterprise. Науковий вісник Національного гірничого університету 3, 159–164 (2014)

51. Vagonova, O., Mormul, T., Zakharchenko, Y., Romaniuk, N., Kasianenko, L.: Topical problems concerning both methods and economy to develop mineral deposits. Min. Miner. Deposits (2018)

52. Tarasov, V., Antoshchenko, M., Rudniev, Y., Zolotarova, O., Davidenko, N.: Metamorphism indicators for establishing the endogenic fire hazard of coal mining plants in mining. Int. J. Environ. Sci. Dev. 12(8), 242–248 (2021). https://doi.org/10.18178/ijesd.2021.12.8.1346

53. Antoshchenko, M., Tarasov, V., Rudniev, Y., Zakharova, O.: Using indices of the current industrial coal classification to forecast hazardous characteristics of coal seams. Min. Miner. Deposits 16(2), 7–13. https://doi.org/10.33271/mining16.02.007

54. Kryvinska, N., Kaczor, S., Strauss, C.: Enterprises' servitization in the first decade—retrospective analysis of back-end and front-end challenges. Appl. Sci. 10, 2957 (2020). https://doi.org/10.3390/APP10082957

55. Kryvinska, N., Bickel, L.: Scenario-based analysis of IT enterprises servitization as a part of digital transformation of modern economy. Appl. Sci. 10, 1076 (2020). https://doi.org/10.3390/app10031076

Assessment of the Efficiency of Decentralization Transformations in the Rural Areas of Ukrainian Western Polissia: Current Trends and Challenges Under the Conditions of Martial Law

Alla Sokolova⬤, Tatiana Ratoshniuk⬤, Iryna Yepifanova⬤,
Yurii Kravchyk⬤, Viktor Ratoshniuk⬤, and Viacheslav Dzhedzhula⬤

Abstract The purpose of the study is to identify modern trends and regional features and assess the effectiveness of decentralization transformations in rural areas of the Western Polissia of Ukraine. The object of the study is the process of identifying regional features and determining the degree of influence of the factors on the effectiveness of the above-mentioned transformations in the United Territorial Communities (UTCs) of the Volyn, Zhytomyr, Rivne, and Chernihiv regions, the analysis of existing challenges in the conditions of martial law. According to the results of the conducted multifactorial correlation-regression analysis, it was established that the general profitability of territorial communities is influenced by the profitability of community lands, local taxes and fees, fiscal return of the territory; expenses of the general fund, expenses for maintenance of the management apparatus from the calculation; capital expenditures and the amount of subsidies (official transfers, basic subsidy). It was found that in recent years, the system of local budgets, as one of the most important tools of the organizational and economic mechanism of ensuring

A. Sokolova
Volyn State Agricultural Experimental Station of the Institute of Potato of the National Academy of Agrarian Sciences of Ukraine, Rokiny Village, Lutsk District, Volyn Region, Lutsk 45626, Ukraine

T. Ratoshniuk · V. Ratoshniuk
Polissia Institute of Agriculture of the National Academy of Agrarian Sciences of Ukraine, 131, Kyivske Shose, Zhytomyr 10007, Ukraine

I. Yepifanova (✉) · V. Dzhedzhula
Vinnytsia National Technical University, 95 Khmelnytske Shose, Vinnytsia 21021, Ukraine
e-mail: yepifanova@vntu.edu.ua

Y. Kravchyk
Khmelnytskyi National University, 11 Instytuts'ka Str, Khmelnytskyi 29016, Ukraine

© The Author(s), under exclusive license to Springer Nature Switzerland AG 2024
A. Semenov et al. (eds.), *Data-Centric Business and Applications*, Lecture Notes on Data Engineering and Communications Technologies 194,
https://doi.org/10.1007/978-3-031-53984-8_16

sustainable rural development in UTCs, has undergone fundamental transformations from its participants and due to the change of financial sources of local self-government bodies of various levels. It was established that the result of the decentralization reform in the rural areas of Western Polissia was an increase in the interest of local self-government bodies in increasing revenues to local budgets, finding reserves to fill them, and improving the efficiency of tax and fee administration.

Keywords United Territorial Communities (UTCs) · Decentralization · Local budgets · Revenue · Regression · Multifactorial correlation · Regression statistics

1 Introduction

Perspectives for sustainable rural development in Ukraine considerably depend on quantitative and qualitative indicators that reflect the effectiveness of decentralization transformations in the regions. For the purpose of detailed research and evaluation of the effectiveness of the decentralization reform, the rural areas of the Western Polissia of Ukraine, such as Zhytomyr, Rivne, Volyn and Chernihiv regions, were selected, which include, which occupy 16.8% of the country's total territory on January 1, 2022. The largest is the Chernihiv region, which has an area of 3,190.3 thousand hectares. There are 50 cities, 110 towns, and 5131 villages in the territory of the Western Polissia of Ukraine, which are united in 17 districts [1].

They should be given one of the decisive places in the national strategy of socio-economic development, since the rural territories of Polissia play a key role in the process of decentralization of power, and their capacity is the main factor in the development of the region. On the basis of the research results, it was found that during the period 2000–2021, the rural areas of the studied region were depopulated, uninhabited settlements were removed from the register, and, as a result, the rural settlement network was degraded. So, for example, in Zhytomyr Region in 2020, there were 1,612 rural settlements; their number decreased by 11 units from the previous survey (2005) [2].

Investigating the economic mechanism of implementing the strategy of sustainable development of the rural areas of Western Polissia, Tymoshenko M. M. notes that the most acute problems of the rural areas of the region are: steady reduction in the number of the rural population and its aging; lack of motivation to work and low incomes of rural residents; external and internal labor migration (especially young people); high level of unemployment; low availability of social infrastructure facilities; insufficient level of competitiveness of agricultural products, most of which are produced by households; deterioration of the ecological situation in the countryside (degradation of land, water and forest resources, non-compliance with scientifically based farming systems, destruction of biodiversity); deformation of the structure of agricultural production; imperfection of the mechanisms of innovation-investment and financial-credit provision of rural development, etc. [3]. The above-mentioned problems intensified with the beginning of the full-scale military aggression of russia, which began on February 24, 2022.

The purpose of the study is to identify modern trends and regional features and assess the effectiveness of decentralization transformations in rural areas of the Western Polissia of Ukraine. The object of the study is the process of identifying regional features and determining the degree of influence of the factors on the effectiveness of the above-mentioned transformations in the United Territorial Communities (UTCs) of the Volyn', Zhytomyr, Rivne, and Chernihiv regions, the analysis of existing challenges in the conditions of martial law.

2 Methods and Technologies

Decentralization is a complex process that affects the social, economic, and environmental spheres of development of united territorial communities, as it brings the processes of management and decision-making closer to the public and, in this way, makes it possible to implement strategies, programs, regional development projects and provide services which would better meet the needs of a certain region [4]. The decentralization reform should increase the financial capabilities of local authorities and increase their potential for socio-economic development.

Investigating the spatial and territorial features of decentralization transformations in the rural areas of Western Polissia, it is worth noting that the network of rural settlements is one of the elements of the strategic potential of the state; therefore the quality of life of the rural population will serve as one of the important indicators of the effectiveness of the decentralization transformations being carried out. The directions and forms of modern transformation of rural settlements (small, medium, large, or great villages) and their main demographic and organizational characteristics form socio-economic prerequisites and opportunities for further development. As you know, on September 17, 2020, the Verkhovna Rada of Ukraine adopted Resolution No. 807-IX "On the formation and liquidation of districts" [5]. As a result of the implementation of this resolution, all 490 administrative districts of Ukraine were liquidated; instead, 136 new ones were created (17 of them are located in the temporarily occupied territories).

Up to January 1, 2022, 17 districts were created on the territory of Western Polissia, in which 241 rural, town, and city UTCs (Table 1).

In Volyn' region, which has the smallest population, only 4 districts were created, and 54 united territorial communities were formed; there are also 4 districts and 64 united territorial communities in the largest—Zhytomyr—region (Table 2).

Summarizing the results of the unification of territorial communities in the rural areas of Western Polissia for the period 2015–2021, it is worth noting that a significant part of them used the provided institutional capacity for effective self-government despite various obstacles and the lack of experience of working in conditions of decentralization. This is evidenced by the results of the annual monitoring of the reform of local self-government and territorial organization of power (Table 3).

Table 1 The state of decentralization transformations in the territory of the regions of Western Polissia up to January 1, 2022

Country/ Region	Area of the territory, km^2	Number of population, persons	Number of districts, units	Number of communities, units	Number of population per 1 community, persons
Ukraine, in general	576,603	38,122,555	119	1439	29,492
Western Polissia, including the regions:	101,888	4,383,888	17	241	18,190
Volyn'	20,144	1,031,421	4	54	19,100
Zhytomyr	29,832	1,208,212	4	66	18,306
Rivne	20,047	1,152,961	4	64	18,015
Chernihiv	31,865	991,294	5	57	17,391
Western Polissia in % of the total	17,7	11,5	14,3	16,7	x

Source Formed according to data [6]

Incomes of local budgets of territorial communities of Western Polissia per 1 inhabitant in 2021 amounted to UAH 6,163.3, which is UAH 996.7 (19.3%) more than in 2020. However, this indicator is UAH 1,845.6 lower than the average in Ukraine. In our point of view the financial aspect in the analysis of the results of decentralization transformations is one of the most essential, which mostly affects the success of the functioning of territorial communities. The presence of economically active business entities, a sufficient number of qualified labor resources, and developed industrial and social infrastructure—all this and much more is the basis for the successful development of the community.

The part of own revenues of the territorial communities of Western Polissia in the total volume of revenues in 2021 in the Volyn' region was 54.4%, in the Zhytomyr region—63.8%, in the Rivne region—53.3%, and in Chernihiv region—69.3%. On average, in Ukraine, this indicator is 70.7%, which indicates the need to provide territorial communities of the region with greater resources and mobilize their internal reserves (Table 4).

In the structure of revenues of the general fund of local budgets of the united territorial communities of Western Polissia, the tax on the income of individuals is the most significant—59.4%; fee for land and real estate—15.6; single tax—16.2; excise duty—6.8; others—2.0%.

It is positive that the local budget expenditures of the studied communities per 1 inhabitant increased by 9.7% (925.1 UAH) during the studied period (Table 5). The part of official transfers from the state budget ranges from 46.7% in the Rivne region to 30.7% in the Chernihiv region.

Table 2 The results of decentralization transformations in the districts of the Western Polissia regions up to January 1, 2022

	Regions/Districts	Number of communities, units	Area, km^2	Number of population, persons
Volyn region				
1	Volodymyr-Volynskyi	11	2556,5	172,947
2	Kamin-kashyrskyi	5	4679,7	131,592
3	Kovel	23	7658,7	269,595
4	Lutsk	15	5249,1	457,287
In general		54	20,144,0	1,031,421
Zhytomyr region				
1	Berdychiv	10	3014,0	161,462
2	Zhytomyr	31	10,508,2	618,111
3	Korosten'	13	10,892,2	258,935
4	Novohrad-Volynskyi	12	5237,3	169,704
In general		66	29,832	1,208,212
Rivne region				
1	Varash	8	3323,5	138,751
2	Dubno	19	3294,2	169,079
3	Rivne	26	7216,6	632,426
4	Sarny	11	6212,7	212,705
In general		64	20,047	1,152,961
Chernihiv region				
1	Koriukivka	5	4600,9	89,920
2	Nizhyn	17	7219,0	224,788
3	Novhorod-Siverskyi	4	4630,4	64,997
4	Pryluky	11	5210,8	155,401
5	Chernihiv	20	10,203,9	456,188
In general		57	31,865	991,294
In general, in Western Polissia		241	101,888	4,383,888

Source Formed according to data [6]

We consider that it will be possible to improve the resource support of rural communities by securing stable sources of income in local budgets and expanding their income base on a legislative basis. After all, as Z. M. Titenko points out, in the conditions of budget decentralization, the main condition for the effective development of Ukraine is the financial support of local self-government bodies, that is, the stability and sufficiency of their sources of funding, which will ensure the effective performance of the functions assigned to local authorities and prompt solution of tasks of a socio-economic nature at the level of the territorial community [8, p. 89].

Table 3 Income dynamics of local budgets per inhabitant for January–November 2020–2021, UAH

Country/Regions	January–November 2020	January–November 2021	2021 to 2020	
			+, −	%
Ukraine, in general	6622,5	8008,9	1386,4	120,9
Western Polissia, including regions:	5166,6	6163,3	996,7	119,3
Volyn	4612,0	5581,8	969,8	121,0
Zytomyr	5506,3	6525,3	1019	118,5
Rivne	4552,0	5551,2	999,2	122,0
Chernihiv	5996,2	6994,7	998,5	116,7
Western Polissia (+, −) to the average in Ukraine	−1455,9	−1845,6	x	x

Source Formed according to data [7]

Table 4 The part of own revenues of local budgets in the total volume of revenues in January–November 2021

Country/Regions	Own revenues of local budgets, UAH million	Transfers from the state budget, UAH million	Total revenues to local budgets (general fund), UAH million	The part of own revenues in the total amount of income, %
Ukraine, in genetal	317,108,6	131,159,5	448,268,1	70,7
Western Polissia, including regions	27,108,8	17,980,0	45,088,8	60,1
Volyn	5774,9	4750,6	10,525,5	54,9
Zhytomyr	7930,7	4500,9	12,431,6	63,8
Rivne	6418,3	5632,4	12,050,7	53,3
Chernihiv	6984,9	3096,1	10,081,0	69,3
Western Polissia in % of Ukraine as a whole	8,5	13,7	10,1	x

Source Formed according to data [7]

At the time of active military operations in Ukraine, which began on February 24, 2022, after the invasion of russian troops on the territory of our country, it is difficult to talk about the implementation of an effective policy in the field of sustainable development of rural areas. In the conditions of martial law, it is extremely important to ensure prompt, proper, and continuous implementation of local budgets. To ensure the effective functioning of the budget sphere and the vital needs of residents of territorial communities during the period of martial law, the Office of the President of Ukraine, the Government, the Verkhovna Rada of Ukraine, the Council of National

Table 5 Dynamics of local budget expenditures per 1 inhabitant for January-November 2020–2021, UAH

Country/Regions	January-November 2020	January–November 2021	2021 to 2020	
			+, −	%
Ukraine in general	10,107,0	11,139,5	1032,5	110,2
Western Polissia, including regions	9514,0	10,439,1	925,1	109,7
Volyn	9623,2	10,397,1	773,9	108,0
Zhytomyr	9714,6	10,571,0	856,4	108,8
Rivne	9065,8	10,252,5	1186,7	113,1
Chernihiv	9652,5	10,535,6	883,1	109,1
Western Polissia (+, −) to the average in Ukraine	−590,0	700,4	x	x

Source Formed according to data [7]

Security and Protection and other central bodies of the executive power make a number of quick, effective, operational decisions.

In order to inform the united territorial communities about the latest key changes in budget legislation, creation of conditions for a timely and prompt response to the needs of financial support for territorial defense measures, protection of public safety and functioning of the budget sphere, communal enterprises during martial law, Support Office of Decentralization Reform of the Ministry of Development of Communities and Territories of Ukraine (Swedish-Ukrainian Project "Supporting Decentralization in Ukraine", SKL International (SALAR)) prepared key changes and moments in the field of local budgets with relevant references to facilitate orientation in the norms of budget legislation [9].

It is worth noting that on July 9, 2022, the Verkhovna Rada of Ukraine adopted in the second reading as a Law the government project of the Law of Ukraine On Amendments to Section VI "Final and Transitional Provisions of the Budget Code of Ukraine according to strengthening the flexibility of local budgets and increasing the expediency of decision-making" No. 7426 dated 01.06.2022 [10]. In accordance with this law, the right is granted to transfer funds from the special fund of the local budget to the general fund of the local budget. It is allowed to carry out expenses not assigned to the relevant local budgets and expenses for the maintenance of budgetary institutions simultaneously from different budgets. Such expenditures are made by providing an interbudgetary transfer from the relevant local budget (proposal of the Verkhovna Rada Committee on Budget Issues).

The right was granted to carry out new local borrowings and provide local guarantees, even if in the process of making payments for the repayment and service of the local debt, the payment schedule was violated, fines were charged, and overdue debt arose. State budget income increased by 80.977 billion hryvnias, in particular, due to [10]: an increase in revenues by UAH 1.546 billion (other assistance provided by the European Union); an increase in funding by UAH 79.431 billion (external borrowing—UAH 75.431 billion, internal borrowing—UAH 4 billion).

These revenues are directed in the following directions:

1. the payment of benefits, compensations, monetary support, and payment of services to certain categories of the population—UAH 5.771 billion;
2. payment of benefits and housing subsidies—UAH 3.330 billion;
3. monthly targeted assistance for internally displaced persons—UAH 32.364 billion;
4. pension payments—UAH 32.304 billion;
5. entrepreneurship development fund—UAH 4.0 billion;
6. subvention to local budgets for the restoration of critical infrastructure facilities— UAH 1.663 billion.

For example, UAH 99,988.791 thousand has been allocated from the reserve fund of the state budget for the Chernihiv Regional Military Administration for the purpose of rebuilding the destroyed single property complex of the communal energy-generating enterprise "Chernihiv Thermal Power Plant" (order of the Cabinet of Ministers of Ukraine from July 13, 2022, No. 597-r).

According to the results of the analysis of the composition and structure of expenses of the budget of the UTCs of the rural territories of Western Polissia in 2021, it was determined that the part of expenses for the maintenance (organizational, information-analytical and material and technical support for the activities of village councils) of local self-government bodies in the communities of the region in 2021 in Volyn' region was 9.7%, Zhytomyr region—11.0%, Rivne region—9.7%, Chernihiv region—11.3%. On average, in Ukraine, this indicator is 10.1% (Table 6).

Table 6 The part of expenses for the maintenance of local self-government bodies in the expenses of the general fund in January–November 2021

Country/Region	The total volume of local budgets, UAH million	Expenses for maintenance of local self-government bodies, UAH million	Part of expenses for maintenance of local self-government bodies in the total amount of expenses, %
Ukraine, in general	349,414,2	35,379,6	10,1
Western Polissia, including regions	38,393,8	4029,8	10,5
Volyn	9053,4	879,3	9,7
Zhytomyr	10,562,9	1163,5	11,0
Rivne	10,106,5	981,6	9,7
Chernihiv	8671,0	1005,4	11,6
Western Polissia in % of Ukraine as a whole	10,9	11,4	x

Source Formed according to data[7]

The analysis of capital expenditures per 1 inhabitant (the ratio of capital expenditures without taking into account own revenues to the number of residents of the UTCs of Western Polissia) indicates a negative trend of their decrease during the studied period (Table 7).

It was determined that in 2020 this indicator was UAH 1,306.8, and in 2021—UAH 1,179.8, which is UAH 640.4 lower than the average for Ukraine. This necessitates the development and implementation of strategic decisions regarding the effective use of the existing potential and diversification of sources of replenishment of the revenue part of budgets of the UTCs located in the rural areas of the Western Polissia of Ukraine. After all, financially capable communities show high and dynamic growth rates of their own incomes.

It is positive that the sectoral reform continues, that is, decentralization involves various spheres of peasant life—school and preschool education, medicine, housing and communal services, etc.; the capabilities of the UTCs to provide high-quality services to citizens are being strengthened, in particular, administrative services center are being created; the material and technical base of schools and kindergartens is being improved, buildings are being repaired, educational reform is continuing; primary medical care centers, etc. are being created (Table 8).

One of the positive results of decentralization transformations in the rural areas of the Western Polissia of Ukraine is an increase in the accessibility of administrative services to rural residents. With the support of the "ULEAD with Europe" program, new administrative service centers continue to be created in the territorial communities of the studied region. Up to January 1, 2022, the highest level of coverage by administrative services centers in Volyn region is 78%, and the lowest is in the Chernihiv region.

Also, according to the results of the analysis, it was found that the number of rural settlements in Western Polissia, which had libraries, clubs or houses of culture, and movie theaters, has significantly decreased in recent years. It is worth noting that the

Table 7 Dynamics of capital expenditures of local budgets per 1 inhabitant for January-November 2020–2021, UAH

Country/Regions	January-November 2020	January-November 2021	2021 to 2020	
			+ , −	%
Ukraine, in general	1839,6	1820,2		98,9
Western Polissia, including regions	1306,9	1179,8	−427,1	90,3
Volyn	1503,6	1268,4	−235,2	84,4
Zhytomyr	1522,2	1384,2	−138	90,9
Rivne	1020,1	1086,8	66,7	106,5
Chernihiv	1181,5	979,8	−201,7	82,9
Western Polissia (+, −) to the average in Ukraine	−532,7	−640,4	x	x

Source Formed according to data [7]

Table 8 Development of the network of administrative services centers (ASC) in territorial communities up to January 1, 2022

Country/ Region	Total number of ASC, including UTCs	Number of UTCs that have ASCs	Number of UTCs that have not ASCs	Number of UTCs that plan to create ASCs in 2022	% UTCs of region which have not ASCs
Ukraine in general	1055	1030	408	182	72,0
Western Polissia, including regions	161	158	83	42	66,0
Volyn	42	42	12	3	78,0
Zhytomyr	44	43	23	10	65,0
Rivne	41	39	25	25	61,0
Chernihiv	34	34	23	4	60,0
Western Polissia in % of Ukraine as a whole	15,3	15,3	20,3	23,1	x

Source Formed according to data [7]

Program "Supporting Decentralization Reform in Ukraine/U-LEAD with Europe: Program for Ukraine on Local Empowerment, Accountability, and Development" since 2016 primarily supports sectoral decentralization in the social sphere: health care, education, and administrative services. The program also advises communities on issues of local finances and human resources management and promotes the development of the so-called "municipal cooperation between UTCs" (Table 9).

According to V. P. Riabokon, it is important to achieve balanced inter-budgetary relations to combine the allocation of financial resources from the state budget and funds from local budgets. Moreover, it is necessary to have such amounts of allocations that will provide all the needs for the development of social infrastructure in the village. First of all, that is the creation of ample opportunities to ensure people's needs in modern education, medicine, social, communal, administrative, and other services. The created new UTCs are able to successfully solve these issues because it is the local self-government bodies that are responsible for school and preschool education, primary health care (polyclinics, public health centers), cultural institutions, landscaping, public order, and many other topical issues [11, p. 9].

Summarizing the results of the unification of territorial communities in the rural areas of Western Polissia for the period 2015–2021, analyzing the current state, achievements and problems of decentralization transformations in the rural areas of Western Polissia, it is possible to declare that the reform has both positive and negative consequences. However, there is a difficult political situation in the country, which is

Table 9 The state of municipal cooperation of territorial communities up to January 1, 2022

Ukraine/ Regions	Total number of cooperation projects	Including					Communities that benefited from municipal cooperation
		Utilities	Landscaping	Fire security	Education, health care, social security	Other	
Ukraine, in general	153	4	3	4	92	50	296
Western Polissia, including regions	23	–	–	2	13	8	52
Volyn	1	–	–	–	1		4
Zhytomyr	2	–	–	–	1	1	5
Rivne	14	–	–	2	7	5	32
Chernihiv	6	–	–		4	2	11
Western Polissia in % of the total	15,0	–	–	50,0	14,1	16,0	17,6

Source Formed according to data [7]

caused by russia's military aggression, and as a result, there is an unfavorable socio-demographic and economic situation in the countryside and therefore there is a small share of the economically active population of working age capable of implementing planned strategic measures to increase the effectiveness of the development of the created UTCs.

The results of studies of the ratio of budget revenues and expenditures of rural territorial communities of Western Polissia in 2021 reflect the possibilities of improving the socio-economic condition of the respective territories and testify to their potential for sustainable socio-ecological and economic development. Considering this, the analysis of the economic efficiency of the activities of the united territorial communities acquires importance for the state and is relevant nowadays [12]. The effectiveness and economic efficiency of the activities of the united territorial community is determined by the amount of income, management expenses, and subsidies received from the state [13].

With the expansion of the powers and capabilities of rural territorial communities, the need to equalize the socio-economic development of settlements and ensure the sustainable development of rural areas is increasingly entrusted to them. In turn, this requires determination of the degree of influence of factors on the level of development of the UTCs in the rural areas of the studied region. To analyze the impact of the above-mentioned factors, an economic-mathematical research method was applied, which made it possible to reveal the relationship between the effective feature and the indicator factors [14]. For this purpose, a statistical correlation-regression model was developed, and a relationship was established between the volume of all revenues to the general fund per 1 resident of the community and a number of financial and economic factors. At the same time, the nature of the specified dependence, according to the results of the conducted research, is direct.

The indicators were determined according to statistical data characterizing the level of development of 54 UTCs of the Volyn' region in 2021, and materials of the author's research and calculations. A number of indicators were calculated based on the information and analytical dashboard "Budgets of Territorial Communities of Ukraine", which visualizes the performance indicators of the budgets of all territorial communities of Ukraine by income, expenditure, and transfers for 2021 (during 4 quarters) and 2022 (1 and 2 quarters) [15]. It is worth emphasizing that this is a new tool for budget analysis at the local level; the indicators are regularly updated, contain minimum, maximum, and average values of the specified indicators, and can be used for detailed economic and financial diagnostics of the implementation of budgets of the local government, identifying shortcomings and developing proposals for improving the financial capacity of communities for the future.

It was established that the relationship and dependence between the effective and factor indicators can be expressed using the following regression equation:

$$y_x = a_0 + a_1 x_1 + a_2 x_2 + \cdots + a_n x_n, \tag{1}$$

where

y_x is the dependent variable (resultant indication—total income to the general fund per 1 inhabitant, UAH);

a_0 is the beginning of the countdown, which has no economic meaning;

a_1, a_2, a_n is the regression coefficients;

x_1, x_2, x_n is the factors.

Factor characteristics were chosen in the following way:

$x1$ is the income of the community's lands per 1 inhabitant, UAH;

$x2$ is the local taxes and fees per 1 inhabitant, UAH;

$x3$ is the fiscal return of the territory, per 1 inhabitant, UAH;

$x4$ is the expenses of the general fund per 1 inhabitant, UAH;

$x5$ is the expense for maintenance of the management apparatus per 1 inhabitant, UAH;

$x6$ is the capital expenditures per 1 inhabitant, UAH;

$x7$ is the amount of subsidy (official transfers, basic subsidy) per 1 inhabitant, UAH;

Based on the results of data processing using the Microsoft Excel application package, the following linear regression model was obtained:

$$y = 467, 61 - 288, 131 x_1 + 1, 668 x_2 - 0, 512 x_3$$

$$+ 0,166x_4 + 1,713x_5 + 0,653x_6 - 1,369x_7 \tag{2}$$

The regression coefficients ($a1 = -288{,}131$, $a2 = 1{,}668$, $a3 = -0{,}512$, $a4 = 0{,}166$, $a5 = 1{,}713$, $a6 = 0{,}653$, $a7 = -1{,}369$) show how much the amount of revenues to the general fund (budget of the UTCs) will change on average per 1 resident in case of change of each factor per unit of its measurement at fixed values of the remaining factors entered into the equation (Table 10). The regression equation shows that the yield of community land per 1 inhabitant has the greatest impact on the profitability of the UTCs.

It is worth noting that there is a close relationship between the selected factors and the resulting feature, as evidenced by the value of the multiple correlation coefficient $R = 0.8844$. Fluctuations in the amount of income to the general fund per 1 inhabitant depend by 78.2% on the factors included in the regression equation (coefficient of multiple determination $R2 = 0{,}7821$). In Table 11 there are other results of the conducted correlation analysis.

Table 10 The results of the multifactorial correlation-regression dependence of incomes to the general fund of the united territorial communities in Volyn' region per 1 inhabitant

Effective and factor characteristics, coefficients and criteria	Economic and statistical indicators, their calculation method and meaning
Effective characteristic—y	Total income to the general fund per 1 inhabitant, UAH
Factor characteristics: x_1	The income of the community's lands per 1 inhabitant, UAH
x_2	Local taxes and fees per 1 inhabitant, UAH
x_3	Fiscal return of the territory per 1 inhabitant, UAH
x_4	Expenditures of the general fund per 1 inhabitant, UAH
x_5	Expenses for maintenance of the management apparatus per 1 inhabitant, UAH
x_6	Capital expenditures per 1 inhabitant, UAH
x_7	Amount of subsidy (official transfers, basic subsidy) per 1 inhabitant, UAH
Regression model	$y = 467{,}61 - 288{,}131x_1 + 1{,}668x_2 - 0{,}512x_3$ $+ 0{,}166x_4 + 1{,}713x_5 + 0{,}653x_6 - 1{,}369x_7$
Multiple correlation coefficient	$R = 0{,}8844$
Coefficient of multiple determination	$R^2 = 0{,}7821$
Standard error	1084,99
Fisher's criterion	$F_{\Phi akm}(5; 10) = 23{,}59$ $F_{meop}(5; 10; 0{,}95) = 3{,}33$ $F_{\Phi akm} > F_{meop}$

Source Author's calculations

Table 11 Indicators of regression statistics obtained from the results of correlation-regression analysis

Effective and factor characteristics	Standard error	t- statistics	P-value	Lower 95%	Upper 95%
y–Effective characteristic	1134,92	0,412,017	0,682,241	−1816,87	2752,083
X_1	130,4741	−2,20,834	0,032,244	−550,762	−25,4999
X_2	0,444,126	3,74,387	0,000,503	0,768,771	2,55,673
X_3	0,427,205	−1,19,928	0,236,561	−1,37,226	0,347,581
X_4	0,082,235	1,91,843	0,061,275	−0,00,777	0,323,291
X_5	0,599,335	2,858,841	0,00,637	0,507,005	2,919,802
X_6	0,332,656	1,963,904	0,055,601	−0,0163	1,322,906
X_7	0,366,124	−3,7135	0,000,551	−2,09,657	−0,62,263

Source Author's calculations

Indicators that reflect the results of variance analysis, predictive indicators of the effective characteristic etc., were also calculated. The correlation matrix of the obtained regression model is shown in Table 12.

Fisher's F-test was used to determine the reliability of the correlation relation. Its calculated actual value (23.59) is higher than the tabulated value (3.33), which ensures the reliability of the correlation relationship [16]. In order to obtain a qualitative regression model and its corresponding description, calculations were made to detect the effect of multicollinearity using the Farrar-Glober algorithm. The existence of such a phenomenon in statistical science as multicollinearity often leads to the selection of incorrect model indicators [17–19]. When they are transferred to the general population, it is possible to get an unreliable picture of the consequences of decentralization transformations taking place in the socio-economic sphere of rural areas. The abovementioned information does not allow us to effectively solve the

Table 12 Correlation matrix of dependence between selected indicators

	X1	X2	X3	X4	X5	X6	X7
X1	1,000,000						
X2	0,412,275	1,000,000					
X3	0,074,172	0,221,921	1,000,000				
X4	-0,09,909	0,198,733	0,236,023	1,000,000			
X5	0,320,009	0,569,928	0,110,368	0,247,187	1,000,000		
X6	0,323,949	0,408,303	0,098,303	0,126,053	0,306,284	1,000,000	
X7	-0,2056	−0,56,711	−0,06,401	−0,30,567	−0,46,398	−0,46,148	1,000,000

Source Author's calculations

identified problems. The calculations confirmed the absence of a multicollinearity effect.

Of course, the situation in separate rural communities of the Volyn region is significantly different, which is connected with natural-climatic and organizational-economic conditions and requires an immediate response. For example, the land yield of the Kamin-Kashyrskyi territorial community in 2021 amounted to UAH 0.093 per 1 resident, and the territorial community of Serehovychi—UAH 5.079. At the same time, the average value in the territorial community of the region is UAH 1,833. However, all communities have shown that they are able to effectively manage and increase the received financial resources, to respond decently to unforeseen challenges and to ensure a high level of service provision in any conditions, as well as to quickly adapt to new conditions. The communities proved that local self-government budgets form the basis of the sustainability of local financial resources and the socio-ecological and economic development of the territory as a whole.

It is worth noting that, in accordance with the requirements of articles 98–100 and clause 24 of chapter VI, "Final and Transitional Provisions" of the Budget Code of Ukraine, in 2022, the budget equalization system continues to operate, which provides for the horizontal equalization of the tax capacity of the territories of Ukraine depending on the sublevel of income per inhabitant. At the same time, horizontal equalization for local self-government budgets is carried out only on the basis of personal income tax. The horizontal equalization mechanism provides that local budgets with a fiscal capacity index below 0.9 of the average indicator for Ukraine receive a basic subsidy to increase the level of their budget security. On the other hand, local budgets with a fiscal capacity index above 1.1 transfer part of their budget resources to support less capable communities.

Up to June 1, 2022, among the partner communities of USAID "HOVERLA" in the Volyn' region, eight communities received a basic subsidy in the total amount of UAH 46.1 million, and two communities (Volodymyr-Volodymyrska and Boratyn) sent a reverse subsidy to the DBU. In 2021, 4 united territorial communities of the region returned funds to the budget: Boratyn UTC—UAH 1,763 per 1 resident; Lypyny UTC—UAH 1,008; Lutsk—UAH 443; Volodymyr-Volynska—UAH 347 per 1 resident.

Therefore, the new scales, structure, and level of powers of the united territorial communities of the Volyn' region, as well as all rural communities in Western Polissia, require a review of the structure and volume of own revenues to the budget, the development of industrial and social infrastructure, the optimization of management costs, the search for additional sources of capital investments and the financing of development programs with in order to ensure the interests of already new, more structured and numerous communities.

As already mentioned, new challenges appeared on the way to ensure the economic efficiency of the united territorial communities in the rural areas of the studied region in wartime conditions. The formation of an effective monitoring system for internally displaced persons and relocated businesses can be conventionally divided into three categories:

1. transit persons (those who settled in new communities for a short period: from a few days to several months, and then go to other communities or regions, or abroad);
2. temporary groups (expected period of residence on the territory of the community—from several months to one year, depending on the prospect of the end of warfare and the possibility of returning home);
3. entities focused on a permanent change of residence, including the irreversible relocation of their business from the East or South of the country to the Center or the West [20].

Each of these three categories of citizens and entrepreneurs has its own needs and peculiarities, which should be taken into account by the local self-government bodies of those territorial communities in which they temporarily live when developing current plans and long-term strategies for the development of the UTC.

Representatives of the third group (internally displaced persons and entrepreneurs who, for various reasons, do not plan to return) obviously need the most attention from the local authorities, because they become new full-fledged members of the UTC. First, if we are talking about representatives of the business environment, the local authorities of communities (and even regions) where representatives of this category of citizens have settled should clearly define the priority areas of entrepreneurial activity that will meet the strategic goals of the development of a certain community or region, and will have priority support from their local self-government bodies. Secondly, for such types of business, work on the formation of adequate production infrastructure should be started today, for example by creating industrial, agricultural, agro-industrial, technological and scientific parks, industrial zones, etc.

It is necessary to create an effective mechanism for coordinated relocation of enterprises and their potential employees. Optimizing the processes of spatial dispersion of relocated enterprises and internally displaced persons is closely related to the need to revise the strategies of socio-economic development of those regions and territorial communities that receive them. In addition, the specified strategies should depict the priority directions for the formation of food reserves on the territory of the community and the main measures for leveling the security risks, in particular, taking into account the experience gained by Ukraine during the war.

In the context of this study, it is worth supporting the scientists of the State institution "Institute of Regional Research named after M. I. Dolishniy of National Academy of Sciences of Ukraine", who believe that it is now necessary to strengthen cooperation with European partners (both with government institutions and with institutions of civil society) regarding the deployment of a number of new programs and projects aimed at preparing our rulers and fellow citizens for the post-war stage of socio-economic development And territorial communities and their local self-government bodies should act as one of the primary initiators in this matter, especially when it comes to communities and regions that are far from the war zone, but located near the border with the EU [21]. The rural area of Western Polissia, which borders Poland [22], belongs to such territories.

We consider that it is now extremely important to establish effective interaction between state authorities and local self-government, taking into account the principle of subsidiarity. After all, as Y. P. Pavlovych-Seneta believes, in the conditions of martial law, all links of the power hierarchy naturally gravitate towards the centralization of management decisions [23]. Therefore, today's absolute priority is to determine the top-priority tasks that must be entrusted to the local self-government bodies of the rural areas of Western Polissia, in order to, on the one hand, ensure their maximum efficiency in managing the resource potential of the respective territories, and on the other hand, to establish the coordination of the activities of all power institutions with the aim of both timely provision of urgent needs of the front and provision of own development on the basis of sustainability.

In the conditions of martial law, it is extremely important to ensure prompt, proper and continuous implementation of local budgets, which will ensure effective economic development of the UTCs of rural areas of Western Polissia. Despite the ban on state authorities to interfere in the process of forming, approving and implementing local budgets, in order to create conditions for timely and prompt response to the needs of financial support for measures of territorial defense, protection of public safety and functioning of the budgetary sphere, the following are allowed for communal enterprises during martial law measures (Fig. 1).

The conducted research shows that the natural and organizational and economic features of the Western Polissia lead to insufficient financial capacity of a significant part of local budgets, especially of united territorial communities located in

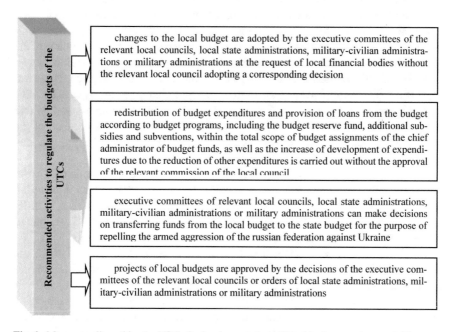

Fig. 1 Measures allowed by the UTCs for implementation of local budgets under martial law

rural areas. The formation of united territorial communities in the rural areas of the region without a sufficient economic base, the creation of new governing bodies in small administrative centers, or with an insufficient population does not contribute to increasing the financial capacity of the relevant local self-government bodies and solving the main socio-economic and environmental issues of development of rural areas in Western Polissia.

Budget and fiscal policy are the most important economic tools of the organizational and economic mechanism of ensuring the sustainable development of the rural areas of Western Polissia, which have a direct impact on the development of the united territorial communities. It is formed and implemented through the system of local and general taxes and fees (revenues) and the corresponding system of expenditures and includes inter-budgetary transfers, contributions to special funds, local loans, rates of taxes and fees, tax benefits, etc.

According to the scientists of the Institute of Agrarian Economics, this is the economic and financial basis for the effective development of each united territorial community [24]. At the same time, we consider that the organizational and economic mechanism for ensuring the sustainable development of rural areas in the UTCs of Western Polissia should ensure the development of social, ecological, and economic-production components, which would not only provide effective development but also preserve historical, cultural and spiritual values of the rural settlement network of the region.

On the basis of the results of the assessment of the progress of the decentralization reform, it was established that up to January 1, 2022, 17 districts were created in the territory of the Western Polissia of Ukraine (Volyn', Zhytomyr, Rivne, and Chernihiv regions), where 241 rural, township, and city united territorial communities are located. It was determined that a significant part of them took advantage of the provided institutional capacity for effective self-governance, despite various obstacles and lack of experience of working in conditions of decentralization. However, there is a difficult political situation in the country, which is caused by russia's military aggression. And as a result, there is an unfavorable socio-demographic and economic situation in the countryside, and therefore, there is a small share of the economically active population of working age, capable of implementing planned strategic measures for development created by UTCs.

It was established that the capacity of the UTCs in the conditions of martial law is its ability to accumulate and rationally use the available financial, material, personnel and other resources, as well as to identify and use their reserves, in order to meet the current needs, the needs of internally displaced persons, the development of the territory and military logistics, formation of competitiveness of the UTCs in the future. It has been established that the local budgets of the regions of the Western Polissia zone of Ukraine, which are currently functioning under martial law and decentralization of the powers of authorities and financial resources for their implementation, have undergone changes in the system of formation of local budgets and inter-budgetary relations. After all, in the conditions of martial law, the UTCs have been given the right to transfer funds from the special fund of the local budget to the general fund of the local budget.

The authors established that currently a significant part of the financial resource, which is transferred from the state budget to local budgets, is directed to the budgets of united territorial communities. First of all, these are financial resources to support the medical and educational sectors. Realization of other expenses from the budgets of the rural UTCs (social organization and social security, housing and communal security, culture and sports, development of tourism, etc.) occurs exclusively at the expense of the revenues of the relevant budgets. The negative thing is that a significant share (more than 30.0%) is spent on state administration—organizational, information-analytical, and logistical support for the activities of village councils. Under the conditions of martial law, the united territorial communities are allowed to make expenditures that are not assigned to the relevant local budgets and expenditures for the maintenance of budgetary institutions at the same time from different budgets. Such expenditures are made by providing an interbudgetary transfer from the relevant local budget.

3 Conclusions

According to the results of the conducted multifactorial correlation-regression analysis, it was established that the general profitability of territorial communities is influenced by the profitability of community lands; local taxes and fees; fiscal return of the territory; expenses of the general fund; expenses for maintenance of the management apparatus from the calculation; capital expenditures and the amount of subsidies (official transfers, basic subsidy).

It was found that in recent years, the system of local budgets, as one of the most important tools of the organizational and economic mechanism of ensuring sustainable rural development in UTCs, has undergone fundamental transformations from its participants and due to the change of financial sources of local self-government bodies of various levels. It was established that the result of the decentralization reform in the rural areas of Western Polissia was an increase in the interest of local self-government bodies in increasing revenues to local budgets, finding reserves to fill them, and improving the efficiency of tax and fee administration.

In order to replenish the revenue part of the budgets of the UTCs located in the rural areas of the Western Polissia of Ukraine, it is considered expedient to diversify the sources of financial resources and centralize them through the distribution of tax on the income of individuals and rent payments for the use of subsoil for the benefit of local budgets. It will be possible to improve the resource support of rural communities by securing stable sources of income in local budgets and expanding their income base on legislative grounds. The main factors of the growth of general fund income: growth of wages and other taxable income of individuals and entrepreneurs; legalization of hired labor; conclusion of new lease agreements for land plots; updating and indexing of the normative monetary valuation of lands; increase in the tax base due to the growth of the living wage and the minimum wage; rise in prices for excise goods, etc.

According to the use of funds in the rural UTCs of Western Polissia, it is worth emphasizing the need to form the most optimal structure of budget expenditures, creation of an effective and not too numerous management apparatus, a constant analysis of the spending of budget funds and prevention of cases of their irrational spending. After all, the presence of sufficient resources in local budgets is a guarantee that the territorial community has the opportunity to provide higher quality and more diverse services to its residents, implement social and infrastructure projects, create conditions for the development of entrepreneurship, attract investment capital, develop local development programs and finance other measures to comprehensively improve the living conditions of community residents. The strategic guidelines for the sustainable development of the rural areas of Western Polissia in the post-war period should be the provision of real self-governance at the level of territorial communities and their material and financial support.

References

1. Number of Administrative and Territorial Units by Regions of Ukraine. Accessed 25 Jan 2023 from https://www.ukrstat.gov.ua/operativ/operativ2016/ds/ator/ator2017_u.htm
2. Savchuk, P.P., Sokolova, A.O., Polishchuk, M.O., Gonta, N.A.: Modern realities and assessment of decentralization transformations in rural areas: regional aspect. In: Development trends of the world agriculture in the XXI-st century: the view of the modern scientific community, pp. 234–257. Baltija Publishing, Riga, Latvia (2022). https://doi.org/10.30525/978-9934-26-203-6-10.
3. Tymoshenko, M.M.: The Strategy of Sustainable Development of Rural Areas of Ukraine and the Economic Mechanism of its Implementation: Monograph. Publisher O.O. Yevenok, Zhytomyr (2018)
4. Khomiuk, N.: Decentralization as a factor diversification of development of rural territories. Econ. J. Lesya Ukrainka Eastern Euro. Natl. Univ. 1(17), 85–91 (2019). https://doi.org/10.29038/2411-4014-2019-01-85-91
5. Resolution of the Verkhovna Rada of Ukraine No. 807-IX dated 17.07.2020 "On the Formation and Liquidation of Districts". Accessed from https://zakon.rada.gov.ua/laws/show/807-20#Text
6. Decentralization: Official state website of Ukraine. Accessed 10 Feb 2023 from https://decentralization.gov.ua/newrayons?area_id=&sort_by_otg_count=&sort_by_villages_count=&sort_by_square=&sort_by_population=
7. Monitoring of the Reform of Local Self-government and Territorial Organization of Authority (2022). Ministry of Development communities and territories of Ukraine. Accessed from https://decentralization.gov.ua/uploads/library/file/800/10.01.2022.pdf
8. Titenko, Z.: Budgetary decentralization as a factor for the rural development of rural areas of Ukraine. Agrosvit 7, 87–91 (2020). https://doi.org/10.32702/2306-6792.2020.7.87
9. Peculiarities of the Budget Process Under Martial Law. Accessed 17 Feb 2023 from https://decentralization.gov.ua/news/14654
10. Parliament Adopted Amendments to the Budget Code Regarding Local Budgets Under Martial Law (9.07.22). Association of Cities of Ukraine. Accessed from https://auc.org.ua/novyna/parlament-pryynyav-zminy-do-byudzhetnogo-kodeksu-shchodo-miscevyh-byudzhetiv-v-umovah-diyi
11. Riabokon, V.P.: Decentralization is the way to rural development in Ukraine. Econ. Agro-Ind. Complex: Int. Sci. Prod. J. 1, 6–17 (2020). https://doi.org/10.32317/2221-1055.202001006

12. Khomiuk, N., Hrytsiuk, P.: Analysis of the united territorial communities' economic activities effectiveness in the regions of Ukraine. Sci. Notes Ostroh Acad. Natl. Univ. Econ. Ser. **14**(42), 45–53 (2019). https://doi.org/10.25264/2311-5149-2019-14(42)-45-53

13. Kachniarz, M.: Bogactwo Gmin–Efekt Gospodarności Czy Renty Geograficznej? Ekonomia **5**(17), 81–94 (2011). Wydawnictwo Uniwersytetu Ekonomicznego we Wrocławiu. https://dbc.wroc.pl/Content/25999/PDF/Kachniarz_Bogactwo_gmin_efekt_gosp odarnosci_czy_renty.pdf

14. Sokolova, A.O., Honta, N.A.: Correlation-Regression Analysis of the Impact of Decentralization Transformations on the Development of Rural Areas. Priority directions of economic development: Scientific Discussions: materials of the All-Ukrainian scientific and practical conference on May 22. Zhytomyr: Polis National University, 17–22 (2020).

15. Municipality budgets in Ukraine: dashboard (2021). Accessed from https://decentralization. gov.ua/news/14281?page=3

16. Semenov, A., Baraban, S., Semenova, O., Voznyak, O., Vydmysh, A., Yaroshenko, L.: Statistical express control of the peak values of the differential-thermal analysis of solid materials. SSP **291**, 28–41 (2019). https://doi.org/10.4028/www.scientific.net/SSP.291.28

17. Engelhardt-Nowitzki, C., Kryvinska, N., Strauss, C.: Strategic demands on information services in uncertain businesses: a layer-based framework from a value network perspective. In: 2011 International Conference on Emerging Intelligent Data and Web Technologies (2011). https:// doi.org/10.1109/EIDWT.2011.28

18. Stoshikj, M., Kryvinska, N., Strauss, C.: Service systems and service innovation: two pillars of service science. Procedia Comput. Sci. **83**, 212–220 (2016). https://doi.org/10.1016/j.procs. 2016.04.118

19. Kryvinska, N., Kaczor, S., Strauss, C., Gregus, M.: Servitization strategies and product-service-systems. In: 2014 IEEE World Congress on Services (2014). https://doi.org/10.1109/SER VICES.2014.52

20. Borshchevskyi, V., Matvieiev, Ye., Kuropas, I., Mykyta, O.: Territorial Communities Under Martial Law: How to Ensure Effective Management in the Context of Post-war Development Priorities. Public space (2022). Accessed from https://www.prostir.ua/?news=terytorialni-hro mady-v-umovah-vojennoho-stanu-yak-zabezpechyty-efektyvne-upravlinnya-v-konteksti-pri orytetiv-povojennoho-rozvytku

21. Borshchevskyi, V., Mahas, V., Tsimbalista, N.: Development Potential and prospects for revitalization of rural areas in the conditions of modern reforms: scientific and analytical report. Lviv: State Institution "Institute of Regional Research named after M.I. Dolishniy of the National Academy of Sciences of Ukraine", p. 44 (2017). Accessed from http://ird.gov.ua/irdp/p20170 202.pdf

22. Rosner, A., Stanny, M.: Socio-economic development of rural areas in poland. Warsaw: EFRWP, IRWiR PAN (2017). Accessed from https://www.irwirpan.waw.pl/dir_upload/site/ files/Lukasz/MROW_en_2017.pdf

23. Pavlovych-Seneta, J.P.: Administrative and legal prerequisites for the economic development of united territorial communities under martial Law. Anal. Comp. Jurisprud. **2**, 222–226 (2022). https://doi.org/10.24144/2788-6018.2022.02.42

24. Patyka, N.I., Bulavka, O.H.: Strategic guidelines and priority directions for sustainable development of rural communities and territories in Ukraine in the context of power decentralization. Econ. Agro-Ind. Complex: Int. Sci. Prod. J. **8**, 91–102 (2021). http://eapk.org.ua/en/contents/ 2021/08/91

The Technological and Environmental Effect on Marketing of Children's Food

Darya Legeza⊙, Yuliia Vlasiuk⊙, Tetiana Kulish⊙, Yana Sokil⊙, Wei Feng, Farhod Ahrorov⊙, and Saule Yessengaziyeva⊙

Abstract Technology, nature, and ecology are increasingly becoming crucial issues in producing foods for children. The major aim of the study is to examine the level of the technological and environmental influence of various factors on the production and marketing of vegetable-based products. We made a hypothesis that there are four alternatives to the new strategy as Variety expansion, eco-intelligent production, implementation of innovative technologies, and entering into the External market. The criteria for selecting factors and alternatives were examined according to the expert survey. We used the AHP indicator (Analytical Hierarchy Process) to evaluate interconnection links between each factor. We investigated that the alternative of organic production has an indicator of 0.341 points and implementation of new technologies has an indicator of 0.255 points, which makes them applicable solutions for the further production of vegetable products for children. The expansion of a product variety and the increase of an entrepreneur's profit in the internal market is still under discussion.

Keywords Marketing macro system · Vegetable consumption · Children's food · Analytical hierarchy process · Eco-intelligent production · Novel technology

D. Legeza (✉) · T. Kulish · Y. Sokil
Dmytro Motornyi Tavria State Agrotechnological University, Zhukovsky Str., 66, Zaporizhzhia 69600, Ukraine
e-mail: darya.legeza@tsatu.edu.ua

Y. Vlasiuk
Khmelnytsky Polytechnic Proffessional College By Lviv Polytechnic National University, Zarichanska Str., 10 Khmelnytsky, Khmelnytskyi 29015, Ukraine

W. Feng
Northwest Agricultural & Forestry University, 3# Taicheng Rd., Yangling 712100, Shaanxi, China

F. Ahrorov
Samarkand Branch of Tashkent State University of Economics, 51, Professorlar Street, Samarkand, Uzbekistan 140147

S. Yessengaziyeva
Al-Farabi Kazakh National University, 71 Al-Farabi Avenue, Almaty 050040, Kazakhstan

1 Introduction

Technology, nature, and ecology are increasingly becoming crucial issues in producing foods for children. The fundamental factors of healthy nutrition are quality control systems, a clean environment, and eco-intelligent production. While environmental pollution negatively affects the quality of products, organic and ecological production offers a new solution to healthful consumption. Primary studies of environmental and technological influence are viewed as separate parts of a process, but a total quantitative effect has not been understood. While technology-oriented investigators propose solutions for precocious production, advocates of ecological direction point out the idea of offering eco-friendly foods with higher costs. Only 50.6% of consumers notice their satisfaction with food nutrition [1].

A sustainable food production system requires reducing energy consumption with the help of organic production [2]. From this side, such a system should take into account each hazard during the formation of value-added products [3]. Current technologies should cover special peculiarities of children's nutrition such as dietary, obesity, and allergies [4]. Emerging processing and preservation technologies (high-pressure processing) provide safe nutrition with the help of key control points of safety management [5]. It should be mentioned that not all consumer may change their regular nutrition and buy new products produced with the use of innovative technologies. While neophobic people prefer familiar food, neophilic people are eager to try some new foods.

What makes children's nutrition friendly to the natural environment and stimulates people to consume ecological products? In the US, children from five years old prefer natural products and evaluate foods by tastiness, safety, naturalness, and desire to consume [6]. However, the total level of children's consumption of healthy foods in the world is not contented because children do not evaluate fresh vegetables as nutrition. From this side, the involvement of children in the cognitive process leads to increasing interest in consuming natural products, which they perceive as not tasty foods. Interactive games, practical cases, and bonuses play an essential role in the development of goal-setting by children to prefer fresh and nutrient-concentrated foods [7]. Interaction with consumers in social networks drives up sales when a producer takes into account the individual image of a food brand [8]. Social environment correlates positively with children's nitration of foods and fresh vegetables when parents' school authority award them [9]. Healthful making decisions for youth should be a key issue for a future marketing strategy for a company and state programs [10]. Due to the absence of the quantitative influence of technological and environmental factors on consumer decisions to buy children's food, the hierarchy-marketing model is set to become the most suitable alternative for making decisions. It allows producers to develop an eco-oriented customer profile and offer demanded products in the children's food market.

2 Literature Review

The focus of the latest research in the marketing of children's food has been on the topic of safety and natural production, which includes new technology of production, packaging, storage, and transportation [11–14]. For several last years, great efforts have been devoted to the study of COVID-19's influence on children's nutrition [15–17]. To solve such problems the literature on the marketing of children's food emphasizes various approaches such as the development of PR companies [18–20], and social media marketing [21, 22]. Moreover, questions about the expansion of a market and product variety are still essential. The central problem of the article is to take into account the cross-factor impact on the selection of different marketing strategies. Significant marketing alternative strategies are variety expansion, eco-intelligent production, implementation of innovative technologies, and entering into a new market.

2.1 Variety Expansion

The pattern of children's nutrition is shaped by the interplay of many marketing factors which may potentially lead to a positive trend of increasing healthy food consumption [23]. Due to increasing interest in natural and safe products, the variety of children's food is extremely growing but still limited. 48% of parents of children under 6 years claim that the choice of organic products is too small [24]. Lack of support and funds in science, slow innovation in logistic chains, and limited variety of fresh vegetables are claimed to be the significant reasons for an insufficient list of foods on shelves. Ilic's group supports the hypothesis about the influence number of meals on a product variety [25]. They claim that 50% of children, who have a complete list of nutrition at a school, consume leafy vegetables. Pickard et al. [26] conclude that children make decisions about food selection according to formed knowledge and script structures of experience, which they have gotten from parents or at school. The more a household has a higher level of education, the more a child eats vegetable products and less pays attention to advertisements about unsafety food in mass media [22]. However, producers prefer to invest in advertisements for unhealthy food to promoting of healthy dietary food [27]. It potentially leads to a decrease in interest and a lack of information about new types of products. To grow the demand for children's food restaurants in the US spend 59% of their marketing budget to buy toys and other games [28]. In such a way, owners stimulate families with children to order meals, which do not meet nutrition standards for children. The findings from this literature review suggest that the promotion of natural products is restricted by producers' wishes to offer unsafety products while the government does not pay enough attention to the promotion of various healthy nutrition among children. Effective practices of public and social promotional programs may increase interest in healthy fresh food, especially vegetables. Trujillo et al. [29] examine that involvement of children in educational activities (market fairs, culinary master-classes, seminars of nutrition, home gardening, and sensor laboratory) increased consumption of fruits and vegetables in three times per week.

2.2 Eco-Intelligent-Production

The ecological model reveals eco-environmental principles based on human behavior. In 1999, [30] estimated that parents prefer environmental consequences more to purchase intentions when they make a decision. Evans et al. [31] underline four factors such as availability, opportunities, social structures, and policies. Health, society and ethical drivers play a vital role in selecting ecological products by parents [32]. Food produced in a local region is healthier for inhabitants' consumption and has natural benefits than non-local products [33]. Cleveland et al. [34] calculated 'food miles' and proved that transportation of fruits and vegetables produced out of a country leads to increasing greenhouse gas emissions. They suggest if producers supply and fill a regional market with local brands, it will promote harmony inside the agro-food system. In 2019–2022 years, [35] studied, that the formation of food circles and food consumers' cooperatives caused demand for local and organic foods.

Packaging influences environmental pollution essentially. Production of metal and glass packages demands energy resources in addition to polyethylene utilization, which causes sulfur dioxide emissions [3]. Moreover, ex-packaging utilization may lead to greenhouse gas emissions, and the decomposition of plastic waste can last for years. If a consumer buys local food, he or she may reduce the greenhouse gas emission by 4–5% [36]. Urban and industrial pollution ranked as significant factors of environmental problems after a lack of knowledge, facilitation skills, and training [37]. At that time, the European market had strict rules for using plastic packages in the food industry. Food packaging may have two antagonist functions eco-intelligent utilization or biodiversity pollution. Organic and natural-friendly products require the appropriate eco-intelligent pack to support the idea of healthy and natural nutrition. It reveals the explicit intentions of a food producer to follow sustainable approaches in the production, distribution, storage, and utilization of a product and its package. On the other side, high-innovated chemical packaging allows traders to prolongate storage. A market requires attractive packaging, while the environment needs rapidly decaying waste.

2.3 Implementation of Innovative Technologies

Gradual population growth with a simultaneous deterioration in the quality of land resources and post-harvest losses leads to food shortages and famine in densely populated countries. A study by Zhang et al. [38] has shown that farmers might lose only about 3% of the harvest in regions where famine occurred, while mortality will get only 1%. Using the difference-in-difference (DID) model, they concluded that farmers who memorize famine in childhood might predict further harvest losses. Farmers lose about 40% of fruits and vegetables after harvesting in developing countries because they do not use machines and equipment for harvesting, shipping, and storage [39]. In India, farmers lose from 30 to 80% of fruits and vegetables because of lack of technologies in clod chain monitoring and packaging [40]. Low producers' profit and restrictions in investments limit opportunities to imply cutting-edge technologies in the agriculture and food industry [40–43]. Therefore, producers are looking for innovative decisions in mass food production with prolonged time for distribution and storage. Advocates of industrial breakthroughs propose innovative technologies in plant breeding, production processes, freezing, storage, shipping, and distribution [44]. Such novelty ideas are claimed to result in the quality level of children's consumption in volume and value indicators [5, 6, 45]. A controlled management system in vertical farming allows optimal land use and boost the harvest of vegetable-based plant. There is still considerable demand for vegetables grown by regular land-using technologies ftp compared with products from vertical farming [46]. While vertical farming is still a challenge in the mass production of vegetables today, it will provide cheaper products without the usage of significant sizes of land resources in the future [47]. Hydroponics and aquaponics improve the results of getting more yield without the use of chemical fertilizers. Kumar's group [48] claimed that hydroponics as a new model system of agriculture allows plant vegetables and greens in regions with limited arable lands and producing eco-friendly products. Shock and dry freezing allow providing vegetables for children during a year, and avoiding the problems of seasonal production. Pandey et al. [49] emphasize that demand for natural antimicrobials and insecticides will increase stably because food companies need quality fresh raw materials, and final consumers require natural products. Technologies with instruments of augmented reality provide permanent monitoring of technological processes and reveal hazard points. Companies in the food industry and retailers maintain software with AI to connect with consumers on issues of diet and nutrition [50]. Suryantari's group [51] concluded that augmented reality instruments assist in exploring information and visualizing data for product ordering, for instance, by using 3-D menu. Lockdown during COVID-19 have launched a virtual ordering and digital search of products with the help of 3-d photo [52]. The scientist emphasizes that companies such as Blippar maintain an application with games or animation to promote a brand. AR motivates consumers to order products online because they pay attention to an attractive and novel package [53]. Masnar's group [54] revealed that AR technologies on packaging might inform the population directly about their potential healthy consumption. Such packages contain educational messages about the consistency of harmful ingredients in a product. Leading food companies use AI for food sorting, product safety monitoring or quality testing, and supply chain management. Ukrainian scientists investigated drone plant-monitoring systems to enhance climate control management and improve the technology of optimal fertilization.

2.4   Entering into a New Market

The intentions and needs of consumers stimulate the development of new business models in the food market. It is paramount for companies to be both up-to-date with consumer behavior and competition in new market niches. See, for example, [55–59]. Peer mental model [60–62], white label [63], and co-creation [64–66] as an example might be the variants to launch business abroad. Communicating with clients online in different ways enhances opportunities to enter an unknown market. According to studies of Dagoudo's group [67], the joint-educated model might increase the sustainability of women's entrepreneurship in organic production. Nowadays, there is an abundance of diverse market approaches to enter the market, either to promote a company or to find new customer bases. Companies are significantly developing numerous services to make distribution affordable for new companies. See examples at [68–72]. Drop shipping, fulfillment, and online marketplace provided in the food market are essential parts of business marketability and the development of Industry 4.0. Miljenović and Beriša [73] accented that drop shipping has stimulated the development of international trade during the lockdown in COVID-14. Such businesses offer a sizable number of services and analytical materials about food requirements that are vital in boosting a company's existence in the market. The better a company knows its customer behavior, the deeper it enters a market, the more purposeful it offers a product, and the more successful it will be in a new market. Online fulfillment services make rural businesses more accessible to an urban population. Abdoellah's group (2023) has concluded that issues such as the remoteness of agricultural lands and urban on-market culture influence significantly a producer's income.

Nowadays, the proposition of a new idea but not a product satisfies customer needs. A current consumer has experience-based knowledge about its pain on the market and looks for a decision. For example, a mother of a child who needs vitamins or minerals searches for any product that consists of it. She does not compare products, but she evaluates products' composition. Buyers of children's food pay for a product's value and its preferences, for instance, organic, natural, vitamin-based, antiallergenic, fat or lactose-free, and zero-kilometer. In consistence with Aizaki's group [74], the location of production affect significantly the consumer decision and preferences to buy local products. Low household income exacerbates the problem of malnutrition and the purchase of low-quality cheap food in families with children. In agreement with studies of Scholz and Kulko [75], the paying ability of consumers and the price sensitivity of a food company influence the decrease of fruit waste on a shelve and total sale income. On this side, the government supports sustainable and healthy consumption in families with children through vibrant programs. Leading companies invest in educational programs for households and schools to interest people in using natural and healthy products on their menus.

3 Aim, Hypothesis and Tasks

The major aim of the study is to examine the level of the technological and environmental influence of various factors on the marketing of fresh and processed vegetables.

We made a hypothesis that there are four alternatives to the new strategy such as variety expansion, eco-intelligent production, implementation of innovative technologies, and entering into the children's food market. This paper is divided into three sections. The first section gives a brief overview of central factors (technological, ecological, and natural) and the main alternatives to develop a marketing strategy. The second section examines a comparison of each factor and each group. A new strategy is described in the third section according to four proposed alternatives.

4 Methods

The criteria for selecting forces and alternatives have been examined according to expert survey. Eleven experts are managers from Ukrainian vegetable factories, marketing cooperatives and farmers. One more expert is a member of Ukrainian Horticulture Business Development Project. The primary survey was conducted in 2018 and prolonged during September and October (2020) to investigate influences of technological and environmental causes on consumer behavior. Previous study has revealed particular technological and natural factors, which might lead to positive and negative changes in market development [76]. Experts estimated all groups, forming marketing macro system in order to reveal the crucial forces according to a 10-point scale. A factor from whatever groups has integer indicator from 0.25 to 1.35 points. We extract the most influenced factors, which have integer indicator more than 0.45 points. The most essential among others have been revealed technological, natural and ecological factors (Fig. 1). For simplicity of modelling, we ignore the dependence of other factors in modelling.

Evaluating effect on food marketing changes have shown that technological forces lead to marketing changings mostly. Based on previous research and analysis of experts' opinion, such technological trends inspire or inhibit development of marketing of vegetable for children:

– low growth of domestic innovation process (Innovation growth);
– lack of government support in science and technological development (Government support);
– low efficiency of plant breeding in science institutions (Breeding);
– outdated manufacturing technologies (Outdated technologies);
– the emergence of novel technologies of vegetable freezing and storage (Freezing and storage technologies);
– modern technologies in quality management control and supply chain management (QMC);

Technological factors
- Slow growth of domestic innovation process;
- Lack of government support in science and technological development;
- Low efficiency of plant breeding in science institutions;
- Outdated manufacturing technologies;
- The emergence of novel technologies of vegetable freezing and storage;
- Modern technologies in quality management control and supply chain management;
- Innovative packaging;
- The requirements of European trade associations to standards for children's food;
- The possibility of non-waste production.

Marketing strategies

1. Expansion of product variety
2. Eco-intelligent production
3. Implementation of innovative technologies
4. Entering into the abroad market

Nature factors
- A limited volume of fresh vegetables;
- Perishability of vegetables;
- The usage of adjuvants in production and storage;
- Limitation of fresh vegetable range in farms.

Ecological factors
- Environmental pollution in a growing place of vegetables;
- Specific hygiene requirements to manufacturing process;
- standards and certification procedures;
- The use of alternative energy sources in manufacturing;
- Foreign investments into organic production.

Fig. 1 Factors and alternatives on the marketing of vegetables for children

- innovative packaging (Packaging);
- the requirements of European trade associations to standards for children's food (Requirements to standards);
- possibility of non-waste production (Non-waste production).

We used a variation of natural forces to eliminate essential influences on children's behavior. The key formed natural factors are:

- a limited volume of fresh vegetables (Limited volume);
- perishability of vegetables (Perishability);
- the usage of adjuvants in production and storage (Adjuvants usage);
- limitation of fresh vegetable range in farms (Limited Variety).

As a third group of forces, we choose ecological influence on children's nutrition and their reaction on the marketing mix. Nevertheless, the ecological group presents rather productive instruments then marketing because conditions of production ground the quality of future value-added products. Furthermore, the questions of environmental safety should be taken into account during research on human behavior. Current clients pay attention to issues of safety and nutrition during operational processes of production. By the whole of the ecological group consists of such factors:

- environmental pollution in a growing place of vegetables (Environmental pollution);
- specific hygiene requirements for the manufacturing process (Hygiene requirements);
- standards and certification procedures (Standards and certification);
- the use of alternative energy sources in manufacturing (Alternative energy sources);
- foreign investments into organic production (Investments into organic production).

To forecast potential future behavior Camerer, [77] use Cognitive hierarchy (Poison CH) and compare it with equilibrium models. The approaches of the model evaluation in the article are the same as those used by [78]. In order to correlate each factor, we used the AHP indicator (Analytical Hierarchy Process). We used the program SoftAnalyzeHierarch because it supports numerous interconnections. This method was chosen to select the outstanding alternatives of marketing strategies for food companies.

The first hypothesis regarding a marketing strategy suggests that a diverse variety of products offered at a market will lead to ensure vegetable and fruit consumption. The more a consumer has choices of proposed products the more products he or she is ready to buy.

Secondly, various approaches have been proposed to ensure marketing alternatives through the implementation of eco-intelligent production. While ecological factors affect directly, the technological group might carry out negative results.

Thirdly, when a farmer uses current technologies to grow products, he might create a new product or even develop a new market. It has now been hypothesized, that implementation of modern technologies might attract customers to demand a product.

Finally, the other point of view reveals solutions connected with food market expansion. Natural factors allow growing only a limited list of vegetables and fruits in a region, while a season of production might hinder farmers' opportunities for a long time of distribution.

5 Results

The market for children's food is changing under the influence of direct or indirect factors that affect the results of marketing activities of enterprises, including fruits and vegetables. Such changes include natural-geographical, ecological, and technological forces, which should be taken into account for developing a marketing strategy for companies. The consumer-oriented strategy is the key to the company's competitive position. Moreover, such a strategy should take into account the specifics of children's food because demand and supply in the market of such products are formed by not only the price and quantity of goods but also their safety. Globalization of economic relations intensifies internal competition among companies, which try to enter the children's food market. Local companies meet competitive barriers built by world leaders such as Gerber and lose their position in local markets in price and variety. The solution for little-known companies is to offer products produced by the national traditional recipes, in ecological regions, or from rare fruit and vegetable varieties. From this side, it is necessary to ground the basic marketing directions of the sale of vegetable products in the children's food market to strengthen its competitive position. The authors have evaluated four possible alternatives of marketing strategy development, which take into account environmental, scientific, and technological forces such as a variety expansion of canned vegetables, eco-intelligent products, implementation of innovative technologies in production and distribution, and development of new marketing channels. Table 1 presents a comparison of the importance of technological factors.

According to the evaluation of the interdependence of technological forces, QMC affects most comprehensively and provides consumer quality on technological, transport, distribution, and other operational processes. The reason is that a consumer, a child's parent, evaluates the quality of the product by safe consumption. Because of improper storage or non-compliance with temperature requirements, the marketable condition of a product might deteriorate, and damage to a product might endanger the child's nutrition.

Priss and Zhukova [79] emphasize that current technologies of storage and freezing of vegetables provide numerous opportunities for further development of vegetable products. The supply of frozen vegetables on the children's food market provides an increase in the season of product consumption. However, such technologies should involve the use of reagents, which are not dangerous for a child. Table 2 presents possible alternatives to the development strategy taking into account technological factors.

Technological processes such as storage (0.458) and non-waste production (0.296) influence directly on expansion of a vegetable variety in the market. For example, a consumer evaluates a vegetable through a possible time of keeping a product edible. A producer might offer it in fresh, dry, frizzing, canned, or baked form according to consumer intention. In the children's food market, parents expect to get healthy eatable food during the year regardless of seasonality. Furthermore, the need for vitamins set for children requires purchasing vegetables not produced in a living

Table 1 A comparison of the group of technological factors

Factor	Innovation growth	Government support	Breeding	Outdated technologies	Freezing and storage technologies	QMC	Packaging	Requirements to standards	Non-waste production	AHP
Innovation growth	1	2	5	1	1/2	4	4	5	4	0.190
Government support	1/2	1	3	1/2	1/3	4	3	4	4	0.139
Breeding	1/5	1/3	1	1/6	1/4	1/3	1/3	1	1/2	0.031
Outdated technologies	1	2	6	1	1	3	2	2	5	0.180
Freezing and storage technologies	2	3	4	1	1	4	3	5	3	0.219
QMC	1/4	1/4	3	1/3	1/4	1	2	6	4	0.934
Packaging	1/4	1/3	3	1/2	1/3	1/2	1	1	1/3	0.050
Requirements to standards	1/5	1/4	1	1/2	1/5	1/6	1	1	1/4	0.035
Non-waste production	1/4	1/4	2	1/5	1/3	1/4	3	4	1	0.062

Source Examined by the authors according to the expert survey

Table 2 AHP of the group of technological factors

Strategy	Variety expansion	Eco-intelligent product	Innovation technologies	External market	AHP
A comparison of strategies by innovation growth					
Variety expansion	1	4	1/5	1	0.165
Eco-intelligent product	1/4	1	1/7	1/4	0.055
Novel technologies	5	7	1	4	0.608
External market	1	4	1/4	1	0.172
A comparison of strategies by government support					
Variety expansion	1	1/7	1/6	1	0.063
Eco-intelligent product	7	1	3	7	0.573
Novel technologies	6	1/3	1	6	0.302
External market	1	1/7	1/6	1	0.062
A comparison of strategies by breeding					
Variety expansion	1	1/2	4	6	0.360
Eco-intelligent product	2	1	4	4	0.461
Novel technologies	1/4	1/4	1	1	0.093
External market	1/6	1/4	1	1	0.086
A comparison of strategies by outdated technologies					
Variety expansion	1	7	1	1/7	0.173
Eco-intelligent product	1/7	1	1/7	1/7	0.034
Novel technologies	1	7	1	3	0.398
External market	7	7	1/3	1	0.395
A comparison of strategies by freezing and storage technologies					
Variety expansion	1	7	1	6	0.458
Eco-intelligent product	1/7	1	1/7	1/6	0.043
Novel technologies	1	7	1	3	0.360
External market	1/6	6	1/3	1	0.139
A comparison of strategies by QMC					
Variety expansion	1	1/6	1/2	1/8	0.055
Eco-intelligent product	6	1	7	2	0.513

(continued)

Table 2 (continued)

Strategy	Variety expansion	Eco-intelligent product	Innovation technologies	External market	AHP
Novel technologies	2	1/7	1	1/5	0.081
External market	8	1/2	5	1	0.352
A comparison of strategies by packaging					
Variety expansion	1	1/3	1/6	1/8	0.046
Eco-intelligent product	3	1	1/3	1/7	0.093
Novel technologies	6	3	1	1/5	0.213
External market	8	7	5	1	0.648
A comparison of strategies by requirements to standards					
Variety expansion	1	1/9	1/9	1/9	0.031
Eco-intelligent product	9	1	1	1/4	0.178
Novel technologies	9	1	1	1/7	0.163
External market	9	4	7	1	0.628
A comparison of strategies by non-waste production					
Variety expansion	1	9	1	9	0.296
Eco-intelligent product	1/9	1	1/9	1	0.044
Novel technologies	4	9	1	9	0.614
External market	1/9	1	1/9	1	0.045

Source Examined by the authors according to the expert survey

areal. Because numerous vegetables are perishable, parents need to look for imported processed products.

Government support (0.573) plays an important role in the presence of eco-intelligent products in the vegetable market. Various state programs, such as culinary schools for children or classes in food preparation, stimulate families to consume eco-intelligent vegetables. They pay attention more to safety, naturalness, and health, which form part of sustainable consumption.

Table 3 presents a comparison of the importance of natural factors.

According to the findings, the usage of adjuvants in production and storage (0.476) is the major group of natural factors, which might influence on consumer intention in the vegetable market. When parents select products for their children, they are looking for safety. Preservatives, sweeteners, and thickeners, for instance, might lead child's allergy. The results of modeling show that such an issue is more valuable

Table 3 A comparison of the group of natural factors

Factor	Limited volume	Perishability	Adjuvants usage	Limited variety	AHP
Limited volume	1	2	1/2	5	0.289
Perishability	1/2	1	1/3	4	0.176
Adjuvants usage	2	3	1	6	0.476
Limited Variety	1/5	1/4	1/6	1	0.059

Source Examined by the authors according to the expert survey

among others to advent marketing decisions. Table 4 presents possible alternatives to the development of marketing strategy taking into account natural factors.

Natural factors influence children's food market besides environmental and technological factors. The level of contamination of the territories where the products were grown, and the possibility of growing organic vegetables become the main criteria for selecting a brand. Each region has its individual regional peculiarities of vegetable production especially the exploitation of machines and technological belts. Therefore, the last studies have been in-depth and prolonged to reveal unaccounted essential ecological factors. Experts who participated in the last studies have evaluated the existing list of forces, such as environmental pollution, hygienic requirements for staff, sanitation, and investment in organic production. They have supplemented the list by factor usage of alternative sources of energy.

Table 5 presents the indicator of the Analytical Hierarchy Process of ecological factors.

The crucial economic forces are environmental pollution (0.292) and hygiene requirements (0.308). The usage of alternative energy sources has a minimal rate, equaling 0.050. It is essential to note, that experts emphasize on vitality of investments in organic production because a vegetable market has a niche for children's food. However, high vested interests and prices on raw materials limit the ongoing development of organic products for children. Costs of organic vegetable production include costs of bio fertilization, usage of which influences harvesting, quality, and price of final fresh goods. Table 6 presents possible alternatives to the development strategy, taking into account ecological factors.

The data show that the alternative production of environmentally friendly food is due to such factors as environmental pollution (0.592), the usage of alternative energy sources (0.737), and the investment attractiveness of organic production (0.613).

The exploitation of innovative technologies is based on taking into account hygienic requirements for production and sales (0.312) and production standards (0.455). Experts note that not all modern technologies help to improve the physical and chemical components of product quality. Producers need to use a certain list of chemicals to increase product life during freezing and transporting vegetables. In this regard, numerous technologies, which include the application of chemical components, cannot ensure healthy and safe products for children's nutrition.

Table 4 AHP of the group of natural factors

Strategy	Variety expansion	Eco-intelligent product	Innovation technologies	External market	AHP
A comparison of strategies by limited volume of fresh products					
Variety expansion	1	8	7	1	0.435
Eco-intelligent product	1/8	1	2	1/8	0.068
Novel technologies	1/7	1/2	1	1/8	0.049
External market	1	8	8	1	0.447
A comparison of strategies by perishability					
Variety expansion	1	3	1/6	1	0.156
Eco-intelligent product	1/3	1	1/4	1/3	0.079
Novel technologies	6	4	1	4	0.596
External market	1	3	1/4	1	0.168
A comparison of strategies by adjuvants usage					
Variety expansion	1	1	4	1	0.358
Eco-intelligent product	1	1	1/3	1/3	0.144
Novel technologies	1/4	3	1	1	0.220
External market	1	3	1	1	0.278
A comparison of strategies by limited variety					
Variety expansion	1	5	3	1	0.411
Eco-intelligent product	1/5	1	1	1/4	0.100
Novel technologies	1/3	1	1	1/2	0.135
External market	1	4	2	1	0.353

Source Examined by the authors according to the expert survey

Calculations show that the expansion of the vegetable variety will not indirectly provide any influence on overcoming environmental factors. Therefore, such an alternative will be less effective. Figure 2 shows the results of the Analytical Hierarchy Modeling.

The most affordable marketing strategy is increasing eco-intelligent production (0.341). Such choice includes the highest AHP indicator because of the high influence of technological and ecological effects. Breeding (0.461) and government support (0.573) are suggested to be the main reasons for the appearance of health products on trade shelves. To meet market demands, a food company should

Table 5 A comparison of the group of ecological factors

Factor	Environmental pollution	Hygiene requirements	Standards and certification	Alternative energy sources	Investments into organic production	AHP
Environmental pollution	1	1	4	5	1	0.292
Hygiene requirements	1	1	3	4	2	0.308
Standards and certification	1/4	1/3	1	4	1/3	0.108
Alternative energy sources	1/5	1/4	1/4	1	1/5	0.050
Investments into organic production	1	1/2	3	5	1	0.242

Source Examined by the authors according to the expert survey

monitor each manufactural and logistic process and reveal critical points in quality (0.513). The main barrier to eco-intelligent marketing could be environmental pollution. Because a consumer wants to be confident in food safety, a company should present comprehensive information about a region of production and distribution. From this side, vegetable companies that use alternative energy sources (0.737) and natural remedies (0.592) will be more attractive in the food market. Consumer confidence in organic-made food (0.613) is gradually growing because people select health-oriented consumption.

Novel technologies (0.255) in marketing may turn around a company's position in vegetable marketing. While numerous farmers still use newspaper ads and email (0.398), modern businesses offer a product in social nets by applying AI and VR (0.608). A company should consider standards (0.455) and consumer requirements to enter a regional market with innovative products.

Expansion of the sales market (0.238) is the third marketing alternative after eco-intelligent production and implementation of novel technologies. Each country has its own regulations for storage, transport, and distribution, reflected in the standards (0.628) and packaging requirements (0.648). A country might have stringent requirements and regulations for vegetables used in children's consumption. Because the global market is limited by the variety of fresh vegetables (0.447) in seasons and regions, trade integration may diversify the presence of natural vegetables in different countries during the year.

Table 6 AHP of the group of ecological factors

Strategy	Variety expansion	Eco-intelligent product	Innovation technologies	External market	AHP
A comparison of strategies by environmental pollution					
Variety expansion	1	1/8	1/5	1/7	0.038
Eco-intelligent product	8	1	7	4	0.592
Novel technologies	5	1/7	1	1/6	0.092
External market	7	1/4	6	1	0.278
A comparison of strategies by hygiene requirements					
Variety expansion	1	1/2	1/5	6	0.238
Eco-intelligent product	2	1	1/3	4	0.233
Novel technologies	5	3	1	1/4	0.312
External market	1/6	1/4	4	1	0.217
A comparison of strategies by standards and certification					
Variety expansion	1	3	1/5	1/5	0.120
Eco-intelligent product	1/3	1	1/3	1/4	0.082
Novel technologies	5	3	1	2	0.455
External market	5	4	1/2	1	0.343
A comparison of strategies by alternative energy sources					
Variety expansion	1	1/9	1	1/2	0.070
Eco-intelligent product	9	1	9	8	0.737
Novel technologies	1	1/9	1	1/2	0.070
External market	2	1/8	2	1	0.123
A comparison of strategies by investments into organic production					
Variety expansion	1	1/8	1/7	1/6	0.038
Eco-intelligent product	8	1	4	7	0.613
Novel technologies	7	1/4	1	3	0.231
External market	6	1/7	1/3	1	0.118

Source Examined by the authors according to the expert survey

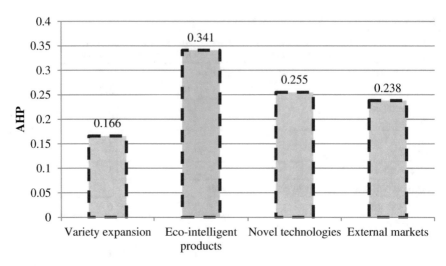

Fig. 2 Output data of a strategy selection

6 Approbation of the Research Results

The results of the study have been presented at the Entrepreneurship Coordination Board of Melitopol Council in Ukraine to facilitate the support of companies that produce food for children. The authors explain the alternative strategies of the on-market development in accordance with requirements for children. The results suggest that it is essential to show the consistency of vegetable products and the location of production on packaging. City authorities should develop educational programs for children and parents to interest families in buying local eco-intelligent products. It could be beneficial to support local farmers to produce zero-kilometer products by maintaining new current technologies in production, processing, and storage.

7 Conclusions

Four potential marketing alternatives include diversification of the product list, adaptation of eco-friendly production, implementation of innovative technologies, and access to the children's food market. Essential factors consist of technological, ecological, and natural aspects, as well as the primary alternatives for developing a marketing strategy. Each factor and its respective groupings are delved into a comprehensive comparison according to the selected marketing strategy.

QMC has the most comprehensive impact, significantly influencing consumer quality across technological, transportation, distribution, and operational processes. It is primarily due to the fact that consumers, especially parents of children, assess product quality based on its safety for consumption. The entrance of frozen vegetables into the children's food market extends the period of product consumption throughout the year. Government support has a fundamental impact on promoting eco-friendly products within the vegetable market. Various state initiatives encourage families to incorporate eco-friendly vegetables into their diets. This emphasis on safety, naturalness, and health aligns with the principles of sustainable consumption and contributes to the increased demand for eco-intelligent products.

The usage of additives in production and storage stands out as the primary natural factor that might impact consumer preferences in the vegetable market. The modeling results indicate that addressing this concern holds greater importance compared to other aspects when making marketing decisions. The degree of contamination in the areas where products are cultivated and the potential for growing organic vegetables are significant criteria for families when selecting a brand.

The most cost-effective marketing approach involves enhancing eco-friendly production, which boasts the highest AHP rating due to its substantial impact on technological and ecological factors. To cater to market demands, food companies must meticulously oversee every aspect of the manufacturing and logistics processes, pinpointing critical quality checkpoints. To instill consumer confidence in the safety of their products, companies should provide comprehensive information about the production and distribution regions. As people increasingly prioritize health-conscious consumption, consumer trust in organic food is steadily rising. In the time of digital marketing, innovative technologies in Marketing 4.0 can significantly alter a company's standing in the vegetable market. Given the limitations of the global market concerning the availability of fresh vegetables across seasons and regions, trade integration can diversify the presence of natural vegetables in different countries throughout the year.

Acknowledgements The study is related to the Ukrainian State Scientific and Technical program No. 0116U002738 "Marketing strategy development of agrarian enterprises". The questions of digital marketing implementation in food children's consumption are part of the ERASMUS+ Digitalization of economics as an element of sustainable development of Ukraine and Tajikistan/ DigEco 618270-EPP-1-2020-1-LT-EPPKA2-CBHE-JP. The authors wish to thank all experts who took part in the evaluation of factors and shared their professional experience to examine major alternatives of further strategies of the development of children's food market of vegetables.

References

1. Villalobos, B., Miranda, H., Schnettler, B.: Satisfaction with food: profiles of two-parent families with adolescent children. Int. J. Environ. Res. Public Health **19**(24), 16693 (2022). https://doi.org/10.3390/ijerph192416693
2. Wood, R., Lenzen, M., Dey, C., Lundie, S.: A comparative study of some environmental impacts of conventional and organic farming in Australia. Agric. Syst. **89**(2–3), 324–348 (2006). https://doi.org/10.1016/j.agsy.2005.09.007

3. Del Borghi, A., Gallo, M., Strazza, C., Del Borghi, M.: An evaluation of environmental sustainability in the food industry through life cycle assessment: the case study of tomato products supply chain. J. Clean. Prod. **78**, 121–130 (2014). https://doi.org/10.1016/j.jclepro.2014.04.083
4. Gutiérrez, T.: Food science. Technol. Nutr. Babies Children (2020). https://doi.org/10.1007/978-3-030-35997-3
5. Torrents-Masoliver, B., Sandjong, D., Jofré, A., Ribas-Agustí, A., Muñoz, I., Felipe, X., Castellari, M., Meurillon, M., den Besten, H.M.W., Engel, E., Bover-Cid, S.: Hazard control through processing and preservation technologies for enhancing the food safety management of infant food chains. Global Pediatrics **2**, 100014 (2022). https://doi.org/10.1016/j.gpeds.2022.100014
6. Wilks, M., Bloom, P.: Children prefer natural food, too. Dev. Psychol. **58**(9), 1747–1758 (2022). https://doi.org/10.1037/dev0001387
7. Bodunde, I.O., Karanja, A., McMullin, S., Mausch, K., Ickowitz, A.: Increasing fruit and vegetables consumption among children: a systematic review of animated nutrition interventions. World Nutr. **13**(4), 29–45 (2022). https://doi.org/10.26596/wn.202213429-45
8. Gómez, P.V., Tamburini, C., García, V.R., Chamorro, V.: Brand social media marketing strategies for foods consumed by children and adolescents in Argentina. Archivos Argentinos De Pediatria (2022). https://doi.org/10.5546/aap.2021-02528.eng
9. Cullen, K.W., Baranowski, T., Rittenberry, L., Cosart, C., Owens, E., Hebert, D., de Moor, C.: Socioenvironmental influences on children's fruit, juice and vegetable consumption as reported by parents: reliability and validity of measures. Public Health Nutr.Nutr. **3**(3), 345–356 (2000). https://doi.org/10.1017/s1368980000000392
10. Edmonds, J., Baranowski, T., Baranowski, J., Cullen, K., Myres, D.: Ecological and socioeconomic correlates of fruit, juice, and vegetable consumption among African-American boys. Prev. Med. **32**(6), 476–481 (2001). https://doi.org/10.1006/pmed.2001.0831
11. Enax, L., Weber, B., Ahlers, M., Kaiser, U., Diethelm, K., Holtkamp, D., Faupel, U., Holzmüller, H.H., Kersting, M.: Food packaging cues influence taste perception and increase effort provision for a recommended snack product in children. Front. Psychol. **6** (2015). https://doi.org/10.3389/fpsyg.2015.00882
12. Guiné, R., Lima, R., Florença S.: Consumer perspectives of innovative technologies in the food sector. In: Florença, S.G. (ed.) n/a. essay (2020)
13. Zainal Abidin, S.A., Ahmad Nizar, N.N., Bujang, A., Shariff, S.S., Rahmat, A.K., Jalil, S.A., Taib, M.N.: Identification of potential hazard and halal control points for perishable food during Food Logistics. J. Contemp. Islamic Stud. **8**(3) (2022) https://doi.org/10.24191/jcis.v8i3.7
14. Wang, H., Ab. Gani, M.A.: Research on emotional design for children food packaging. Idealogy J. **7**(2), 8–24 (2022). https://doi.org/10.24191/idealogy.v7i2.342
15. Islam, N.A., Tanni, F.T., Akter, A., Shahrin, S.: Food consumption pattern among children in an English medium school during COVID-19. J. Prev. Soc. Med. **40**(2), 32–37 (2022). https://doi.org/10.3329/jopsom.v40i2.61794
16. Sandeep, R.S., Prakash, O.M., Gupt, D.K., Zaman, T., Kumari, V., Tomar, R., Mertia, S.: Virtual class and children food patterns during pandemic: a review. J. Educ. Soc. Res. **12**(4), 315 (2022). https://doi.org/10.36941/jesr-2022-0116
17. Loth, K.A., Hersch, D., Trofholz, A., Harnack, L., Norderud, K.: Impacts of covid-19 on the home food environment and eating related behaviors of families with young children based on food security status. Appetite **180**, 106345 (2023). https://doi.org/10.1016/j.appet.2022.106345
18. Vergeer, L., Vanderlee, L., Potvin Kent, M., Mulligan, C., L'Abbé, M.R.: The effectiveness of voluntary policies and commitments in restricting unhealthy food marketing to Canadian children on Food Company websites. Appl. Physiol. Nutr. Metab.Nutr. Metab. **44**(1), 74–82 (2019). https://doi.org/10.1139/apnm-2018-0528
19. Mulligan, C., Potvin Kent, M., Christoforou, A.K., L'Abbé, M.R.: Inventory of marketing techniques used in child-appealing food and beverage research: a rapid review. Int. J. Public Health **65**(7), 1045–1055 (2020). https://doi.org/10.1007/s00038-020-01444-w
20. Jensen, M.L., Dillman Carpentier, F.R., Adair, L., Corvalán, C., Popkin, B.M., Taillie, L.S.: TV advertising and dietary intake in adolescents: A pre- and post- Study of Chile's Food Marketing Policy. Int. J. Behav. Nutr. Phys. Act. **18**(1) (2021). https://doi.org/10.1186/s12966-021-01126-7

21. Olstad, D.L., Raman, M., Valderrama, C., Abad, Z.S., Cheema, A.B., Ng, S., Memon, A., Lee, J.: Development of an artificial intelligence system to monitor digital marketing of unhealthy food to children: research protocol. Curr. Dev. Nutr. **6**(Supplement_1), 1151–1151 (2022). https://doi.org/10.1093/cdn/nzac072.023
22. Valderrama, C., Olstad, D., Lee, Y., Lee, J.: What factors shape whether digital food marketing appeals to children? Curr. Dev. Nutr. **6**(Supplement_1), 406–406 (2022). https://doi.org/10. 1093/cdn/nzac054.061
23. McGinnis, J.M., Gootman, J., Kraak, V.I.: Food marketing to children and Youth (2006). https:// doi.org/10.17226/11514
24. Woś, K., Dobrowolski, H., Gajewska, D., Rembiałkowska, E.: Organic food consumption and perception among Polish mothers of children under 6 years old. Int. J. Environ. Res. Public Health **19**(22), 15196 (2022). https://doi.org/10.3390/ijerph192215196
25. Ilić, A., Rumbak, I., Karlović, T., Marić, L., Brečić, R., Colić Barić, I., Bituh, M.: How the number and type of primary school meals affect food variety and dietary diversity? Hrvatski Časopis Za Prehrambenu Tehnologiju, Biotehnologiju i Nutricionizam **17**(1–2), 27–33 (2022). https://doi.org/10.31895/hcptbn.17.1-2.4
26. Pickard, A., Thibaut, J.-P., Philippe, K., Lafraire, J.: Poor conceptual knowledge in the food domain and food rejection dispositions in 3- to 7-year-old children. J. Exp. Child Psychol. **226**, 105546 (2023). https://doi.org/10.1016/j.jecp.2022.105546
27. Folkvord, F., Hermans, R.C.: Food marketing in an obesogenic environment: a narrative overview of the potential of healthy food promotion to children and adults. Curr. Addict. Rep. **7**(4), 431–436 (2020). https://doi.org/10.1007/s40429-020-00338-4
28. Otten, J.: (rep). Food Marketing: Using Toys to Market Children's Meals. P.4 (2014). Available at http://www.healthyeatingresearch.org
29. Trujillo, L.M., Tiboni Oschilewski, O., Hernández-Garbanzo, Y.: P112 fruit and vegetable fairs as a food education strategy for Chilean children. J. Nutr. Educ. Behav. **54**(7) (2022). https:// doi.org/10.1016/j.jneb.2022.04.153
30. Follows, S., Jobber, D.: Environmentally responsible purchase behaviour: a test of a consumer model. Eur. J. Mark. **34**(5/6), 723–746 (2000). https://doi.org/10.1108/03090560010322009
31. Evans, A., Dowda, M., Saunders, R., Buck, J., Hastings, L., Kenison, K.: The relationship between the food environment and fruit and vegetable intake of adolescents living in Residential Children's Homes. Health Educ. Res. **24**(3), 520–530 (2008). https://doi.org/10.1093/her/cyn053
32. Haftenberger, M., Lehmann, F., Lage Barbosa, C., Brettschneider, A.-K., Mensink, G.B.M.: Consumption of organic food by children in Germany—Results of EsKiMo II. J. Health Monit. **5**(1), 19–26 (2020). https://doi.org/10.25646/6399
33. Edwards-Jones, G.: Does eating local food reduce the environmental impact of food production and enhance consumer health? Proc. Nutr. Soc.Nutr. Soc. **69**(4), 582–591 (2010). https://doi. org/10.1017/s0029665110002004
34. Cleveland, D., Radka, C., Müller, N., Watson, T., Rekstein, N., Van M. Wright, H., Hollingshead, S.: Effect of Localizing Fruit and Vegetable Consumption on Greenhouse Gas Emissions and Nutrition, Santa Barbara County. Environ. Sci. Technol. **45**(10), 4555–4562 (2011). https:// doi.org/10.1021/es1040317
35. Korhonen, K., Muilu, T.: Characteristics and stability of consumer food-buying groups: the case of food circles. Rev. Agric. Food Environ. Stud. **103**(3), 211–245 (2022). https://doi.org/ 10.1007/s41130-022-00172-4
36. Weber, C., Matthews, H.: Food-miles and the relative climate impacts of food choices in the United States. Environ. Sci. Technol. **42**(10), 3508–3513 (2008). https://doi.org/10.1021/es7 02969f
37. Hajkowicz, S.: Supporting multi-stakeholder environmental decisions. J. Environ. Manag. **88**(4), 607–614 (2008). https://doi.org/10.1016/j.jenvman.2007.03.020
38. Zhang, K., Luo, Y., Han, Y.: The long-term impact of famine experience on harvest losses. Agriculture **13**(6), 1128 (2023). https://doi.org/10.3390/agriculture13061128

39. Kogan, F.: Global warming crop yield and food security. Remote Sens. Land Surf. Changes 217–275 (2022). https://doi.org/10.1007/978-3-030-96810-6_8
40. Poonam, J.R.K., Kumawat, P., Prajapat, A., Kumawat, S.: Post-harvest management of fruits and vegetables. Worldwide Trends Sustain. Agric. 148–160 (2022)
41. Muhaimin, A.W., Toiba, H., Retnoningsih, D., Yapanto, L.M.: The impact of technology adoption on income and food security of smallholder cassava farmers: empirical evidence from Indonesia. Int. J. Adv. Sci. Technol. 29(9), 699–707 (2020)
42. Sissoko, P., Synnevag, G., Aune, J.B.: Effects of low-cost agricultural technology package on income, cereal surplus production, household expenditure, and food security in the drylands of Mali. AIMS Agric. Food 7(1), 22–36 (2022). https://doi.org/10.3934/agrfood.2022002
43. Brenya, R., Zhu, J., Sampene, A.K.: Can agriculture technology improve food security in low- and middle-income nations? A systematic review. Sustain. Food Technol. (2023). https://doi.org/10.1039/d2fb00050d
44. Burdina, I., Priss, O.: Effect of the Substrate Composition on Yield and Quality of Basil (*Ocimum basilicum L.*). J. Hortic. Res. 24(2) 109–118 (2016). https://doi.org/10.1515/johr-2016-0027
45. Yang, J., Kuang, H., Xiong, X., Li, N., Song, J.: Alteration of the allergenicity of cow's milk proteins using different food processing modifications. Crit. Rev. Food Sci. Nutr. 1–21 (2022). https://doi.org/10.1080/10408398.2022.2144792
46. Son, J., Hwang, K.: How to make vertical farming more attractive: effects of vegetable growing conditions on consumer assessment. Psychol. Mark. (2023). https://doi.org/10.1002/mar.21823
47. Ion, M.: Vertical Farms: A business model built for the future? Biotechnol. Entrepreneurship 1–52 (2022). https://doi.org/10.13140/RG.2.2.32585.83048
48. Kumar, S., Kumar, S., Lal, J.: Assessing opportunities and difficulties in hydroponic farming. Bhartiya Krishi Anusandhan Patrika (Of) (2023). https://doi.org/10.18805/bkap556
49. Pandey, A.K., Sanches Silva, A., Chávez-González, M.L., Dubey, N.K.: Editorial: Novel tools to improve food quality and shelf life: advances and future perspectives. Front. Sustain. Food Syst. 7 (2023). https://doi.org/10.3389/fsufs.2023.1148977
50. Chai, J.J.K., O'Sullivan, C., Gowen, A.A., Rooney, B., Xu, J.-L.: Augmented/mixed reality technologies for food: a review. Trends Food Sci. Technol. 124, 182–194 (2022). https://doi.org/10.1016/j.tifs.2022.04.021
51. Suryantari, P.A., Wijanarko, R.P., Wati, S.F., Kom, S., Kom, M., Deviyanti, I.G.: Application of augmented reality in food ordering system. IJEEIT : Int. J. Electr. Eng. Inf. Technol. 6(1), 25–36 (2023). https://doi.org/10.29138/ijeeit.v6i1.2112
52. Styliaras, G.D.: Augmented reality in food promotion and analysis: review and potentials. Digital 1(4), 216–240 (2021). https://doi.org/10.3390/digital1040016
53. Gu, C., Huang, T., Wei, W., Yang, C., Chen, J., Miao, W., Lin, S., Sun, H., Sun, J.: The effect of using augmented reality technology in takeaway food packaging to improve young consumers' negative evaluations. Agriculture 13(2), 335 (2023). https://doi.org/10.3390/agriculture13020335
54. Masnar, A., Hidayah, F.O., Ar Rachmah, I., Nurbaya, N.: Combating excessive food consumption through augmented reality packaging: An explorative study of Generation Z. Jurnal Kesehatan Manarang 9(1), 34 (2023). https://doi.org/10.33490/jkm.v9i1.895
55. Nopparat, N., Motte, D.: Business model patterns in the 3D food printing industry. Int. J. Innov. Sci. (2023). https://doi.org/10.1108/ijis-09-2022-0176
56. Felicetti, A.M., Volpentesta, A.P., Linzalone, R., Schiuma, G., Ammirato, S.: Business model innovation in the food sector: towards a dimensional framework for analyzing the value proposition of digital platforms for food information services. Eur. J. Innov. Manag.Innov. Manag. (2023). https://doi.org/10.1108/ejim-10-2022-0563
57. Pellegrini, G., de Mattos, C.S., Otter, V., Hagelaar, G.: Exploring how EU Agri-food smes approach technology-driven business model innovation. Int. Food Agribus. Manag. Rev. 1–20 (2023). https://doi.org/10.22434/ifamr2022.0122
58. Ciccullo, F., Fabbri, M., Abdelkafi, N., Pero, M.: Exploring the potential of business models for sustainability and big data for food waste reduction. J. Clean. Prod. 340, 130673 (2022). https://doi.org/10.1016/j.jclepro.2022.130673

59. Lanciano, E., Lapoutte, A., Saleilles, S.: What business models for food justice? Systèmes Alimentaires—Food Syst. **4**, 159–183 (2019). https://doi.org/10.15122/isbn.978-2-406-09829-4.p.0159
60. Oliver, T.L., McKeever, A., Shenkman, R., Diewald, L.K.: Successes and challenges of using a peer mentor model for nutrition education within a food pantry: a qualitative study. BMC Nutr. **6**(1) (2020). https://doi.org/10.1186/s40795-020-00352-9
61. Dhanjal, R., Dine, K., Gerdts, J., Merrill, K., Frykas, T.L., Protudjer, J.L.: An online, peer-mentored food allergy education program improves children's and parents' confidence. Allergy, Asthma Clin. Immunol. **19**(1) (2023). https://doi.org/10.1186/s13223-023-00800-8
62. Zhou, M., Bian, B., Huang, L.: Do peers matter? unhealthy food and beverages preferences among children in a selected rural province in China. Foods **12**(7), 1482 (2023). https://doi.org/10.3390/foods12071482
63. Gabison, G.A.: White label: the technological illusion of competition. Antitrust Bull. **67**(4), 642–662 (2022). https://doi.org/10.1177/0003603x221126160
64. Vargas, C., Brimblecombe, J., Allender, S., Whelan, J.: Co-creation of health-enabling initiatives in food retail: academic perspectives. BMC Public Health **23**(1) (2023). https://doi.org/10.1186/s12889-023-15771-z
65. de Koning, J.I.: Exploring co-creation with agri-food smallholders in Vietnam. Int. J. Food Des. (2023). https://doi.org/10.1386/ijfd_00056_1
66. Qian, J., Lin, P., Wei, J., Liu, T., Nuttavuthisit, K.: Cooking class travel in Thailand: an investigation of value co-creation experiences. SAGE Open **13**(2), 215824402311769 (2023). https://doi.org/10.1177/21582440231176994
67. Dagoudo, B.A., Ssekyewa, C., Tovignan, S.D., Ssekandi, J., Nina, P.M.: Agroecological business model: A pillar stone for women's entrepreneurship in Agroecology and Sustainable Food Systems. Int. J. Curr. Sci. Res. Rev. **06**(01) (2023). https://doi.org/10.47191/ijcsrr/v6-i1-31
68. Rico, C., Wong, P., Etrata, Jr., A.E.: The impact of modern business disruption on Heritage Brands: A resiliency model for the packaged food product industry. Millennium J. Humanit. Soc. Sci. 1–13 (2023). https://doi.org/10.47340/mjhss.v4i2.1.2023
69. Sodero, A.C., Namin, A., Gauri, D.K., Bhaskaran, S.R.: The strategic drivers of drop-shipping and retail store sales for seasonal products. J. Retail. **97**(4), 561–581 (2021). https://doi.org/10.1016/j.jretai.2021.09.001
70. Menaouer, B., Khalissa, S., Belayachi, M.E., Amine, B.: The role of drop shipping in e-commerce. Int. J. E-Bus. Res. **17**(4), 54–72 (2021). https://doi.org/10.4018/ijebr.2021100104
71. Abdoellah, O.S., Suparman, Y., Safitri, K.I., Mubarak, A.Z., Milani, M., Margareth, Surya, L.: Between food fulfillment and income: Can urban agriculture contribute to both? Geogr. Sustain. **4**(2), 127–137 (2023). https://doi.org/10.1016/j.geosus.2023.03.001
72. Obikhod, S., Legeza, D., Nestor, V., Harvat, O., Akhtoian, A.: Digitization of business processes and the impact on the interaction of business entities. Econ. Affairs (New Delhi) **68**(1), 115–121 (2023). https://doi.org/10.46852/0424-2513.1s.2023.14
73. Miljenović, D., Beriša, B.: Pandemics trends in e-commerce: Drop Shipping Entrepreneurship during COVID-19 pandemic (2022). Pomorstvo. https://hrcak.srce.hr/279300
74. Aizaki, H., Sato, K., Nakatani, T.: Hometown effect on consumer preferences for food products. Int. Food Agribus. Manag. Rev. **26**(2), 309–323 (2023). https://doi.org/10.22434/ifamr2021.0164
75. Scholz, M., Kulko, R.-D.: Dynamic pricing of perishable food as a sustainable business model. Br. Food J. **124**(5), 1609–1621 (2021). https://doi.org/10.1108/bfj-03-2021-0294
76. Dunn, J., Brunner, T., Legeza, D., Konovalenko, A., Demchuk, O.: Factors of the marketing macro system effecting children's food production. Econ. Ann.-XXI, **170**(3–4), 49–56 (2018). https://doi.org/10.21003/ea.v170-09
77. Camerer, C., Ho, T., Chong, J.: A Cognitive hierarchy model of games. Q. J. Econ. **119**(3), 861–898 (2004). https://doi.org/10.1162/0033553041502225

78. Van de Water, H., van Peet, H.: A decision support model based on the analytic hierarchy process for the make or buy decision in manufacturing. J. Purchasing Supply Manag. **12**(5), 258–271 (2006). https://doi.org/10.1016/j.pursup.2007.01.003
79. Priss, O.P., Zhukova, V.F.: Optimized concentration of exogenous antioxidants for the storage of Zucchini Fruit. J. Chem. Technol. **27**(1), 40–47 (2019). https://doi.org/10.15421/081904

Model for Universal Classification of Social Agents' Activity/Behavior in Hierarchical Systems

Anatolii Shyian⦿ **and Liliia Nikiforova**⦿

Abstract Today, social systems consist of people and artificial agents (technical objects, learning, expert systems, bots, etc.). They carry out activities in a hierarchical environment. The paper develops a model for the universal classification of social agents' activity/behavior in hierarchical systems. The activities of social agents are carried out in real space, while the analysis, decision-making, choosing of behavior and verification of the result, and the transfer of experience are carried out in the model space. The universal description of the model space for hierarchical structures is initially built. The set of 64 operators is introduced into the model space, which can describe changes in the space. It is proved that the minimum set of these operators is 16 classes. It is suggested that this minimal set describes the activity/behavior of the human component of agents. As a result, the activity of people and artificial agents can be described within the framework of a universal/unified theoretical apparatus. This makes it possible to increase the efficiency of management both in social systems and in systems with artificial social agents. An example of determining the effectiveness of a person's activity at a given workplace is given. To apply the method, the functional duties of a person must be expressed in terms of the information space. Possible areas of application of the obtained results and areas of further research are discussed.

Keywords Social agent · Activity · Behavior · Classification · Hierarchy

1 Introduction

Today, a person is forced to exist in a social environment that consists of natural (people as a social component) and artificial (robots, technical and software components, educational systems, bots, etc.) agents. Human is gradually transitioning to activities in the world of social agents, some of which are artificial [1]. Artificial

---

A. Shyian (✉) · L. Nikiforova
Vinnytsia National Technical University, Khmelnytske Shose 95, Vinnytsia 21021, Ukraine
e-mail: anatoliy.a.shyian@gmail.com

© The Author(s), under exclusive license to Springer Nature Switzerland AG 2024
A. Semenov et al. (eds.), *Data-Centric Business and Applications*, Lecture Notes on Data Engineering and Communications Technologies 194,
https://doi.org/10.1007/978-3-031-53984-8_18

agents are increasingly integrated into human activity. The number of technical systems that operate fully or partially autonomously is growing.

Modeling the development and behavior of such communities is complicated because these two components are described by fundamentally different models. Further development of artificial social actors can go in three directions. The first direction is to imitate human activity. This requires a powerful model for human behavior. Unfortunately, so far, such models are rather probabilistic [2], which makes it impossible to effectively move in this direction. The second direction is when a person will adapt to technology. This will lead to stress and ineffective results of joint activities [3]. Finally, the third way is the formation of unified models for the activity of technical systems and people (see, for example, [4]). It allows the most comfortable humanity to carry out activities in the technical environment [5]. In addition, part of the ethical problems, in this case, will be solved effectively [6].

The advantage of using unified models to jointly describe the behavior of artificial agents and humans is that they provide an opportunity to describe them as a single system [7]. As a result, the artificial social agent will be programmed into typical human behavior. This will allow people to effectively make decisions and be active in joint teams [8].

Thus, the problem arises of developing models for the formation of behavior classifications that could describe both the characteristics of a person and the characteristics of technical objects in the frame of "one language" [9]. This will allow transferring results between different types of objects. In particular, to organize training of both people and technical objects [10].

The purpose of the paper is to develop a model for the universal classification of social agents' activity/behavior in hierarchical systems.

2   Methodology

The activity/behavior $a_R(g_i)$ is carried out by the subject in the Real Space (World) R under a certain (given) goal g_i. To analyze activity/behavior, it is necessary to proceed to the consideration of the models. Models are formed in the Model Space (World) M. At the same time, the goal of activity g_i is also set, as a rule. The transition to the model space is carried out with the help of the projection (transition) operator Pr, which also depends on the goal of activity g_i. In the model space, the activity/behavior is described by the operator $a_M(g_i)$.

In the general case, the analysis of activity/behavior will be described by the diagram of Fig. 1.

One model space M can correspond to several real worlds R. In other words, one model can correspond to several (different) implementations of activity in real situations. As a result, education and communication between the entities that carry out activities are implemented in the model world.

Fig. 1 Transition from
activities in the real world to
activities in the model space

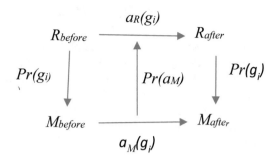

Essentially, the projection operator Pr translates the characteristics of the real world into terms, that is, into the names of abstract classes that characterize a sufficiently large number of real situations/entities. This operator is an algorithm that discards "insignificant" information for a given goal.

Here, $Pr(a_M)$ means the transition from the model description of the activity to the implementation of the activity in the real world. If the operator Pr of designing the Real World into the Model World is described as compression of information (discarding many data and/or characteristics that do not relate to the purpose of the activity), the operator $Pr(a_M)$ specifies the transition from the image of the activity to its implementation. It is, in general, a transition from a sequence of transitions (processes) between states in the Model World to their "analogs" in the Real World.

Thus, meaningful activity is carried out like this.

1. The goal of the activity is determined. As a rule, the goal is defined as the transfer of the Real World from the existing state/process to the desired state/process.
2. Projection/transition from the Real World to the Model World is carried out.
3. The solution to the problem of achieving the goal in the Model World is carried out.
4. The design of the solution to the model achievement of the goal in the set of methods and technologies of activity and their implementation in the Real World is carried out.
5. Projection/transition of the Real World into the Model World is carried out and it is determined whether the goal has been achieved or not. If the goal is not achieved, the cycle repeats.

The most telling example of this approach to realization is physics as a science. It starts with the Real World. The apparatus of experimental physics here performs the role of the projection operator Pr, which allows you to compress reality to the Model World, that is, to the set of terms M. And the activity in the Model World is carried out with the help of the apparatus of theoretical physics. To verify the activity in the Model World, appropriate experiments are carried out during which the planned activity is carried out in the Real World. The results of this activity, again, using the apparatus of experimental physics, are reduced to the set of terms M.

3   The Structure of the Model Space

Both the Real World R and the Model Space (World) M have a distinct hierarchical structure. Therefore, this feature should be reflected in the projection operator's Pr. This feature is that each of the hierarchical levels must be reduced to a set of terms (certain abstract structures) in a single, universal way.

We present a variant of the formation of such a projection operator Pr, which allows us to describe universally one hierarchical level of an arbitrary level and an arbitrary nature.

1. Each separate hierarchical level consists of separate elements, each of which should, as a rule, have specific (special) characteristics. The set of these characteristics will be denoted as Ob (from "object").
2. All elements at the hierarchical level must be connected by certain links, which must also have specific (special) characteristics. The set of these characteristics will be denoted as L (from "links").
3. The hierarchical level must be separated from its environment by a certain boundary, the elements of which are characterized by certain special characteristics (which distinguish them from other elements in the hierarchical level as a whole). We denote the set of these characteristics as B (from "boundary").
4. Finally, the hierarchical level should have a structure that consists of a certain set of elements that will have special characteristics (which distinguish them from other elements in the hierarchical level as a whole). We denote the set of these characteristics as St (from "structure").

The two sets Ob and L describe the characteristics that detailed the composition of the hierarchical level. The two sets B and St describe characteristics that refer to the entire hierarchical level as a whole.

To complete the construction of the operator Pr, it is necessary to describe each of the four introduced sets within the framework of dynamics D (process, change in time, variability) and state S (immutability, identity in time).

Thus, we get 8 sets (information components) that make up the Model Space for each fixed problem.

Briefly, the above is shown in Table 1.

Let's introduce such definitions.

Definition 1 A set of parameters and characteristics that allow to description of the separate hierarchical level in the frame of the considered problem will be named the Model Space M for the considered goal.

Definition 2 A set of characteristics of the elements that relate to a certain component of the information space will be named the "component of the Model Space".

Definition 3 The names for eight components of the information space M construct the "basis (backbone) of the Model Space" (or a "set of Model components").

Table 1 The structure of the component of the model (information) space

Data for the description of the activity/ behavior in Model Space	Data about the class of similar objects (*generalizing* components)	Supporting elements of the level (*structure*, topology)	Static	St-S
			Dynamism	St-D
		The *boundary* between this level and the environment	Static	B-S
			Dynamism	B-D
	Data about this particular object (*detailing* components)	The *object* itself is unique	Static	Ob-S
			Dynamism	Ob-D
		The *links* of this object with other concrete, similar to him	Static	L-S
			Dynamism	L-D

Thus, the model space M can be represented by the vector $(m_1, m_2,..., m_8)$, where $m_k, k = 1...8$ are the characteristics of the Model Space.

Note that the method of building the Model Space described above can also be considered a universal method of building ontologies within the framework of information technologies and databases [11].

Arbitrary Model Space can be written in such form using the constructed basis of Model components m_k.

$$M = \sum_{k=1}^{k=8} m_k \cdot \overrightarrow{e_k} \tag{1}$$

Here e_k is the basic vectors of the space of Model Space components that define the names of Model components (they are listed in Table 1). m_k is a set (database) of characteristics that can be attributed to this Model component (that is, the filling of these Model components).

Relation (1) is understood in the sense that m_k is a specific database that belongs to this class of Model components. For example, if e_1 denotes the Model components of *St–S* (see Table 1), then m_1 will denote the entire set of parameters and characteristics (assigned to the database as a whole) that is included in this particular Model component. In this sense, the "point" in the Model space is a set of eight databases that do not intersect with each other.

Note that m_k is not a number, so the "additional components" operation should be defined as the union of two homogeneous (which describe the same information components) databases in one. Similarly, "subtraction of components" is defined. In some cases, the information space can also be equipped with an appropriate metric (for example, similar to [12, 13]).

4   Modeling Activity/Behavior in Model Space

Definition 4   The activity/behavior is described as the change in one or more Model components.

Definition 4 can be rewritten as the following formula.

$$a_M(g_i) = M_{afterr}(g_i) - M_{before}(g_i) \tag{2}$$

In the future, the dependence on the goal g_i and the index M will be omitted.

If such changes are present (i.e., $a \notin \emptyset$), then the "act of activity/behavior" took place.

Note that if such a change has not existed (i.e., $a = \emptyset$), then the "act of activity/ behavior" was completed without result in the frame of the selected methods for constructing the Model space. It is possible that for other methods of constructing a Model Space (for example, for other goals and/or other methods of projection Pr to a Model space), an act of activity/behavior can be fixed.

The method of interconnection between Model Spaces "before" and "after" can be described as follows. In the first Model Space, created before the action (by which the considered object is programming to activities). The second Model Space is built (within the same goal) to decide whether an activity/behavior effect has been achieved. These Model Spaces are considered at different times.

Definition 5   An object that perceives (assimilates) Model components from the Model space and is capable of transforming (changing) Model components from it, i.e., which realizes the operator $a_M(g_i)$, will be named a "Model activity operator" (MAO).

Note. For the correct definition of the term MAO, it is necessary that the researched object can implement the diagram in Fig. 1.

From formula (2) and definition 5, the MAO will have the following structure.

$$< input | output > \tag{3}$$

Or, in a different form, like.

$$< perception(programming)block | activity(behavior)block > \tag{4}$$

In (3) and (4) the vector <input| describes those components M_{before} of the Model Space, by which the object under study is programmed for activity/behavior. The |output> vector, accordingly, describes those components M_{after} of the Model Space that the researched object changed in the process of activity/behavior. We note that the goal of g_i for carrying out activities within the framework of the proposed approach can be described in terms of "filling with specific values" of operators (3) or (4). The representation form of the operator (4) describes g_i exactly.

Notation (3) or (4) corresponds to the form of operator notation proposed by Dirac for quantum mechanics [14]. Essentially, this corresponds to Definition 5, where the operator $a_M(g_i)$ is the transition operator between two information spaces (i.e., environmental states).

Let us emphasize that, for the sake of ease of further use, in (3) and (4) the standard order of following "input" and "output" in mathematics has been changed. The logical sequence for activity is precisely the transition from "input" (motivation, setting the task, programming of action) to "output" (results of activity/behavior, analysis of goal achievement).

The a_M operator in the form (3) and (4) is a separate class for a set of ways, methods, technologies, algorithms, etc., to perform activity/behavior within the Model World. In essence, it is a projection of the corresponding class for ways to implement an activity/behavior in the Real World.

Thus, the activity of MAO in the information space can be represented in the form of an operator A (which will be written in the operator form $<input|output>$), which transforms information space M_{before} into M_{after}. This can be written as follows:

$$M_{after} = A \cdot M_{before} \tag{5}$$

It is easy to see that the operator A has the following property: if the information space of the problem can be divided into two subspaces M_{b1} and M_{b2}, then $A \cdot (M_{b1} + M_{b2}) = A \cdot (M_{b1}) + A \cdot (M_{b2}) = M_{a1} + M_{a2}$. This property is a consequence of the fact that the solution of the set of problems, each of which is obtained by expanding the basic (complex) problem into additional parts, each of which is solved separately, is equivalent to solving the initial complex problem. Of course, this is done in the absence of the effect of synergy and nonlinearity. But this, in fact, means that the subspaces M_{b1} and M_{b2} from the information space simply do not intersect (i.e., have no common (joint) points).

Thus, the information space M_{before} is divided into a direct sum of subspaces.

$$M_b = \bigcup_k M_b^k, \forall k, n : M_k \cap M_n = \emptyset \tag{6}$$

As a result, operator A acts as follows:

$$M_a = \bigcup_k M_a^k = A \cdot \bigcup_k M_b^k = \bigcup_k A \cdot M_b^k \tag{7}$$

In general, it follows from (7) that operator A can be represented as a tensor operator $A_{(n)}^{(m)}$ with n "lower" and m "upper" indices. Moreover, because of the presence of a basis in the information space, the number of "upper" and "lower" components in the tensor $A_{(n)}^{(m)}$ is bounded by 8: $n, m \leq 8$.

Using properties (6) and (7), the conclusion is that the action of any tensor operator $A_{(n)}^{(m)}$ can be expressed as the action of the sum of "binary" operators having the form a_{ik} (here the first subscript describes M_b, and the second subscript describes M_a).

This assertion can be formulated in the form of such a theorem.

Theorem 1 To implement any activity/behavior in the information space of a problem, only such MAOs are necessary and sufficient that are programmed by one component of the input information (from the M_{before} information space) and whose activities are also expressed in a change of one output information component from the information space M_{after} the problem (i.e., the resulting change in the transition from M_{before} to M_{after} means that the information space M_{after} is simply different one component as compared to the information space M_{before}).

Prove. The validity of this theorem is based on the fact that the information space is a database of characteristics, parameters, etc., which relate to the problem under consideration.

In this sense, any MAO operator acts as $M_{after} \rightarrow M_{before}$, that is, it does not change the design of the information space: only the numerical (or values) of its components change. For other "two-component" MAO, the result of the "previous" MAO action is the programming information space for other MAOs. The chain can continue as long as necessary.

Thus, there is an equality $M_{before} \rightarrow M_{after} = M_{after}^1 \rightarrow M_{after}^2 \rightarrow \ldots \rightarrow M_{after}$ (in the sense of filling the components of the information space), where the "intermediate" information spaces are different, each of them differs only one information component.

The sufficiency of the theorem follows from the fact that we can only compare the "initial" information space and the "final" information space with each other; it is impossible to determine that the activity/behavior was performed by the operator $A_{(n)}^{(m)}$, and it was not realized in a set of successive applied "two-component" operators a_{ik}.   ∎

In other words, one action/behavior, which is based on a set of information space components, can be replaced by the number of sequentially applied acts of activity/behavior, each of which "uses" only one component from the information space M_{before}, and the result, which can be expressed in changing only one component in the information space M_{after}. This statement is a standard method for breaking down a complex problem into successive intermediate stages.

Thus, from Theorem 1, each operator corresponding to MAO can be expressed as the sum of some "binary" operators that connect only two information components: one of the space M_{before}, and the other from the space M_{after}.

The following theorems will be valid for introduced operators a_{ik}.

Theorem 2 The operator g_{ik} is a commutative $a_{ik} + a_{nm} = a_{nm} + a_{ik}$ and an associative $a_1 + (a_2 + a_3) = (a_1 + a_2) + a_3$.

The proof follows from Theorem 1 for the properties of the operators a_{ik}.

Theorem 3 The total number of a_{ik} operators is 64 different variants.

Prove. The binary operator a_{ik} can have only one of the eight components of the information space M_{before} and only one of the eight components of the information space M_{after}. The number of different possible options: $8 \times 8 = 64$.   ∎

The operators a_{ik} can be used as a general method for creating automated process control systems based on abstract information machines (see, e.g., [15]). With the help of the introduced above operators a_{ik}, it is possible to describe the automatic control systems of arbitrary complexity. For example, operators *<process|state>* (programmed by the process, the result of the activity is the achievement of a given state) will describe the negative feedback. Accordingly, the operators *<state|process>* will describe negative feedback. Note that for many problems, it will be necessary to create complex control systems that include several components from M_{before} and M_{after}.

Thus, with the use of MAO, the arbitrary activity of implementing automatic control can be described as "constructing a control design from standard elements". That is, as an option for assembling a given structure from the elements of the Lego constructor.

5   Minimum Number Quantity of MAO (Activity/Behavior Classes)

Consider a simple set of MAO, which will describe any activity/behavior.

Definition 6   MAO is named an "activity/behavior class" (ABC, or "ABC operator") when it satisfies the following conditions:

1. For each MAO, one information component is generalizing, and the other is detailing.
2. For each MAO, one information component is static, and the other is dynamism.

In Definition 6, only the common dichotomies from which the information spaces were constructed are used (see Table 1).

The correct definition for ABC operators as an object that implements a certain class of activity/behavior can be performed only in the above way. Let's justify the Definition 6.

Suppose we define ABC in this way when both components (input and output) are either generalizing or detailing. In this case, defined in this way, the set ABC will be divided into two unrelated subsets: one is generalizing, and the second is detailing. So, this second condition is necessary. For the second condition, the argument is the same.

Thus, for any other definition of ABC, their number will be greater. Consequently, the set of ABCs defined above has the minimum necessary number of elements (ABC operators) for describing the activity/behavior in the arbitrary information space.

According to ABC's definition, one of its components corresponds to the hierarchical level as the whole, and its second component corresponds to the specific

Fig. 2 ABC as objects that
perform the transformation
of information components
into the information space

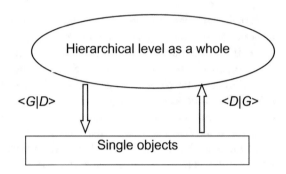

unitary objects from which this level is built. This is schematically shown in Fig. 2, where the generalizing and detailing of the information components are labeled G and D, respectively. From Fig. 2, one can see that class ABC creates feedback loops.

The number of different ABS can be calculated as follows. First, any information component can be selected as an "input" component. Thus, there are eight different options. For the second component of information, it is already necessary to abandon many options for information components. Thus, all those components that describe the same action time (the same pole in the "static–dynamic" dichotomy), that is, the four information components, must be rejected (for example, if the input information component is "static", then the output component cannot be "static"). Further, information components that describe the same pole of the "generalizing—detailing" dichotomy, as for the input information component, should be discarded (for example, if the input information component is "generalizing", the output information component must be "detailing"). There are no other limitations in the definition of ABC. Thus, the two information components can be used as output (for each fixed input information component). For example, if the input component is static and dynamic, then the output can accept any two informational components: static and detailing. The result is $8 \times 2 = 16$ classes for ABC. The result has followed the theorem.

Theorem 4 An arbitrary activity/behavior in arbitrary information space can be carried out in general form with a set of 16 classes of a_M in representation form as ABC, which have the following form:

$$< St - S|L - D >, < St - S|Ob - D >, < St - D|L - S >, < St - D|Ob - S >,$$
$$< B - S|L - D >, < B - S|Ob - D >, < B - D|L - S >, < B - D|Ob - S >,$$
$$< Ob - S|St - D >, < Ob - S|B - D >, < Ob - D|St - S >, < Ob - D|B - S >,$$
$$< L - S|St - D >, < L - S|B - D >, < L - D|St - S >, < L - D|B - S > .$$

The names of the information components shown in Table 1 were used to record the naming of ABC. The first component (input) of the information component corresponds to the input block (component, function) in ABC, with the help of which this ABC is programmed for activity/behavior. The second component (output) describes the activity/behavior of ABC.

The Theorem is obtained as a result of the above.

Theorem 5 For realization (or description, modeling, etc.) of an arbitrarily complex activity/behavior using the Model space, it is necessary and sufficient to have only the 16 defined above ABC operators. Every ABC operator is one class for realizations of activity/behavior in the Model Space is built.

6 An Example of the Application of Behavior Classification

Here we will use the names of four dichotomies [16], by which the class of human activity can be identified.

The calculation of the effectiveness of a person's activity is formed based on the analysis of the functional duties that he must perform.

We carry out such operations for each ith person.

1. The interior of the $as^i_e \in AS_e$ activity in which it functions is selected.
2. The purpose of activity $g^i_e \in G_e$ is selected for a person in this interior.
3. The database of functional responsibilities for this interior and this purpose of activity is imported (the database of functional responsibilities of this "workplace"). Each kth functional duty is assigned a certain weight h^i_k, which is a natural number (if not assigned, the weight is chosen equal to "1").
4. Each k-th functional duty is matched (by an expert or a corresponding computer program) with the corresponding pole of the dichotomy $p^i_{k,j}$ from the list of those 4 dichotomies that determine the class 2AIA for a person (j denotes the corresponding pole of the dichotomy, $j = 1,\ldots,8$; here $j = 1,2$ are the corresponding poles of the first dichotomy and so on). One functional duty may correspond to several poles of different dichotomies, or none.
5. For a given workplace, the following value is calculated for each pole of the dichotomy j using the formula.

$$P^i_j = \frac{1}{\sum_k h^i_k} \sum_{k=1}^{K} p^i_{k,j} \cdot h^i_k \qquad (8)$$

Summation in the formula is carried out for all functional duties, $k = 1..K$.
The class of activity for a person at a given workplace is selected as

$$t^i_e = \left\{ \max_{j=1,2} P^i_j, \max_{j=3,4} P^i_j, \max_{j=5,6} P^i_j, \max_{j=7,8} P^i_j \right\} = \left\{ a^i_e, b^i_e, c^i_e, d^i_e \right\}. \qquad (9)$$

We import the real class of behavior for the ith person

$$t_r^i = \left\{ a_r^i, b_r^i, c_r^i, d_r^i \right\} \tag{10}$$

The effectiveness of the activity of a real ith person at a given workplace is calculated according to the following formula:

$$E^i = \begin{cases} 0, & \text{if } a_e^i \neq a_r^i \\ 3 - \left| b_e^i - b_r^i \right| - \left| c_e^i - c_r^i \right| - \left| d_e^i - d_r^i \right|, & \text{if } a_e^i = a_r^i \end{cases} \tag{11}$$

The creation of a theoretical database for calculating the quantitative correspondence "workplace—class of human behavior/activity" is finished. A detailed example is presented in [12].

7 Discussion

One of the important issues is the possibility of applying the activity/behavior classification to people. This is the most promising direction of application of ABC classification for a_M. Humans and automatic control systems, as well as the activities of robots, artificial intelligence, and swarms of drones, can be described within a unified approach. Development in this direction will allow to development of more effective models of management and behavior prediction for socio-technical systems.

At the same time, people will act as objects that implement certain ABC classes. An important feature is that the classification can be applied only to that activity/behavior of a real person, which he performs purposefully and as a result of his analysis (that is, meaningfully and realized by him). The activity that a person performs under the normative rules (e.g., which is in culture, social norms, etc.) should not be subject to classification. The classification developed in the article differs from psychological classifications in several features. The first feature is that the division into classes is unambiguous: a person cannot belong to the ABC class by less than 100% (in contrast to psychological or social classifications—see, for example, [2]). The second feature is that the person can show activity/behavior outside of his ABC class. But such behavior occurs only after training and it will be effective only for clearly defined situations. Such unusual behavior is formed through special training and is one of the components of a person's personality (see, for example, [3]).

In [16], one of the possible variants of construction of projection operators Pr, oriented not on the construction of the Model Space, but on the identification of the class of activity/behavior for a real person, is given. Since the markers for identifying the activity/behavior class are described by objective characteristics, this allows the verification of the class for a real person. And although these results can be considered only preliminary, they indicate new possibilities that open up the results obtained above. Note that one class is assigned to ¼ of all information components that determine human activity/behavior. This corresponds to the well-known "Ebinghaus effect", which is characteristic of human perception of new information [17].

The use of the obtained ABC operators allows for mathematical modeling and prediction of the behavior of real people in given conditions (some examples are given in the Supplement). The use of projection operators Pr allows the use of transitions both $Pr : a_R \rightarrow a_M$ and $Pr^{-1} : a_M \rightarrow a_R$. Note that if the goal of activity/ behavior cannot be specified in the form of one class of ABC, then the effectiveness of a real person's activity will have a random component. In this case, to improve the efficiency of the activity/behavior, it is necessary to involve people who belong to other classes. The method of such involvement is given in Theorem 1.

It should also be emphasized that the ABC classification for a_M will not be able to describe the individual characteristics of a person. If people are evenly distributed by classes of activity/behavior, each class will make up only $1/16 \approx 6\%$ of the number of people under study. However, for many areas of activity, it is possible to focus not only on individual classes of behavior but also on certain combinations of classes. For example, as can be seen from Fig. 2, the division into detailing ($a_M = \, < G|D >$) and generalizing ($a_M = \, < D|G >$) behavior classes will already be $\frac{1}{2} = 50\%$. For example, such a division will be important for significantly increasing the effectiveness of learning, as detailing classes ABC better perceives specific, differentiating characteristics, and generalizing classes ABC better perceives general, collective (abstract) characteristics.

Also, the obtained results open up new possibilities for defining the concept of "mind". In particular, Humankind acts as an example of one of the possible realizations (one of the possible variants of the "carrier of the mind"). An important feature is that the definition of "mind" does not refer to one person but to at least 16 people. Only in this case can we talk about the possibility of optimal activity/behavior or management. Research in this direction is important for the development of artificial intelligence, which emphasizes the importance of human orientation [1]. There are certain indications that the class of human activity/behavior can be manifested in the philosophical concepts developed by him or in theories of human personality (personal observations).

Figure 1 shows that a person $h \in H$ can be characterized by a tuple

$$h \in H =< G, RW, MW, \text{Pr}(G), \text{Pr}(A) > \tag{12}$$

A person's personality is manifested through the individualization of almost every element of the tuple. At the same time, the personality includes both the elements of the tuple that characterize the class of activity/behavior H_c, and the elements of the tuple H_n that are set by the normative, learned activity of a specific/real person.

$$h_i \in H_c + H_n, \, H_c \cap H_n = \emptyset \tag{13}$$

Let's emphasize that the procession of normative behavior can consist of professional functional duties, social and/or legal norms, rules, religious norms, etc.

The obtained results can be applied to the formation of artificial intelligence, as well as to the design of control systems for technical objects of various levels of complexity (for example, robots and drones). Also, the classification of the components of the Model World makes it possible to form new approaches to training artificial intelligence systems. At the same time, it is important to establish the fact that a certain number of effector objects is required to manage complex objects and to solve management problems in a changing environment. At the same time, such effector objects must satisfy the conditions described in the article. It should also be taken into account that the technical component of social systems is described in a different way than the human component. For example, the total number of MAOs is 64 classes (Theorem 3), while the total number of ABCs is 16 classes (Theorem 4). This feature must be taken into account when developing models for the effective joint activity of social systems consisting of humans and artificial agents. Also, the use of MAO classification makes it possible to develop such algorithms of robot activity based on artificial intelligence, which are understandable to people. This will increase the level of trust in socio-technical systems.

By using robots/drones that use activities/behaviors as 64 MAO classes, the possibilities of activities in the environment can be greatly expanded. This will allow the social systems consisting of the people and such robots/drones to set the goals and objectives that previously could not be set before purely human systems. For example, it allows modeling the logic and the behavior that lies beyond the limits of the human mind. This may be important for understanding the nature of the non-human mind.

Artificial social systems can be built based on artificial objects that implement the described classes of activity/behavior. These social systems may exhibit characteristics that correspond to human social systems. In particular, they can have a hierarchical structure, which will reveal new topological patterns. This will be discussed in detail in a separate article.

The obtained classification can be used to develop the functionality of bots (for example, in social networks), which will more effectively imitate human behavior in communication. It may also offer new approaches to computer systems passing the Turing Test [18].

The obtained results open opportunities for scientific research in various fields of science. For example, the possible presence of the ABC classes described in the article on activity/behavior for humans raises the question of the presence of relevant gene systems that encode the formation of such a class. The structure of the experiments is obvious: first, a genetic study of people who belong to the same class to find "common", and then a comparison of "common genes" for different classes to find "genes that determine the difference between classes". Moreover, these sets of genes will act as a basis for the formation of "mind", thereby distinguishing Man from Animals. Some empirical human typologies based on body structure [19, 20] or obtained as a result of personal observations [21, 22] may be related to certain genetic patterns. There are also indications (personal observations) that the behavior class of children may imitate some characteristics of the behavior class of parents.

Also, the existence of differentiation of activity/behavior classes should be reflected in the organization of the human brain. It can also lead to new directions of experimental activity. For example, it may be promising to link the class of human activity/behavior with the presence of a specific brain structure. Probably, we are talking about some topological features of its structural and functional construction.

8 Conclusion

An apparatus has been built that allows modeling the activity/behavior of social systems consisting of people and artificial agents. For the first time, the activity/behavior of a person is described within the framework of those environmental changes that a person makes. It is shown that there is a finite number of classes of operators that can be used to describe human activity/behavior. It is shown that artificial systems of various origins, such as technical systems, robots, drones, expert and learning systems, computer bots, etc., can be described in the same terms as human activity/behavior. In this approach, a person's personality is manifested in the choice of a set of goals of activity/behavior and a set of those tools, methods, algorithms, and technologies of activity that are beyond the limits of the class of activity/behavior inherent in it. An example of determining the effectiveness of a person's activity at a given workplace is given. To apply the method, the functional duties of a person must be expressed in terms of the information space.

The results obtained in the article can provide a new impetus for scientific research in various branches of science and technology.

References

1. Dignum, V.: Responsible Artificial Intelligence: How to Develop and Use AI in a Responsible Way. Springer, Berlin (2019)
2. Almlund, M., Duckworth, A.L., Heckman, J.J., Kautz T.D.: Personality psychology and economics. National Bureau of Economic Research. NBER Working Paper No. 16822 (2011). https://doi.org/10.3386/w16822. Last accessed 17 Jan 2023
3. Hjelle, L.A., Ziegler, D.: Personality Theories. McGraw-Hill Inc., US (1992)
4. Dignum, F.P.M.: Social Practices: a Complete Formalization. Preprint at arXiv (2022). https://doi.org/10.48550/arXiv.2206.06088. Last accessed 17 Jan 2023
5. Dignum, V.: Responsible artificial intelligence–from principles to practice. ACM SIGIR Forum. 56, No.1 June 2022. Preprint at https://sigir.org/wp-content/uploads/2022/07/p03.pdf. Last accessed 17 Jan 2023
6. Verdiesen, I., Dignum, V.: Value elicitation on a scenario of autonomous weapon system deployment: a qualitative study based on the value deliberation process. AI Ethics (2022). https://doi.org/10.1007/s43681-022-00211-2,lastaccessed2023/01/17
7. Stoshikj, M., Kryvinska, N., Strauss, C.: Service systems and service innovation: two pillars of service science. Procedia Comput. .Sci 83, 212–220 (2016). https://doi.org/10.1016/j.procs.2016.04.118
8. Rauer, J.N., Kroiss, M., Kryvinska, N., Engelhardt-Nowitzki, C., Aburaia, M.: Cross-university virtual teamwork as a means of internationalization at home. Int. J. Manag. Educ. 19, 100512 (2021). https://doi.org/10.1016/j.ijme.2021.100512

9. Bashir, M.F., Arshad, H., Javed, A.R., Kryvinska, N., Band, S.S.: Subjective answers evaluation using machine learning and natural language processing. IEEE Access **9**, 158972–158983 (2021). https://doi.org/10.1109/ACCESS.2021.3130902

10. Kopetzky, R., Günther, M., Kryvinska, N., Mladenow, A., Strauss, C., Stummer, C.: Strategic management of disruptive technologies: a practical framework in the context of voice services and of computing towards the cloud. IJGUC **4**, 47 (2013). https://doi.org/10.1504/IJGUC.2013.054490

11. Staab, S., Studer, R. (eds.).: Handbook on ontologies. In: International Handbooks on Information Systems. Springer, Berlin (2009)

12. Dekel, E., Fudenberg, D., Morris, S.: Topologies on Types. Theor. Econ. **1**, 275–309 (2006)

13. Dekel, E.: Interim rationalizability. Theor. Econ. **2**, 15–40 (2007)

14. Dirac, P.A.M.: The Principles of Quantum Mechanics. University Press, Oxford (1948)

15. Dorf, R.C., Bishop, H.R.: Modern Control Systems. Prentice Hall, New York (2010)

16. Shyian, A., Nikiforova, L.: Model of human behavior classification and class identification method for a real person. Supplement. Available at SSRN (2022). https://ssrn.com/abstract=4035369. Last accessed 17 Jan 2023

17. Ebbinghaus, H.M.: A Contribution to Experimental Psychology. Original Work Published 1885, New York, Dover (1964)

18. Turing, A.M.: Computing machinery and intelligence. In: Mind LIX, pp. 433–460 (1950)

19. Sheldon, W.H., Stevens, S.S., Tucker, W.B.: The Varieties of Human Physique: An Introduction to Constitutional Psychology. Harper & Brothers, New York (1940)

20. Kretschmer, E.: Körperbau und Charakter Gebundene Ausgabe. Verlag, Springer (1955)

21. Jung, C.G.: Psychological Types, The Collected Works of C.G. Jung, vol. 6. Princeton University Press, Princeton (1976)

22. Leonhard, K.: Classification of Endogenous Psychoses and their Differentiated Etiology. In: Beckmann, H. (ed.). Wien, Springer (1999)

Printed in the United States
by Baker & Taylor Publisher Services